Materials and Contact Characterisation IX

WIT*PRESS*

WIT Press publishes leading books in Science and Technology.
Visit our website for new and current list of titles.
www.witpress.com

WIT*eLibrary*

Home of the Transactions of the Wessex Institute.
Papers published in this volume are archived in the WIT eLibrary in volume 124 of WIT Transactions on Engineering Sciences (ISSN 1743-3533).
The WIT eLibrary provides the international scientific community with immediate and permanent access to individual papers presented at WIT conferences.
http://library.witpress.com

Ninth International Conference on
Computational Methods and Experiments in
Materials and Contact Characterisation

MATERIALS CHARACTERISATION 2019

Organised by

Wessex Institute, UK
University of Groningen, The Netherlands
University of Windsor, Canada
Lisbon University, Portugal

Sponsored by

WIT Transactions on Engineering Sciences
International Journal of Computational Methods and Experimental Measurements

WIT Transactions

Wessex Institute
Ashurst Lodge, Ashurst
Southampton SO40 7AA, UK

Editorial Board

Materials and Contact Characterisation IX

Editors

S. Hernández
University of A Coruña, Spain
Member of WIT Board of Directors

J. De Hosson
University of Groningen, The Netherlands

D. O. Northwood
University of Windsor, Canada

R. Vilar
Lisbon University, Portugal

WIT*PRESS* Southampton, Boston

Editors:

S. Hernández
University of A Coruña, Spain
Member of WIT Board of Directors

J. De Hosson
University of Groningen, The Netherlands

D. O. Northwood
University of Windsor, Canada

R. Vilar
Lisbon University, Portugal

Published by

WIT Press
Ashurst Lodge, Ashurst, Southampton, SO40 7AA, UK
Tel: 44 (0) 238 029 3223; Fax: 44 (0) 238 029 2853
E-Mail: witpress@witpress.com
http://www.witpress.com

For USA, Canada and Mexico

Computational Mechanics International Inc
25 Bridge Street, Billerica, MA 01821, USA
Tel: 978 667 5841; Fax: 978 667 7582
E-Mail: infousa@witpress.com
http://www.witpress.com

British Library Cataloguing-in-Publication Data

A Catalogue record for this book is available
from the British Library

ISBN: 978-1-78466-331-5
eISBN: 978-1-78466-332-2
ISSN: 1746-4471 (print)
ISSN: 1743-3533 (on-line)

Preface

The contents of this Volume reflect the rapid advances that have taken place in materials science and engineering, prompted by the demand for high quality performance materials by industry.

The papers in this book were presented at the ninth International Conference on Computational Methods and Experiments in Materials Characterisation held in Lisbon in 2019. It follows the success of the eighth previous meetings in the series, which started in Santa Fe, New Mexico in 2003, followed by conferences in Portland, Maine (2004); Bologna (2007); the New Forest, home of the Wessex Institute of Technology (2009); Kos, Greece (2011); La Certosa di Pontignano of the University of Siena (2013), Valencia (2015) and Tallinn (2017).

The conference was organized by the Wessex Institute in collaboration with the Lisbon University, the University of Groningen and the University of Windsor.

The contributions represent some of the latest developments in the field of Materials Characterisation, describing procedures for the assessment of mechanical, thermal, acoustic and chemical properties of materials by experimental, computational of combined methods. Advances in micro and macro-structural approaches were presented. The range of material described in the presentations contained metallic substances, recycled materials, composites, biomaterials, polymers, ceramics and cementitious materials. The effectiveness of various surface treatments to enhance materials behaviour was discussed and techniques for designing performance base material were shown. The wide range of topics includes interaction between disciplines, which is sometimes essential to achieving a proper understanding of material behaviour.

All papers published in this Volume, as well as those of the previous conferences are permanently archived at www.witpress.com/elibrary, where they are easily available to the international community.

All volumes, including this one, have been published in paper as well as digital format and are being widely distributed throughout the world. As with other papers presented at Wessex Institute conferences, they are part of the WIT Transactions, a collection that appears in notable reviews, publications and databases.

The Meeting was sponsored by WIT Transactions on Engineering Sciences and the International Journal of Computational Methods and Experimental Measurements.

The Editors are also grateful to all authors for their contributions and to the members of the International Scientific Advisory Committee of the conference and other colleagues who reviewed the papers to ensure the quality of this book.

The Editors
Lisbon, Portugal, 2019

Contents

Section 4: Mechanical properties

Section 5: Mechanical and thermal properties

Section 6: Composite materials

SECTION 1
EXPERIMENTAL APPROACHES

EFFECT OF CURRENT MODE ON THE PLASMA DISCHARGE, MICROSTRUCTURE AND CORROSION RESISTANCE OF OXIDE COATINGS PRODUCED ON 1100 ALUMINUM ALLOY BY PLASMA ELECTROLYTIC OXIDATION

RIYAD O. HUSSEIN[1], XUEYUAN NIE[2] & DEREK O. NORTHWOOD[2]
[1]Natural Sciences Division, Higher Colleges of Technology, United Arab Emirates
[2]Department of Mechanical, Automotive and Materials Engineering, University of Windsor, Canada

ABSTRACT

Oxide coatings were produced on an 1100 aluminum alloy using a plasma electrolytic oxidation (PEO) process in an alkali-silicate electrolyte. Three different current modes were investigated, namely DC, unipolar and bipolar. Optical emission spectroscopy (OES) was used to characterize the plasma in terms of the plasma electron concentration and temperature. SEM and EDS were used to characterize the surface morphology and the cross-sectional structure of the coatings. Potentiodynamic polarization corrosion testing was performed in 3.5 wt% NaCl solution for both uncoated and coated samples. Plasma electron temperatures were in the range of 4,500 K to 10,000 K. The plasma characteristics, including sparking, have been related to the surface morphology of the coating. The coating morphology, microstructure and corrosion resistance varied with current mode. A bipolar current mode produced a higher quality (surface morphology, compactness and corrosion resistance) coating than either a DC or unipolar mode.

Keywords: PEO coatings, process parameters, plasma characteristics, soft sparking, coating quality, corrosion.

1 INTRODUCTION

Aluminum and its alloys are important structural materials in the automotive, aerospace and transportation industries [1], [2]. In order to improve surface properties, such as wear and corrosion resistance, various surface treatments have been employed including electrochemical treatments, chemical treatments and coatings [3], [4]. Of the electrochemical treatments, anodizing, where the metal in converted to an oxide coating, has found widespread usage [5]. However, conventional anodized coatings do not provide sufficient wear and corrosion resistance for specific applications [6].

Plasma electrolytic oxidation (PEO), also known as micro-arc oxidation (MAO), is a surface coating technology that is related to anodizing, but produces coatings that are superior in metal protection to those formed by anodizing or chemical treatment [7], [8]. PEO is operated at a voltage (400–600 V) exceeding the dielectric breakdown of the anodic oxide first formed at the surface. Sparks move across the surface, form a plasma envelope, and oxidize the surface As sparking proceeds the oxide grows and consolidates, and its adhesion to the surface is improved [7]. The mechanisms of coating growth are complex but essentially involve electrochemical reactions, plasma-chemical reactions and thermal (oxygen) diffusion [9]–[13].

The formation and structure (microstructure; phases; porosity; stress state) of the coatings depend on the PEO processing parameters, including electrical parameters (current density, frequency, duty cycle; type of current (DC, AC, pulsed-unipolar and bi-polar), electrolyte composition, deposition time and substrate composition [6], [10], [11], [14], [15]. Changes in the PEO processing parameters give rise to changes in the microdischarge characteristics,

WIT Transactions on Engineering Sciences, Vol 124, © 2019 WIT Press
www.witpress.com, ISSN 1743-3533 (on-line)
doi:10.2495/MC190011

which determine the thermal and chemical conditions on the oxidizing surface [16], and thus the structure of the resultant oxide coating. To produce the desired structure, it is now recognized that it is necessary to prompt a soft sparking transition where there is a sudden decrease in light and acoustic emissions, and drop in the anodic voltage when operating under a controlled current mode [7].

Several approaches have been employed for PEO process diagnostics including optical, spectral, electrical, acoustic, frequency response and electrochemical impedance spectroscopy [17]. Optical emission spectroscopy (OES) can provide valuable information on the plasma characteristics including electron number density, ionization temperature and elemental composition. This information is useful in understanding the coating formation process, and in the development of PEO and associated processes [18].

In the present study, the effect of current mode (DC, unipolar or bipolar) was investigated for the PEO processing of 1100 aluminum alloy in an alkali-silicate electrolyte. OES was used to characterize the plasma (electron concentration and temperature). The plasma characteristics, including sparking, are correlated both to the coating surface morphology, microstructure and composition, and to the corrosion resistance of the coated and uncoated alloy.

2 EXPERIMENTAL PROCEDURES

2.1 Material

A 1100 aluminium alloy was used for the study. The composition was Al-0.10% Cu-0.15% Fe-0.05% Si-0.05% Zn. Specimens with dimensions 25 mm x 25 mm x 5 mm were cut from the sheet. They were manually ground and polished with 240, 400, 600 and 1200 grid SiC papers to a surface finish of 0.1±0.02 µm. After cleaning with water, they were degreased with acetone prior to PEO processing.

2.2 PEO processing

A schematic of the PEO system is given in Fig. 1. The electrolyte used was an alkaline solution containing sodium silicate (Na_2SiO_3) at two levels of concentration (4 and 8 g/l) and potassium hydroxide (1 g/l KOH). The electrolyte was kept below 25°C by a water cooling system. The system was operated under a constant current control function in three different current modes (DC, unipolar, bipolar). Since the conductance of the oxide film decreases with coating thickness, the input voltage between anode and cathode must be increased to maintain the current density. Voltage versus PEO treatment time was recorded for each sample. Further details of the PEO system can be found in [19], [20].

2.3 OES and diagnostics

Since the discharges occur randomly, the OES measurements were made using an integrated signal over the whole sample surface. An Ocean Optics PC2000-UV-VIS fiber optic spectrometer, with a 200–1100 nm range was used. Six different spectral lines were recorded. Prior to the experimentation, the spectrometer was calibrated using two different laser beams: He-Cd (441.6 nm) and He-Ne (632.8 nm). Further details of the instrumentation can be found in [19].

The electron density can be determined through the Stark broadening of the Hβ line. Stark broadening is the broadening due to interaction with charged particles. The hydrogen lines

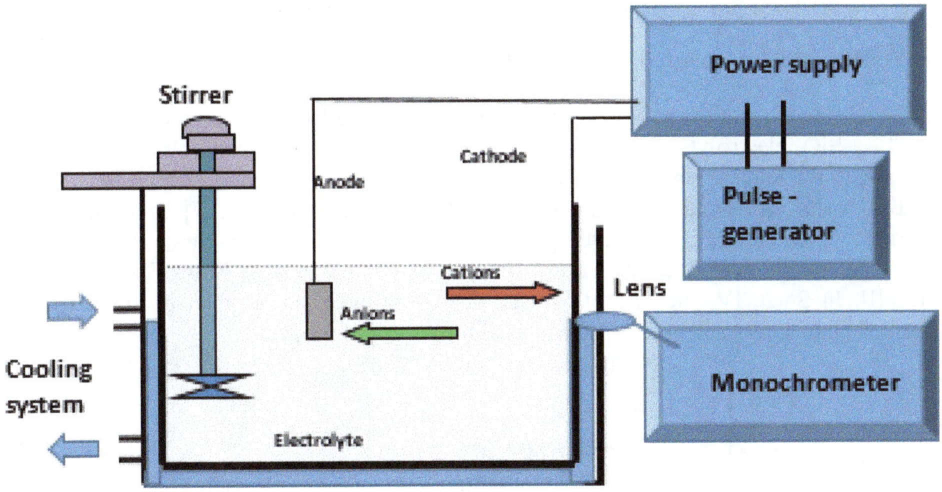

Figure 1: Schematic diagram of the PEO coating system.

Hβ are wide so that they are easy to record and process without perturbing the plasma. The PEO plasma falls into the category of an optically thin plasma, i.e. none of the observed lines emitted from the plasma were affected by self-absorption [21], [22]. For a hydrogen line profile, the relationship between full half-widths FWHM (full width at half maximum intensity) $\Delta\lambda_{1/2}$ in nm and electron concentration Ne in m^{-3} is given by [23]:

$$\Delta\lambda_{1/2} = 4.80\left(\frac{N_e}{10^{23}}\right)^{0.68116}. \qquad (1)$$

As a cross check, and to include the effect of electron temperature on the electron density calculation, the formula by Czernikowski and Chapelle [24], using the Stark broadening in the line at half-width ($\Delta\lambda_{1/2}$) was used where [25]:

$$\log Ne = c_0 + c_1 \log \Delta\lambda_{1/2} + c_2 (\log \Delta\lambda_{1/2})^2 + c_3 \log Te, \qquad (2)$$

with c_0=22.758; c_1=1.478; c_2= -0.144; and c_3 = -0.1265, $\Delta\lambda_{1/2}$ being given in nm, Te in K and electron density in cm^{-3}.

The plasma electron temperature Te, was determined from the relative intensities of spectral lines of the same atomic species from the same ionization stage as detailed in [19], [26].

2.4 Materials characterization

A FEI Quanta 200 FEG SEM equipped with a Kevex 5100c energy dispersive x-ray analysis detector (EDS) was used to characterize the coated specimens for surface morphology, coating thickness, integrity (cross-sectional samples) and chemical composition.

A Solartron 1285 Potentiostat with a Corrware Software was used for the potentiodynamic polarization corrosion testing. Corrosion testing was conducted in a 3.5% NaCl solution.

3 RESULTS AND DISCUSSION

3.1 Voltage variation during PEO processing

Fig. 2 presents plots of the anodic voltage vs treatment time for the three current modes with a 4 g/l Na_2SiO_3 electrolyte. Four stages can be seen in the PEO process [19]. In Stage I, the rapid rise in the voltage corresponds to a conventional aluminium anodizing process. Above a critical voltage, the anodic film undergoes dielectric breakdown [27] small sparks are generated which facilitates continued growth of the oxide film. This is Stage II, and the rate of voltage change with time decreases. The rate of voltage change slows down further in Stage III. In Stage IV, the rate of voltage change is even slower than in Stage III, and remained almost constant at 1 volt/min for all three current modes.

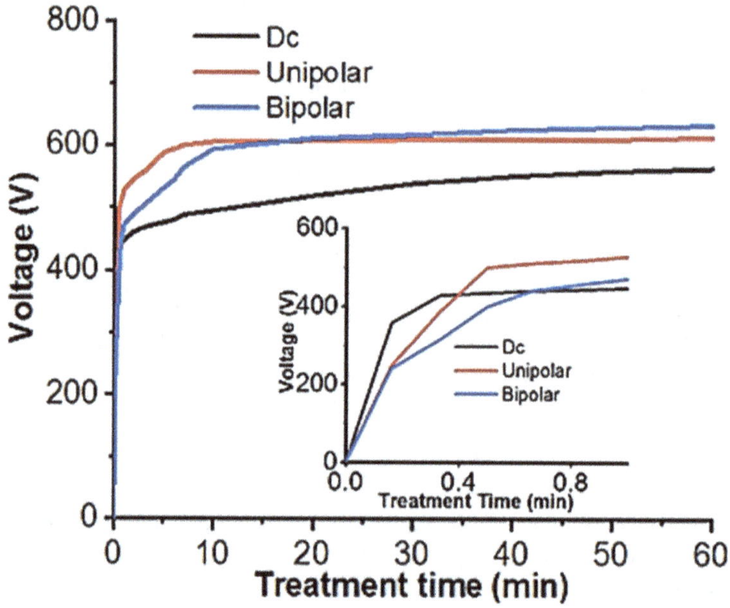

Figure 2: Plots of anodic voltage (V) vs. treatment time (t) for 4 g/l Na_2SiO_3 electrolyte using three different current modes. (Insert is an enlargement for the first minute.)

Table 1 summarizes the breakdown voltages and the time periods for the four stages. Stage 1, before breakdown, is very short (<20 seconds). Stage II, occurs for times up to 10 minutes. Stage III for times up to 40 minutes. Stage IV occurs for processing times up to 60 minutes, the maximum processing time investigated. The breakdown voltage was the highest (358 V) for the DC current mode: over 100 V higher than the unipolar or bipolar current. The magnitude of the breakdown voltage has conventionally been correlated to the conductivity of the electrolyte but recent work by Simchen et al. has shown that it depends on the physico-chemical properties of the anodic film, which is defined by the substrate material [28]. In the present tests, the substrate material was always 1100 aluminium alloy. If Simchen et al. [28] are correct, then it is suggested that the change in current mode changes "the physico-chemical properties of the anodic film".

Table 1: Breakdown voltages and durations of the four PEO stages for different current modes and a 4 g/l Na_2SiO_3 electrolyte.

Current mode	Stage I		Stage II	Stage III	Stage IV
	Breakdown voltage	Period (min)	Period (min)	Period (min)	Period (min)
DC	358	0–0.3	0.3–3.0	3–40	40–60
Unipolar	247	0–0.2	0.2–5	5–22	22–60
Bipolar	240	0–0.2	0.2–10	10–28	28–60

3.2 Plasma electron density and temperature

Fig. 3 presents a typical Lorentz profile of the experimental data of H_B for the unipolar current mode at a specific time (15 minutes from the start of the process), with a peak width (FWHM) of 1.45 nm. In the current experiments, FWHM values between 1.3 and 1.8 nm were recorded. Using both eqns (1) and (2), these half-widths correspond to a plasma electron concentration in a range of approximately $1.5 \pm 0.23 \times 10^{22}$ m^{-3} to $2.4 \pm 0.36 \times 10^{22}$ m^{-3} which are in good agreement with the plasma electron concentration results of Klapkiv et al. [29]. The error in measuring the electron concentration ($\delta N_e / N_e$) is determined to be ≈ 15% [26], which is a reasonable value in such experiments [26].

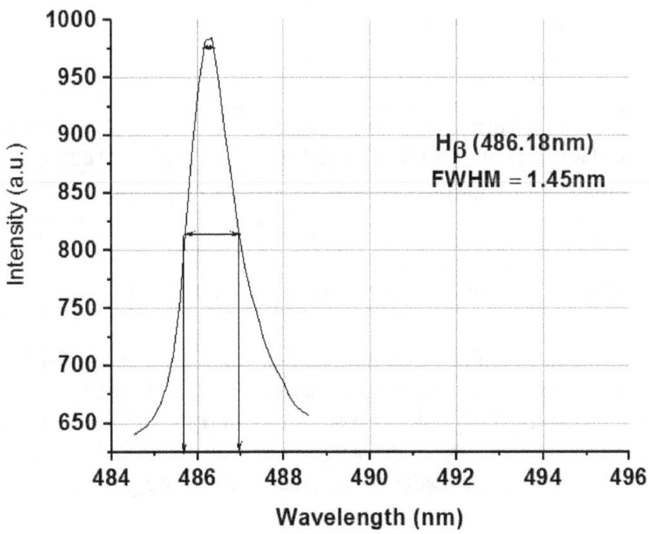

Figure 3: Typical results of a Lorentz profile of H_β.

When using spectral information to calculate parameters of the discharge, two assumptions are usually made: (1) The plasma is optically thin, which is the case for PEO [21], [22] and (2) There is local thermodynamic equilibrium (ETL). A partial LTE was expected where the electron density yield is given by [26]:

$$N_e \geq 7.0 \times 10^{24} \frac{Z^7}{n^{17/2}} (\frac{kT}{Z^2 E_H})^{1/2} (m^{-3}), \quad (3)$$

where kT is the electron temperature in (eV), E_H (eV) is the ionization potential for the corresponding element, z is the ionization stage (z =1 for H) and n is the principle quantum number of the lowest quantum level included in the partial LTE (n=2). The plasma electron concentration obtained in this work (1.5–2.4 $\times 10^{22}$ m^{-3}) is higher than the minimum plasma electron concentration for partial LTE, Ne > 0.4$\times 10^{22}$ m^{-3}, and therefore partial LTE exists, and the electron temperature estimated by Boltzmann assumption and Stark broadening effect is valid.

The intensity ratio of 396.2 nm (Al I) to 309.1 nm (Al I) was used to measure the plasma temperature (Te). Fig. 4 is a plot of the plasma temperature vs PEO processing time for the three current modes with an electrolyte concentration of 4 g/l Na_2SiO_3. Table 2 summarizes the baseline, minimum and maximum temperatures for the three current modes at the two electrolyte concentrations. Temperatures ranged from about 4,000 K to 10,000 K. The low temperature represent the early stage discharges. The strong peaks (6,000–10,000 K) correspond to relatively strong discharges. Application of a bipolar current mode, virtually eliminates the very strong discharges and the electron temperature rarely exceeds 6,000 K. Comparing that data in Table 2 for the two Na_2SiO_3 concentrations in the electrolyte (4 g/l and 8 g/l), it can be readily seen that the baseline temperatures are lower for 8 g/l electrolyte. This is likely related to the higher electrical conductivity of the electrolyte.

Table 2: Electron temperatures for different electrolyte concentrations (4 and 8 g/l Na_2SiO_3) and current modes.

Current mode	4 g/l Na_2SiO_3			8 g/l Na_2SiO_3		
	Base line Temp(K)	Minimum Temp (K)	Maximum Temp(K)	Base line Temp(K)	Minimum Temp(K)	Maximum Temp(K)
DC	5,200	3,900	8,500	5,000	3,900	8,600
Unipolar	5,600	3,900	10,000	5,100	3,950	8,300
Bipolar	5,400	4,600	6,000	4,800	4,200	5,700

It is interesting to compare the electron densities and temperatures for PEO to those in a lightning step leader, where the discharge channel is also optically thin and is in local thermodynamic equilibrium [30]. Electron densities of 10^{24} m^{-3} [30] and temperatures of ~14,000 K [30], [31] have been reported for lightning. The electron densities and temperature decrease as the lightning stepped leader propagates towards the ground.

3.3 Morphology, microstructure and chemistry of coatings

Fig. 5 presents SEM micrographs of the surface morphology of the coatings produced after 60 minutes treatment time. All samples showed a distinct "pancake" structure in which the hole at the center of the "pancake" is the discharge channel through which molten aluminium was ejected, sintered, and rapidly solidified by electrolyte. Fig. 5 are also shows micropores and cracks of different size, which helps to release residual stress in the coating. The size and density of the pores reflecting the strength of the discharges that produced the pores.

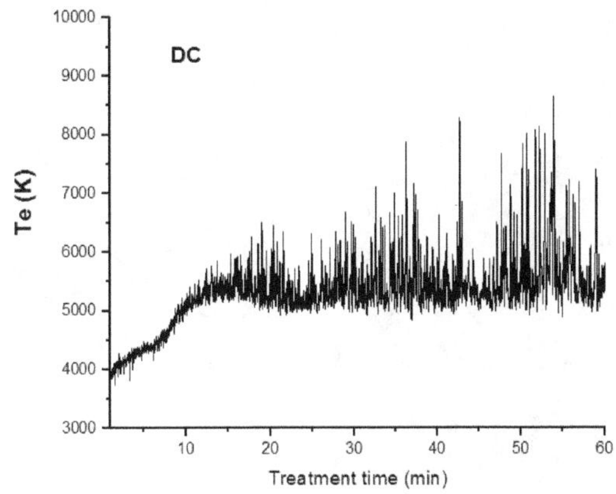

(a)

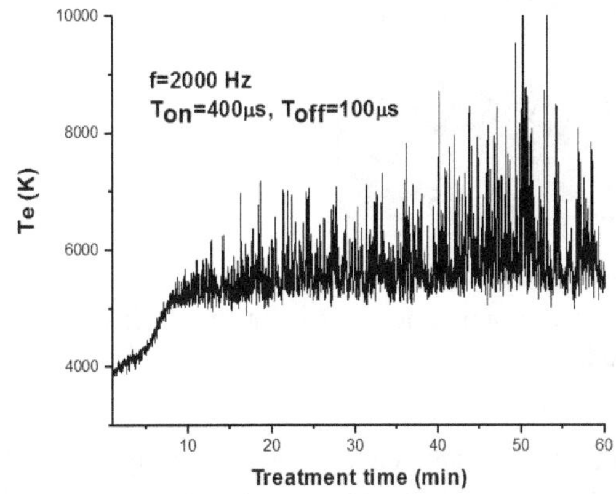

(b)

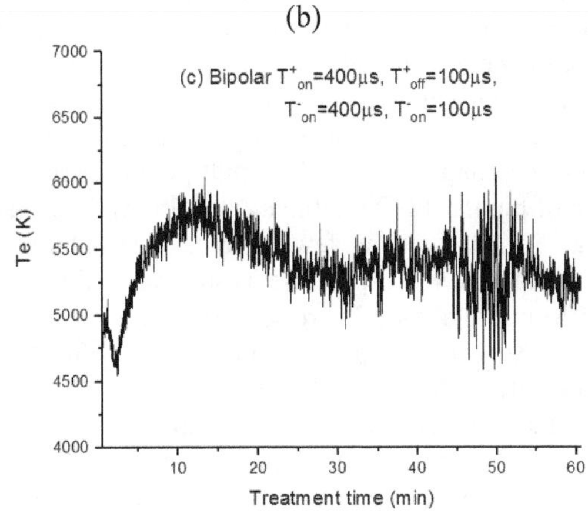

(c)

Figure 4: Plasma temperature as a function of treatment time for: (a) DC; (b) Unipolar; and (c) Bipolar current modes for 4 g/l Na_2SiO_3 electrolyte.

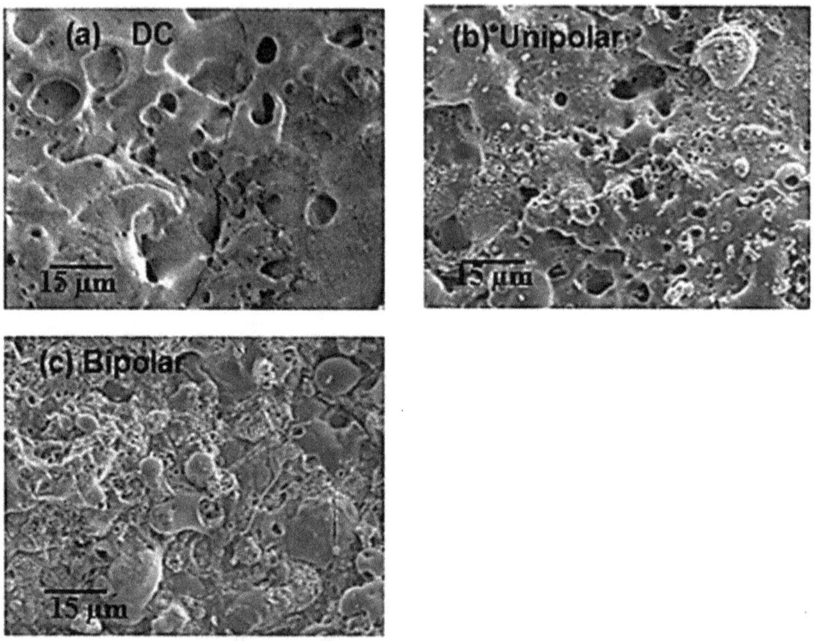

Figure 5: SEM micrographs showing surface morphology of oxide coating after 60 minutes treatment time on Al for: (a) DC; (b) Unipolar; and (c) Bipolar current modes.

The high spike temperatures generated by strong discharges melts the oxide and then traps gas into the melt pool, resulting in some porosity at the interface. SEM micrographs shown in Fig. 5(a) for DC and Fig. 5(b) for the unipolar mode show a number of relatively large pores in the coating. The diameter and features of the pancakes and pores can be considered as a guide to the strength of individual discharges and whether they were from the coating surface or from the interface. The deep holes suggest that the discharges are strong and penetrate deeply through the coating while, the small pores reflect relatively small surface discharges. For the case of bipolar PEO treatment, the flower-like shapes and smaller holes diameter shown in Fig. 5(c) as well as fewer pancake structures, indicates less strong discharges and a lower average base temperature as summarized in Fig. 4 and Table 2.

Fig. 6 is a typical SEM micrograph of a cross section of a coated Al alloy. PEO coatings consist of a three-layered structure: a porous outer layer with several large-sized, deep pores/cavities, an intermediate layer that is relatively dense that can provide good mechanical properties (e.g. wear) and a thin inner dense layer (coating/substrate interface barrier layer) that is well adhered to the substrate. It is the inner dense layer that eventually provides the best corrosion protection. However, due to the thickness of the interface barrier layer, only two layers are often reported when presenting SEM cross-sectional and micrographs [32].

Fig. 7 presents EDS spectra from positions 1 (porous layer), 2 (inner dense layer) and 3 (substrate) shown in Fig. 6. Due to the voids and pores in the coating outer layer, the silicon content is much higher than that in the inner region since the porous layer can trap the silicate electrolyte and allow silica to be formed after the plasma discharging. The inner layer

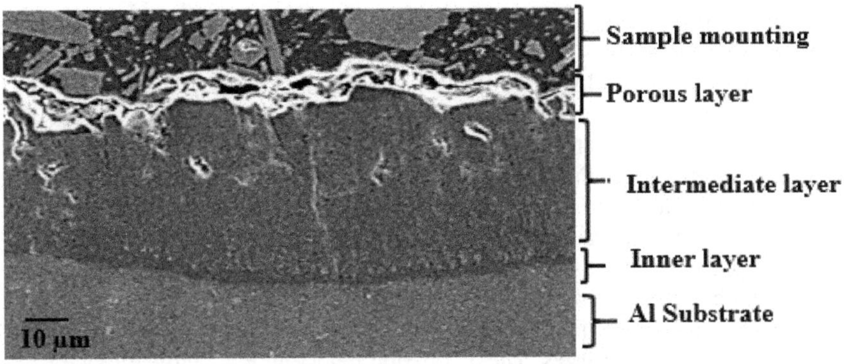

Figure 6: SEM micrograph of cross section of PEO coating on Al alloy produced using a bipolar current mode.

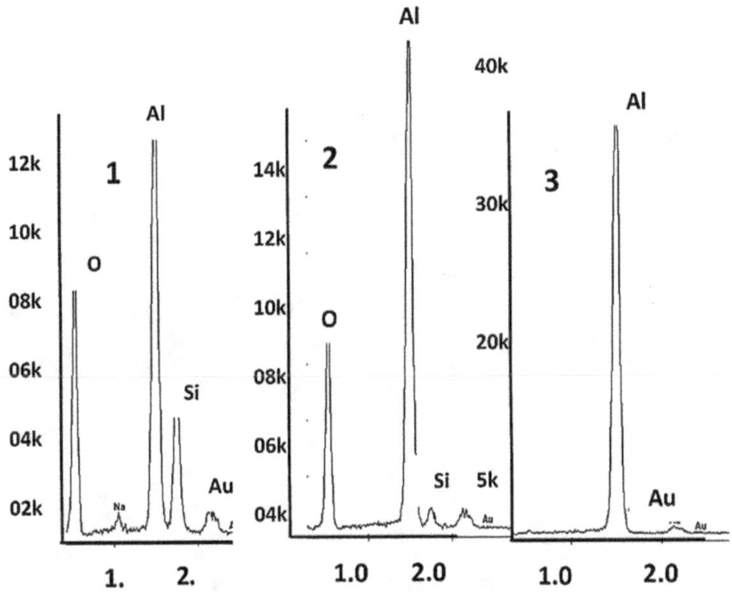

Figure 7: EDX analysis from (1) outer and (2) inner regions of the coated Al samples and for the substrate (region 3).

contains very little Si and no Na from the electrolyte. Aluminum and oxygen are the two major constituents of the inner layer of the coatings. An enhanced concentration of elements from the electrolyte in the outer, porous layer of the coating is commonly observed in the PEO processing of aluminium alloys [33].

Fig. 8(a)–(f) present cross-sectional SEM micrographs of the coatings for the three current modes, and treatment times of 15 and 60 minutes. The coating thicknesses after 60 minutes was much higher for the DC current mode (77.0±4.0 μm) than for the unipolar (47.0±2.3 μm) or bipolar (48.0±2.4 μm) modes. The higher oxide growth rate for the DC current mode comes at the expense of higher porosity within the coating, including near the oxide-metal interface (Fig. 8(a)). The coating for the unipolar current mode appears to be well bonded to the substrate but there is still significant porosity within the coating (Fig. 8(c)). The bipolar current mode, although it results in a similar thickness coating as the unipolar mode, produces a more dense coating with a third layer (coating substrate interface) clearly visible (Fig. 8(e)). These cross-sectional observations are consistent with what was observed both with the coating surface morphology (Fig. 5) and plasma temperatures (Fig. 4 and Table 2). The very high-energy (temperature) discharges produced with the DC current mode, lead to the "pancake" structure and discharge channels that extend to the oxide metal interface. This also leads to higher porosity. The discharges are moderated by use of a unipolar or bipolar current mode, giving higher quality coatings.

A bipolar current mode is particularly effective in preventing the development of long-lived, very large micro-discharges. The T_{off} period helps the local molten oxide cool down before another pulse is initiated. The T_{on} period provides sufficient time for sintering, and a hard coating with minimum porosity is produced [34].

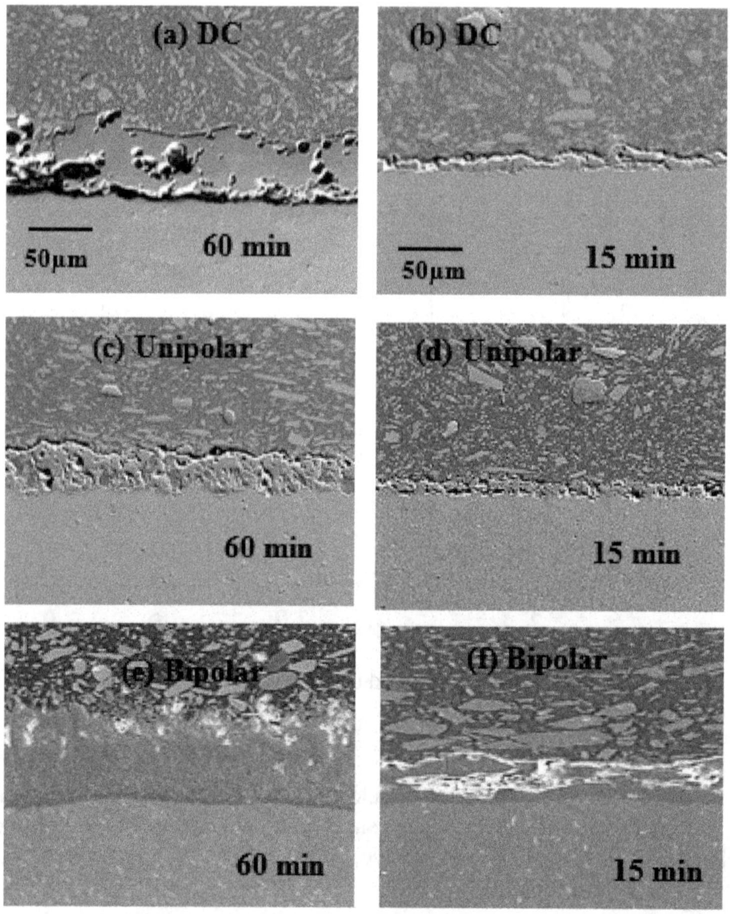

Figure 8: Cross-section micrographs of the oxide ceramic surface layers formed after 15 and 60 minutes treatment time and different current modes (electrolyte concentration 4 g/l Na_2SiO_3).

3.4 Corrosion testing

The potentiodynamic polarization curves of the uncoated 1100 alloy (S0) and the samples coated using DC (S1), unipolar (S2) and bipolar (S3) current modes are shown in Fig. 9. Values for the corrosion potential (E_{corr}), corrosion current density (i_{corr}), and polarization resistance (Rp) obtained from the curves in Fig. 9 are summarized in Table 3.

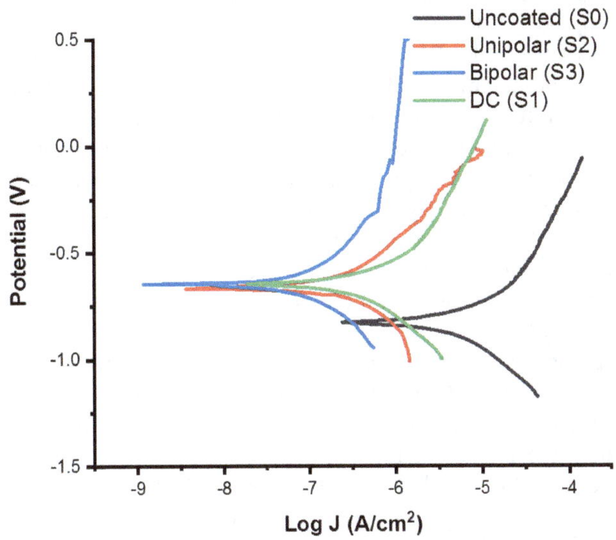

Figure 9: Potentiodynamic polarization curves of the uncoated (S0) and coated samples using DC (S1), unipolar (S2) and bipolar (S3) current modes.

Table 3: Potentiodynamic polarization corrosion test results in 3.5%wt NaCl solution.

Samples	**E_{corr} V**	**I_{corr} µA/cm²**	**R_p MΩ/cm²**
S0 Uncoated	-0.875	3.98	15.9
S1 DC	-0.723	0.177	166.0
S2 Unipolar	-0.72	0.098	367.5
S3 Bipolar	-0.675	0.036	1987.2

The data clearly show the enhanced corrosion resistance provided by the PEO coatings: more noble E_{corr}, reduced i_{corr} and higher Rp. Ranked in terms or their effectiveness as corrosion prevention coatings, they are: Bipolar > Unipolar > DC. This can then be related back to the ability to form a less porous, and more compact, microstructure as seen in the cross-sectional SEM micrographs in Fig. 8. The structure produced by a DC current mode gives the least corrosion resistance since there are pores at the coating-substrate interface (Fig. 8(a)). A bi-polar current mode produces a particularly effective corrosion resistant coating since there is a clear evidence for a thin interface barrier layer (Fig. 8(e)). As noted earlier, the structure of the PEO coatings can be related back to the nature and intensity of the micro discharges as seen in Fig. 4 and summarized in Table 2.

4 CONCLUSIONS

Three different current modes (DC, unipolar, bipolar) were used to produce oxide coatings on an 1100 aluminum alloy using PEO in an alkaline silicate electrolyte. Measurement and analysis of the visible and NUV wavelength emissions from plasmas using OES provided a valuable diagnostic tool for the understanding the processes in the PEO plasma. The plasma electron concentration was calculated to be from $1.5 \pm 0.23 \times 10^{22}$ m^{-3} to $2.4 \pm 0.36 \times 10^{22}$ m^{-3}, while plasma electron temperatures were found to be in the range of $4{,}500 \pm 450$ K to $10{,}000 \pm 1{,}000$ K.

The application of the pulsed bipolar current mode, compared to the DC or unipolar modes, reduced both the magnitude of the high temperature spikes and the average plasma temperature. The magnitude and frequency of these temperature spikes (micro-sparking) affected the microstructure, morphology and chemistry of the oxide coating.

In general terms, the PEO coatings on aluminum consisted of a three-layer structure: an outer porous layer, an inner dense layer, and a thin interface barrier layer. This interface layer was most evident for coatings produced using bipolar current mode. EDS studies showed that the Si, and to a lesser extent Na, originating from the electrolyte, were concentrated in the porous layer. The porous outer layer and the dense inner layer have been referred to as the "technological layer" and "working layer", respectively, to reflect the fact that it is the compact oxide in the inner region that provides the wear and corrosion resistance [33]. The DC current mode, and associated long-lived, very large microdischarges, produced a thicker porous outer region and, even, porosity at the metal-coating interface that would, and did, adversely affect the corrosion resistance.

As noted, the use of the bipolar current mode resulted in what is known as "softer sparking" [7] producing a less porous, more compact coating with smother surface and the highest corrosion resistance of the coatings fabricated using the three different current modes. In terms of their effectiveness in producing corrosion resistant coatings, the three current modes are ranked: Bipolar > Unipolar > DC.

REFERENCES

[1] Miller, W.S. et al., Recent development in aluminum alloys for the automotive industry. *Materials Science and Engineering,* **A280**, pp. 37–49, 2000.

[2] Rambabu, P., Prasad, N.E., Kutumbarao, V.V. & Wanhill, R.J.H., Aluminium alloys for aerospace applications. *Aerospace Materials and Material Technologies*, eds N. Eswara Prasad & R.J.H. Wanhill, Springer Science and Business Media: Singapore, pp. 29–52, 2017.

[3] Total Materia, Surface treatment of aluminium and aluminium alloys. www.totalmateria.com/Article67.htm. Accessed on: 16 Jan. 2019.

[4] Hanshan, D. (ed), *Surface Engineering of Light Alloys*, Woodhead Publishing: Cambridge, UK, 2010.

[5] Yerokhin, A. & Khan, R.H.U., Anodising of light alloys. *Surface Engineering of Light Alloys*, ed. H. Dong, Woodhead Publishing: Cambridge, UK, pp. 83–109, 2010.

[6] Yerokhin, A.L., Nie, X., Leyland, A., Matthews, A. & Dowey, S.J., Plasma electrolysis for surface engineering. *Surface and Coatings Technology*, **122**, pp. 73–93, 1999.

[7] Tsai, D.-S. & Chou, C.-C., Review of the soft sparking issues in plasma electrolytic oxidation. *Metals*, **8**(2), pp. 1–22, 2018. DOI: 10.3390/met8020105.

[8] Jedrusik, J., Debowska, A. & Kopia, A., Characterisation of oxide coatings produced on aluminium alloys by MAO and chemical methods. *Archives of Metallurgy and Materials,* **63**(1), pp. 125–128, 2018.

[9] Hussein, R.O., Nie, X. & Northwood, D.O., An investigation of ceramic coating growth mechanisms in Plasma Electrolytic Oxidation (PEO) processing. *Electrochimica Acta*, **112**, pp. 111–119, 2013.

[10] Jiang, B.L. & Wang, Y.M., Plasma electrolytic oxidation treatment of aluminium and titanium alloys. *Surface Engineering of Light Alloys*, ed. H. Dong, Woodhead Publishing: Cambridge, UK, pp. 110–146, 2010.

[11] Hussein, R.O., Nie, X. & Northwood, D.O., Influence of process parameters on electrolytic plasma discharging behaviour and aluminum oxide coating microstructure. *Surface and Coatings Technology*, **205**, pp. 1659–1667, 2010.

[12] Al Bosta, M.M.S. & Ma, K.-J., Suggested mechanism for the MAO ceramic coating on aluminium substrates using bipolar current mode in the alkaline silicate electrolytes. *Applied Surface Science*, **308**, pp. 121–138, 2014.

[13] Hussein, R.O., Northwood, D.O. & Nie, X., Coating growth behavior during the plasma electrolytic oxidation process. *Journal of Vacuum Science and Technology A*, **28**(4), pp. 766–773, 2010.

[14] Egorkin, V.S., Gnedenkov, S.V., Sinebryukhov, S.L., Vyaliy, L.E. & Gnedenkov, R.G.C.A.S., Increasing thickness and protective properties of PEO-coatings on aluminum alloy. *Surface and Coatings Technology*, **334**, pp. 29–42, 2018.

[15] Dehnavi, V., Surface modification of aluminum alloys by plasma electrolytic oxidation. Electronic Thesis and Dissertation Repository, 2311, 2014. https://ir.lib.uwo.ca/etd/2311.

[16] Rakoch, A.G., Khokhlov, V.V., Bautin, V.A., Lebedeva, N.A., Magurova, Y.V. & Bardin, I.V., Model concepts on the mechanism of microarc oxidation of metal materials and the control over this process. *Protection of Metals and Physical Chemistry of Surfaces*, **42,** pp. 158–169, 2006.

[17] Yang, K. & Cao, B., Electrical characteristics identification of dielectric film breakdown during plasma electrolytic oxidation process. *Materials Letters*, **143**, pp. 177–180, 2015.

[18] Maximov, A.I. & Khlustova, A.V., Optical emission from plasma discharge in electrochemical systems applied for modification of material surfaces. *Surface and Coatings Technology,* **201**, pp. 8782–8788, 2007.

[19] Hussein, R.O., Nie, X., Northwood, D.O., Yerokhin, A. & Matthews, A., Spectroscopic study of electrolytic plasma and discharging behaviour during the plasma electrolytic oxidation (PEO) process. *Journal of Physics D: Applied Physics*, **43**, p. 105203, 2010.

[20] Northwood, D.O., Hussein, R.O. & Nie, X., Environmentally friendly surface treatment of light alloy materials for automotive applications. *Proceedings of the 18th International Corrosion Congress*, Perth, Australia, pp. 825–836, 2011.

[21] Dunleavy, C.S., Golosnoy, I.O., Curran, J.A. & Clyne, T.W., Characterisation of discharge events during plasma electrolytic oxidation. *Surface and Coatings Technology*, **203**, pp. 3410–3419, 2009.

[22] Sullivan, S.A., Experimental study of the absorption in distilled water, artificial sea water, and heavy water in the visible region of the spectrum. *Journal of the Optical Society of America*, **53**, pp. 962–968, 1963.

[23] Gigososa, M., Gonzalez, M. & Cardenoso, V., Computer simulated Balmer-alpha, -beta and -gamma Stark line profiles for non-equilibrium plasmas diagnostics. *Spectrochim. Acta Part B*, **58**, pp. 1489–1504, 2003.

[24] Czernikowski, A. & Chapelle, J., Experimental-study of Stark-broadening of the argon-I 430.01 nm line. *Acta Phys. Pol. A,* **63**, pp. 67–75, 1983.

[25] García, M.C., Rodero, A., Sola, A. & Gamero, A., Spectroscopic study of a stationary surface-wave sustained argon plasma column at atmospheric pressure. *Spectrochimica Acta Part B: Atomic Spectroscopy*, **55**, pp. 1733–1745, 2000.
[26] Griem, H.R., *Principles of Plasma Spectroscopy*, Cambridge University Press: Cambridge, 1997.
[27] Hussein, R.O., Zhang, P., Northwood, D.O. & Nie, X., Improving the corrosion resistance of magnesium alloy AJ62 by a plasma electrolytic oxidation (PEO) coating process. *Corrosion and Materials*, **36**, pp. 38–49, 2011.
[28] Simchen, F., Sieber, M. & Lampke, T., Electrolyte influence on ignition of plasma electrolytic oxidation processes on light metals. *Surface and Coatings Technology*, **315**, pp. 205–213, 2017.
[29] Klapkiv, M.D., Nykyforchyn, H.M. & Posuvailo, V.M., Spectral analysis of an electrolytic plasma in the process of synthesis of aluminum oxide. *Material Science*, **30**(3), pp. 333–343, 1995.
[30] Cen, J. et al., Electron density measurement of a lightning stepped leader by oxygen spectral lines. *AIP ADVANCES*, **8**, p. 085019, 2018.
[31] Chang, X., Yuan, P., Gen, J. & Wand, X., Variation of the channel temperature in the transmission of lightning leader. *Journal of Atmospheric and Solar-Terrestrial Physics*, **159**, pp. 41–47, 2017.
[32] Safa-Yilmaz, M. & Sahin, O., Applying high voltage cathodic pulse with various pulse durations on aluminium via micro-arc oxidation (MAO). *Surface and Coatings Technology*, **347**, pp. 278–285, 2018.
[33] Sieber, M., Simchen, F., Morgenstern, R., Scharf, I. & Lampke, T., Plasma electrolytic oxidation of high-strength aluminium alloys substrate effect on wear and corrosion performance. *Metals,* **8**(5), p. 356, 2018.
[34] Hussein, R.O., Northwood, D.O. & Nie, X., The influence of pulse timing and current mode on the microstructure and corrosion behaviour of a plasma electrolytic oxidation (PEO) coated AM60B magnesium alloy. *J. Alloys and Compounds*, **541**, pp. 41–48, 2012.

EXAMINATION OF THE CARBONIZATION PROCESS USING KAOLIN AND SAWDUST

EMESE KUROVICS, BELLA UDVARDI, KRISZTINA ROMÁN,
JAMAL ELDIN F. M. IBRAHIM & LÁSZLÓ A. GÖMZE
Institute of Ceramic and Polymer Engineering, University of Miskolc, Hungary

ABSTRACT

Ceramics reinforced with carbon and silicon carbide particles are of great importance in the field of technical ceramics and ceramic composites. In this research, the possibilities of producing advanced ceramics from traditional, relatively inexpensive materials were studied. The authors have successfully developed mullite ceramic materials reinforced with carbon particles using kaolinite powder and sawdust. Powder mixtures with different compositions were prepared in a planetary ball mill using silica balls. The produced mixtures were compacted in a 100 kN pull-press to produce pellets using a pressure of 140 MPa. The prepared specimens were sintered in a laboratory chamber kiln under oxygen deficient conditions at 1250°C. After sintering of the specimens, considerable amounts of carbon content were incorporated into the material structure of produced ceramics, which were confirmed by energy-dispersive X-ray analysis. Using 10–20 m% sawdust in the mixtures, the carbon (C) content of sintered specimens was found to be more than 8 m%.

Keywords: carbon, carbonization, ceramics, kaolinite, mullite, sawdust.

1 INTRODUCTION

Mullite and mullite-containing ceramics play an important role in both the traditional and technical ceramics industries. In the case of traditional ceramic products, the mullite phase is usually formed by the thermal decomposition of kaolin, while in the case of technical ceramics the mullite phase is formed by the reaction of the free quartz (SiO_2) and aluminium oxide (Al_2O_3) contained in the product. The phase purity and morphology of the mullite depend on the precursor materials and the methods of processing [1], [2]. Fig. 1 shows the SiO_2-Al_2O_3 binary phase diagram.

Pure mullite with high density cannot be produced by conventional sintering technologies. Usually it contains up to 10 m% of glass phase material and about 10% average porosity. Mullite ceramics that contain a small amount of glass phase usually have relatively high strength, low thermal expansion, so they have good heat shock resistance [1]–[6]. They have a better creep resistance than aluminium oxide ceramics. The properties of mullite ceramics can be advantageously influenced using metal oxides produced by mullite activation mechanism [7].

Based on the SiO_2-Al_2O_3 two-component phase diagram, the mullite is formed at 1087°C. Changing the ratio of oxides can highly influence the melting point of the product [3]. Starting from kaolinite minerals, tridymite is also formed from the metakaolin in the well-known thermal decomposition process along with the mullite [3], [8], [9], further increasing of the temperature results on conversion of tridymite into silicon carbide in the presence of carbon [10].

$$SiO_2 + 3C \rightarrow SiC(s) + 2\,CO(g). \tag{1}$$

Stoichiometric silicon carbide consists of one silicon and one carbon per unit compound. Silicon carbide has more than 100 known polytypes, the most important and best-known ones are hexagonal, rhombohedral and cubic crystal types as shown in Fig. 2 [11]. The SiC is often used as a reinforcing phase of various composites due to its high melting point, low density

WIT Transactions on Engineering Sciences, Vol 124, © 2019 WIT Press
www.witpress.com, ISSN 1743-3533 (on-line)
doi:10.2495/MC190021

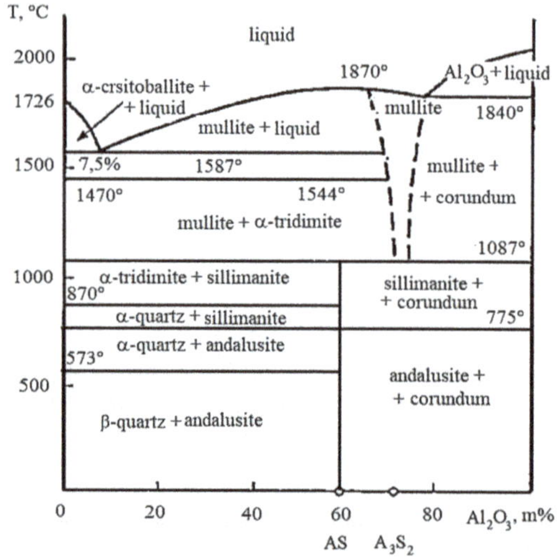

Figure 1: SiO_2-Al_2O_3 system at normal pressure [3].

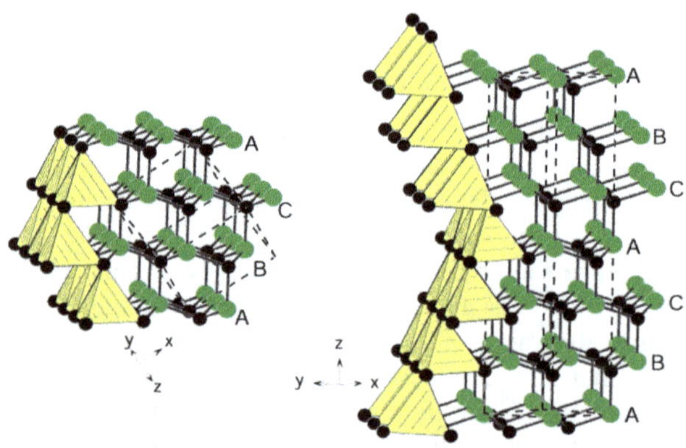

(a) (b)

Figure 2: (a) 3C cubic β-SiC; and (b) 6H hexagonal α-SiC [18].

and high mechanical strength [12]. The SiC particles incorporated into the ceramic matrix properly improve the mechanical properties of the material [13]. SiC has excellent mechanical strength, good chemical resistance, high thermal conductivity, low thermal expansion coefficient and good heat shock resistance. It is therefore ideally used as a material for abrasives, filters, catalysts, acoustic and thermal insulation equipment, as well as a structural material or interior coating for high temperature furnaces [14]–[18].

The future aim of the authors is to develop ceramic composites with increased physical and mechanical properties using relatively inexpensive processing technologies and raw materials like conventional kaolinite or other clay minerals. The goals of this research are to prepare mullite reinforced with SiC or carbon nanoparticles and understand the influence of vegetable organic additive (sawdust) on properties of ceramic products including colors, microstructures, chemical compositions as well as shrinkages.

2 MATERIALS AND EXPERIMENTS

Zettlitz Sedlecky ml kaolin and fine grain oak sawdust (additive) were used as starting raw materials. Fig. 3 shows their microstructures. A suitable amount of kaolin and sawdust were prepared according to Table 1. The produced mixtures were milled in Retsch PM 400 planetary ball mill for 20 min at 150 rpm. The milled mixtures were uniaxially compacted at 140 MPa pressures to produce cylindrical disc specimens with diameters of 25 mm using 100 kN mechanical pull-press. Ten specimens of each mixture were made with a weight of 10 g.

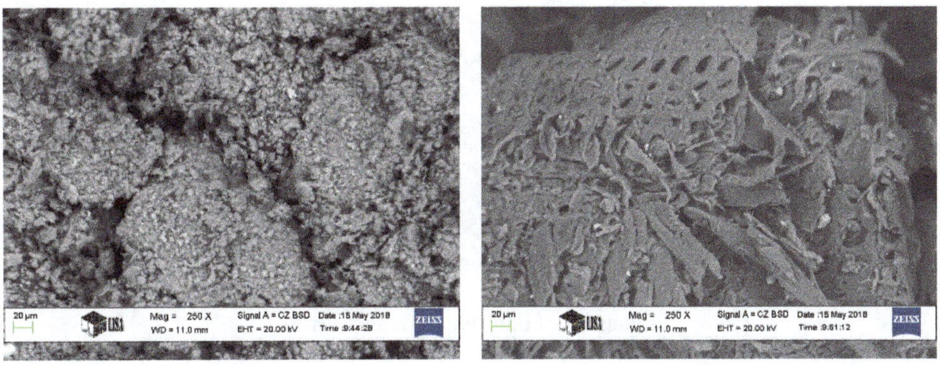

(a) (b)

Figure 3: The microstructure of the raw material powders. (a) Kaolinite; and (b) Sawdust.

Table 1: Composition of the mixtures used in the research in weight percent.

Mark of mixture	Quantity, m%	
	Kaolinite	Sawdust
A	100	0
B	95	5
C	90	10
D	85	15
E	80	20

The thermo-analytical properties of the powder mixtures were examined with a MOM Derivatograph-C. Raising the temperature, firstly leads to the drying of the sample, when all the moisture content of the mixtures leave. The next step is inflammation and burning of the additive (sawdust), which is most intensively carried out at 350–375°C, then, the thermal degradation of kaolin begins at 430°C. The kaolinite mineral ($Al_2O_3 \cdot 2SiO_2 \cdot 2H_2O$) produces metakaolin ($Al_2O_3 \cdot 2SiO_2$) when lose its crystal water [8], [9], [19]. This process is mostly occurred at 538°C. The process of mullite formation begins between 1000–1100°C.

After the thermo-analytical analysis and the compaction process the specimens were sintered in a laboratory chamber kiln at maximum temperature 1250°C and were kept at this temperature for 1 h. The sintering process of half of the specimens were made in a reduction environment in a sealed container.

3 RESULTS AND DISCUSSION

The geometrical parameters of the ceramic products are strongly depending on the material compositions and sintering conditions, like temperature, and sintering environment [5], [20], [21]. Due to sintering in different atmospheres, the colour of the ceramic specimens has changed as shown in Fig. 4.

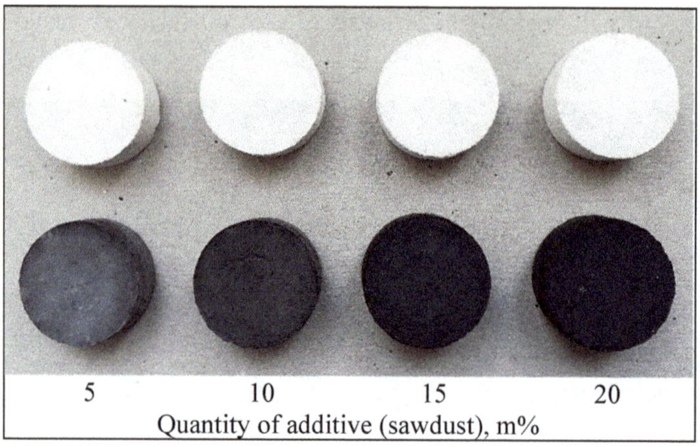

Figure 4: The specimens sintered in oxidation atmosphere (white) and reduction atmosphere (black).

The effect of the applied heat treatment methods can also be observed by examining the volume shrinkage and the weight loss due to sintering (Fig. 5). Volume shrinkage of the pure sintered mullite specimens in oxidation atmosphere was found to be about 30 V% which slightly decrease with increasing the amount of sawdust. The weight loss of the sintered product has increased proportionally with the sawdust content of the green product. Volume shrinkage of the sintered samples in the reducing atmosphere is significantly reduced by increasing the amount of sawdust. However, the change of weight loss varies according to the curve with a lower slope than in the case of oxidation sintering. The reason is that, the additive (sawdust) has not completely burned out from the product. In case of 20% sawdust, the weight loss of the product in reducing sintering decreased by 5% less than that of oxidation sintering.

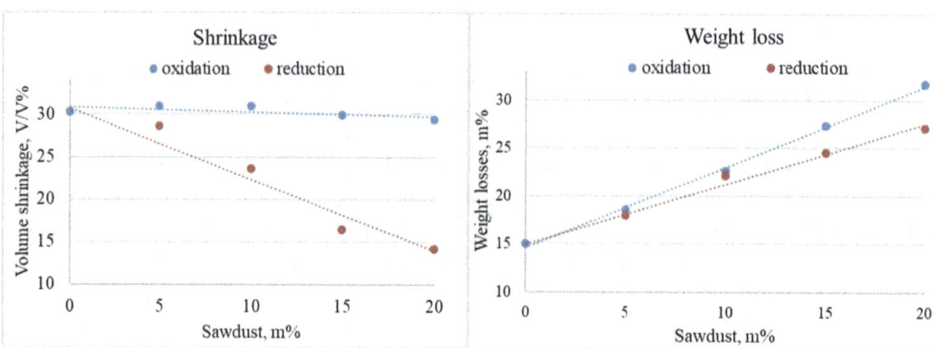

Figure 5: The volume shrinkage and the weight loss due to sintering.

The density of green-ceramic specimens was approximately 1.75 g/cm^3 which increased with oxidation sintering. Increasing sawdust content in all the cases reduced the density of the sintered product. The density of the produced ceramic samples can be seen in Fig. 6.

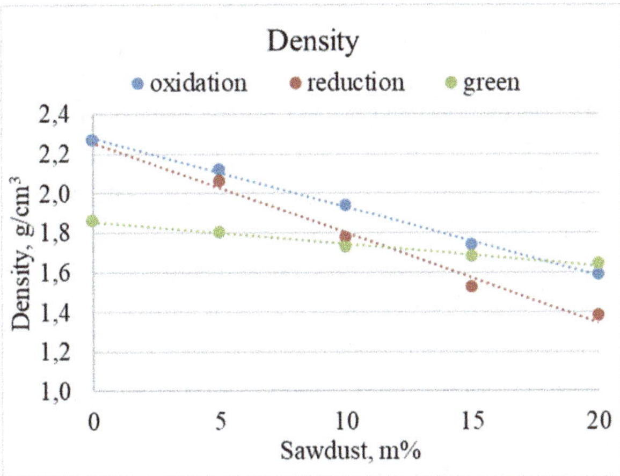

Figure 6: The densities of the produced mullite ceramics.

The microstructure of the powders and the sintered specimens were examined by Hitachi TM-1000 scanning electron microscopy. From the fracture surface investigation of the reduction sintered specimen made with 20 m%, it can be clearly seen that, sawdust is highly incorporated into ceramics structure (Fig. 7) [21].

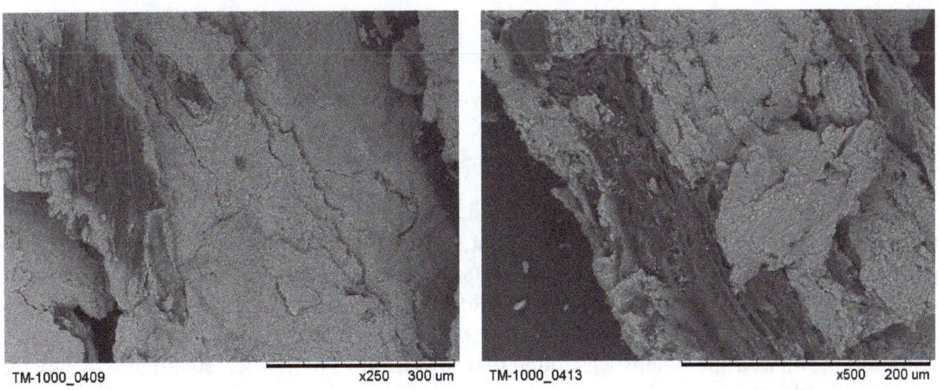

Figure 7: Fracture surface of the reduction sintered specimens with 20 m% sawdust.

The chemical compositions of the raw materials and the final ceramic products were examined by ZEISS EVO MA10 serviced with energy dispersive X-ray analysis (EDXA) micro-switch. Table 2 and Fig. 8 show the measured elemental compositions of the used raw materials and the fracture surface of the reduction sintered specimens. According to the data from EDXA, it can be observed that, increasing the amount of sawdust leads to increase the carbon content of the samples.

Table 2: Elemental composition of the used powders and fracture surface of the final products sintered in reduction atmosphere.

Element, m%	Powder		Fracture surface		
	Kaolinite	Sawdust	0 m% sawdust	10 m% sawdust	20 m% sawdust
C	–	62.15	–	8.58	14.89
O	45.36	37.25	38.80	34.56	28.85
Al	23.39	–	25.55	23.93	24.65
Si	29.14	–	32.18	28.80	27.81
K	0.83	–	1.04	1.05	1.38
Ca	0.47	0.60	0.49	0.61	0.47
Fe	0.81	–	1.94	2.46	1.95

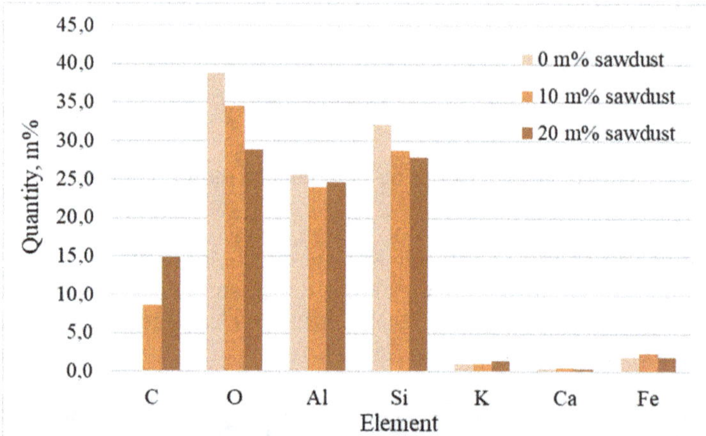

Figure 8: Elemental composition of the fracture surface of specimens of different mixtures sintered in reduction atmosphere.

4 CONCLUSIONS

From this research work we can conclude that, during the reduction sintering process a large amount of the vegetable organic additive (sawdust) was incorporated into the ceramic specimens. With increasing the amount of additive, large difference between the properties of ceramic specimens made by two sintering methods was noticed. The carbon (C) content of reduction sintered specimens was found to be more than 8 m% when 10–20 m% of sawdust was used in the mixtures. In this research work, the authors could successfully produce ceramic composites with increased properties using traditional kaolin and vegetable additives.

ACKNOWLEDGEMENTS

The described article was carried out as part of the EFOP-3.6.1-16-00011 "Younger and Renewing University – Innovative Knowledge City – institutional development of the University of Miskolc aiming at intelligent specialization" project implemented in the framework of the Szechenyi 2020 program. The realization of this project is supported by the European Union, co-financed by the European Social Fund.

REFERENCES

[1] Schneider, H., Schreuer, J. & Hildmann, B., Structure and properties of mullite: A review. *Journal of the European Ceramic Society,* **28**, pp. 329–344, 2008. DOI: 10.1016/j.jeurceramsoc.2007.03.017.

[2] Duval, J., Risbud, S.H. & Shackelford, J.F., Mullite. *Ceramic and Glass Materials*, eds J. Shackelford & R.H. Doremus, Springer, pp. 27–39, 2008. www.springer.com/978-0-387-73361-6.

[3] Bobkova, N.M., *Fizicheskaya Himiya Tugoplavkih Nemetallicheskih i Silicatnyh Materialov, Yvyshej Shaya,* Minszk, pp. 88–90, 2007.

[4] Moreno, E.A.X. & Scian, A.N., High purity mullite by slip casting method from calcined alumina and kaolinitic clay. *Procedia Materials Science,* **8**, pp. 245–250, 2015. DOI: 10.1016/j.mspro.2015.04.070.

[5] Kurovics, E., Shmakova, A., Kanev, B. & Gömze, L.A., Development ceramic composites based on Al_2O_3, SiO_2 and IG-017 additive., *IOP Conference Series: Materials Science and Engineering,* **175**, 012013, 2017. DOI: 10.1088/1757-899X/175/1/012013.

[6] Heimann, R.B., *Classic and Advanced Ceramics,* Wiley-VCH Verlag: Weinheim, 2010.

[7] Vakalova, T.V., Govorova, L.P., Reshetova, A.A., Tokareva, A.Y. & Shvagrukova, E.V., Activation of synthesis and sintering of mullite aluminosilicate ceramics based on natural raw materials. *Advanced Materials Research,* **1040**, pp. 268–271, 2014. DOI: 10.4028/www.scientific.net/AMR.1040.268.

[8] Gömze, L.A., Liszátzné Helvei, Á., Simonné Odler, A. & Szabó, M., *Ceramic Yearbook I,* ÉTK and MÉASZ: Budapest, pp. 30–85, 2001.

[9] Varga, G., The structure of kaolinite and metakaolinite. *Építőanyag – JSBCM*, **59**(1), pp. 6–9, 2007. DOI: 10.14382/epitoanyag-jsbcm.2007.2.

[10] Krstic, V.D., Production of fine, high-purity beta silicon carbide powders. *J. Am. Ceram. Soc.,* **75**(1), pp. 170–174, 1992. DOI: 10.1111/j.1151-2916.1992.tb05460.x.

[11] Izhezskyi, V.A., Genova, L.A., Bressiani, J.C. & Bressiani, A.H., Review article: Silicon carbide. Structure, properties and processing. *Cerâmica,* **46**, p. 297, 2000. DOI: 10.1590/S0366-69132000000100002.

[12] Gömze, A.L. & Lipusz, D., Korszerű SiC fejlesztése bioanyagok felhasználásával. *Anyagmérnöki Tudományok*, **37**(1), pp. 451–458, 2012. (In Hungarian.)

[13] Sternitzke, M., Review: Structural ceramic nanocomposites. *Journal of the European Ceramic Society,* **17**, pp. 1061–1082, 1997. DOI: 10.1016/S0955-2219(96)00222-1.

[14] Eom, J.-H., Kim, Y.-W. & Raju, S., Processing and properties of macroporous silicon carbide ceramics: A review. *Journal of Asian Ceramic Societies,* **1**(3), pp. 220–242, 2013. DOI: 10.1016/j.jascer.2013.07.003.

[15] Wu, S. & Claussen, N., Reaction bonding and mechanical properties of mullite/silicon carbide composites. *Journal of the American Ceramic Society*, **77**, pp. 2898–2904, 1994. DOI: 10.1111/j.1151-2916.1994.tb04521.x.

[16] Akpinar, S., Altun, I.A. & Onel, K., Effects of SiC addition on the structure and properties of reticulated porous mullite ceramics. *Journal of the European Ceramic Society*, **30**, pp. 2727–2734, 2010. DOI: 10.1016/j.jeurceramsoc.2010.05.005.

[17] Akpinar, S., Kusoglu, I.M., Ertugrul, O. & Onel, K., Silicon carbide particle reinforced mullite composite foams. *Ceramics International*, **38**, pp. 6163–6169, 2012. DOI: 10.1016/j.ceramint.2012.04.067.

[18] López-Álvarez, M., Serra, J., de Carlos, A. & González, P., Marine-based carbon and silicon carbide scaffolds with patterned surface for tissue. *Engineering Applications*, 2011. DOI: 10.5772/22046.

[19] Štubňa, I., Varga, G. & Trník, A., Investigation of kaolinite dehydroxylations is still interesting. *Építőanyag – JSBCM*, **58**(1), pp. 6–9, 2006. DOI: 10.14382/epitoanyag-jsbcm.2006.2.

[20] Kurovics, E., Buzimov, A.Y. & Gömze, L.A., Influence of raw materials composition on firing shrinkage, porosity, heat conductivity and microstucture of ceramic tiles. *IOP Materials Science and Engineering*, **123,** 012058, 2016. DOI: 10.1088/1757-899X/123/1/012058.

[21] Kurovics, E., Kulkov, S.N. & Gömze, L.A., Investigation of ceramic brick rods with blackened materials inside. *Építőanyag – JSBCM*, **70**(1), pp. 3–7, 2018. DOI: 10.14382/epitoanyag-jsbcm.2018.1.

STUDY OF CARBONATION DURABILITY OF SEVERAL ETTRINGITE-ENRICHED PASTES

BAO CHEN[1,2], MATTHIEU HORGNIES[2], VINCENT MORIN[2], MOUNA BOUMAAZA[2], BRUNO HUET[2], EDOUARD GENGEMBRE[2], ISABELLE BACO[2], KEVYN JOHANNES[1] & FREDERIC KUZNIK[1]
[1]Université de Lyon, France
[2]LafargeHolcim R&D, France

ABSTRACT

Recent reported materials of high energy density for thermochemical heat storage are mostly inorganic rare elements salts, such as $SrBr_2$ and $LaCl_3$. However, their wide use may be hindered due to the high material cost. An affordable cement-based compound, named ettringite, of which energy density (~500 kWh/m^3) is comparable, and even higher, than rare elements salts and therefore attracts people's sight. As an alkaline phase in hydrated cement pastes, ettringite could be carbonated by CO_2 in the air. The change of this mineral structure upon carbonation could reduce its capacity to store energy. This phenomenon is known to be unstoppable until there is a complete depletion of reactants. Consequently, it is necessary to investigate the carbonation rate of ettringite-based materials for energy storage. In the present work, a large range of characterization methods – such as optical microscopy analysis, mercury intrusion porosimetry (MIP), X-ray fluorescence (XRF), X-ray diffraction (XRD), Thermogravimetric analysis (TGA) – have been carried out to analyse phase structure and phase assemblage. Three series of cement pastes containing different contents of ettringite were firstly prepared. Then, crushed samples of a size of 1–2mm were exposed to an accelerated carbonation environment (composed of 1% CO_2 by volume, and 90% relative humidity (RH) at 20°C and ambient pressure) for different durations. The results have highlighted that the carbonation of ettringite-based granules under set conditions went very fast. The XRD patterns confirmed that the calcium carbonates were vaterite and aragonite accompanied with gypsum. Moreover, a bigger content of ettringite seemed to yield a higher carbonation resistance and a less foamy structure.

Keywords: carbonation, durability, ettringite-based material, TGA, XRD, energy storage.

1 INTRODUCTION

Nowadays, the primary energy consumption in building or industry sector is more significant than petrol depletion in transport sector all over the world. Besides, a local/individual secondary energy supply system is assumed to be efficient to reduce the operating burden on national grids during heating consumption peaks in winter. Therefore, thermal energy storage systems are developed to reduce the load peak of consumption of electricity. Recent investigations [1]–[5] on a thermochemical material named ettringite proposed a good solution to reduce the cost of construction of backup system. The energy density can reach up to 500 kWh/m^3 for a working temperature as low as 60°C, which guarantees its adaption to most renewable heat sources [1]. However, as an alkaline material from hydrated cement paste, its reactivity with CO_2 in air is unavoidable in heat storage systems. Indeed, carbonation can affect the crystal structure of ettringite to several simple chemical compounds thereby losing its capacity to store heat.

Robl et al. [6] reported that dry ettringite samples did not react significantly with CO_2 while humid ettringite reacted rapidly. Pajares et al. [7] reported the harmful effect of carbonate ions on ettringite's stability, from $CaCO_3$ and $MgCO_3$ silica gel saturated solution, at ambient temperature. However, the presence of $Ca(OH)_2$ could evidently delay the carbonation of ettringite, since the capacity of Portlandite ($Ca(OH)_2$) to buffer the carbonation reaction prevents significant deterioration of the remaining phases. The same phenomenon caused by carbonate ions at different temperatures was equally investigated [8].

WIT Transactions on Engineering Sciences, Vol 124, © 2019 WIT Press
www.witpress.com, ISSN 1743-3533 (on-line)
doi:10.2495/MC190031

Grounds et al. [9] described the carbonation rate of a synthetic ettringite powder in atmospheric CO_2. The carbonation reaction was inferred by X-ray diffraction (XRD) as:

$$3CaO{\cdot}Al_2O_3{\cdot}3CaSO_4{\cdot}32H_2O + 3\ CO_2 \rightarrow 3\ CaSO_4{\cdot}2H_2O + 3\ CaCO_3 + 2\ Al(OH)_3 + 23\ H_2O \quad (1)$$

Nishikawa et al. [10] confirmed this carbonation reaction of ettringite powder at high relative humidity (RH). Detailed works by Chen et al. [11] stated that the increase of CO_2 pressure, water vapour and temperature could accelerate the carbonation of ettringite. Zhou and Glasser [12] also reported the acceleration of carbonation with both increasing relative humidity and temperature via carbonation experiments on ettringite powder and pellets (2.5mm thick, pressed to about 80% theoretical density). Differing from powder cases, AFm phase was detected due to a possible condensation of water in the pores and inter-crystalline space in pellet samples. Ndiaye et al. [4] demonstrated that the carbonation of ettringite in CSA paste powder yielded to bassanite ($CaSO_4{\cdot}0.5H_2O$) instead of gypsum accompanied by aragonite and vaterite as $CaCO_3$ at 65% RH.

Most previous research about the carbonation of ettringite were based on synthetic powders which may cause agglomeration during energy storage processes. Consequently, our investigations were carried out on grains (1–2mm) of hydrated pastes that were confirmed to contain high quantities of ettringite. In this paper, we detail our carbonation experiments surveying accelerated carbonation (1 vol. % CO_2) on three mixtures from Belite-Calcium Sulfoaluminate-Ferrite Cement (BCSAF), Calcium Aluminate Cement (CAC) and Ordinary Portland Cement (OPC) via main methods like optical microscopy, mercury intrusion porosimetry (MIP), X-ray fluorescence (XRF), X-ray diffraction (XRD) and Thermogravimetric analysis (TGA). Meanwhile, the evolution of components and according quantity as a function of carbonation time is discussed.

2 MATERIALS AND METHODS

2.1 Raw materials

The tested cement pastes were firstly characterized by XRF and XRD to determine the initial existing phases and quantities. The results are demonstrated in Table 1. The sums of different columns inferior to 100 are related to unlisted minor phases.

2.2 Preparation of the tested samples

Three binders used result in around 565 g anhydrous mixture respectively from pure BCSAF (abbreviated as BCSAF), 80% CAC and 20% OPC (abbreviated as C80P20), and 60% CAC and 40% OPC (abbreviated as C60P40). Then the mixtures were particularly hydrated by demineralized water of 1.1 × anhydrous mixture, a little of Chryso® Premia 180 (0.05 wt. % of binder) and Kelco-Crete® DG-F (0.1 wt. % of water). After a correct agitation by a mixer machine for 10 min at 1000 r/min, the pastes were sealed in polystyrene moulds (4 × 4 × 16cm) by a glass plate for 24 h. At 1 d, harden pastes were then stored in sealed plastic bags at about 22°C for 27 d to avoid carbonation. At last, the prims were crushed down to 1–2mm grains. Meanwhile, several pieces of harden pastes were characterized by MIP (Autopore IV of Micromeritics, USA). The mean size of grains was analysed by optical microscopy (Stereoscopic magnifier STEMI SV 11).

Table 1: Chemical and mineralogical composition of cements by XRF and XRD[1].

	OPC	CAC	BCSAF		OPC	CAC	BCSAF
SiO_2	20.3	1.9	12.4	C_3S	70.7	-	-
Al_2O_3	5.4	23.4	16.9	C_2S	7.3	-	30.9
Fe_2O_3	2.5	0.5	6.3	C_4AF	7.4	-	16.4
CaO	64.2	43.2	47.3	C_3A	8.5	-	-
MgO	1.2	0.3	1.4	$C\bar{S}H_{0.5}$	3.2	-	3.7
K_2O	0.2	0.1	0.7	$C\bar{S}H_2$	2.6	-	1.2
Na_2O	0.2	0.1	0.1	$C\bar{S}$	-	52.5	10.7
SO_3	3.3	28.8	9.9	$C_{12}A_7$	-	3.5	-
TiO_2	0.2	1.1	0.9	$C_4A_3\bar{S}$	-	-	31.1
Mn_2O_3	0.1	<0.1	0.1	MgO	-	-	0.5
P_2O_5	0.3	0.1	0.1	Fero-perovskite	-	-	3.6
Cr_2O_3	<0.1	<0.1	<0.1	Ettringite	-	-	1.5
ZrO_2	<0.1	<0.1	<0.1	Monosulfoaluminate	-	-	0.3
SrO_2	0.1	<0.1	0.1	Amorphous phase	-	44.0	-
Total	98.2	99.9	96.4		99.7	100.0	99.9

[1]Cement notions: A = Alumina (Al_2O_3), C = Lime (CaO), H = Water (H_2O), F = Iron oxide (Fe_2O_3), S = Silica (SiO_2), $\bar{S}$ = Sulphur trioxide (SO_3).

2.3 Accelerated carbonation and solid analysis

About 4 g grains of each carbonation date were evenly paved in an aluminium dish of 8cm. The samples carbonated in a CO_2 climate chamber (Pharma 600 of Weisstechnik®) where RH, CO_2 concentration and temperature were maintained at 90%, 1 vol. % and 20°C for 0, 2 h, 6 h, 1 d, 11 d and 28 d. For each carbonation period, the samples were analysed by XRD (Philips / PANalytical X'Pert Pro-MPD coupled with X'Celerator detector of incident CuKα radiation beam by 40 kV and 40 mA, 2θ = [5–65°]) and TGA (Metller TGA-DSC 3^+, from 30 to 1000°C at 10°C/min under protection of 20 mL/min nitrogen) to determine carbonation products and quantity of absorbed CO_2.

2.4 TGA and XRD data processing

The temperature of releasing CO_2 is supposed to be in the range 550–1000°C. The maximum of bonded CO_2 was called the CO_2 binding capacity. For the variation of mass caused by water loss during carbonation, seeing eqn (1), the indicator of carbonation was defined as:

$$I = \frac{m_{CO2}}{m_{cement}} = \frac{m_{initiale\,TGA} * (a_{550\,°C} - a_{1000\,°C})}{m_{initiale\,TGA} * a_{1000\,°C}}, \tag{2}$$

where I is the carbonation degree, m is the weight from TGA and a is the weight percentage from TGA.

Carbonation was hard to avoid during preparation of samples. The absolute carbonation degree for accelerated carbonation was calculated as the following:

$$I_r = I_t - I_0, \tag{3}$$

where I_t is the carbonation degree at time t, I_0 is the initial carbonation degree before accelerated carbonation.

The analysis results of XRD were equally recalculated with TGA results, seeing eqn (4)

$$X = \frac{X_{XRD} * m_{initial\,TGA}}{m_{cement}} = \frac{X_{XRD} * m_{initial\,TGA}}{m_{initial\,TGA} * a_{1000\,°C}}, \tag{4}$$

where X is the recalculated percentage of components, X_{XRD} the original weight percentage from XRD, m and a are respectively the weight and weight percentage in TGA.

2.5 Maximum CO_2 binding capacity

According to the XRF results, the maximum carbonation degree (CO_2 binding capacity) of cement paste could be computed using eqn (5) under the following assumptions:

- $CaSO_4$ is the only final formula of sulphates;
- CaO and MgO are the main phases to be carbonated

$$I_{max} = \sum \frac{m_{CO2}}{m_{mixture}} = \sum \frac{\left(\frac{a_{CaO}}{M(CaO)} + \frac{a_{MgO}}{M(MgO)} - \frac{a_{SO3}}{M(SO3)}\right) * m_{cement} * M(CO_2)}{m_{anhydrous\,solid}}, \tag{5}$$

where I is the carbonation degree, m is the weight (g), a is the percentage and M is molar mass of compounds (g/mol). Along with eqn (5), the theoretical CO_2 binding capacities are around 25% for C80P20, 31% for C60P40 and 33% for BCSAF.

3 RESULTS AND DISCUSSIONS

3.1 Characterizations of reference binders

The grains from different mixtures generally range in 1–2mm by sieving. More precisely, the mean sizes are 1.04mm, 0.98mm and 1.13mm for C80P20 C60P40 and BCSAF respectively given by optical microscopy (Fig. 1).

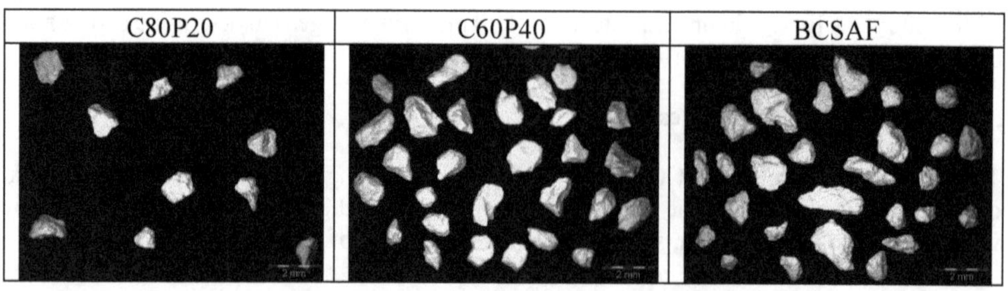

Figure 1: Optical microscopy analysis of grain size (1–2mm) for three binders' pates.

The grains of three hydrated binders were also characterized by MIP to determine their porosity and pore size distribution before exposure to carbonation. As expected, due to their high water to binder ratio, results from Fig. 2 show high porosities of the tested samples (above 30%). Generally, they are in the order of BCSAF (~53%) > C60P40 (~38%) > C80P20 (~34%). This order is in contrast with their ettringite contents indicated by XRD. Actually, C80P20 contains ~73% ettringite while ~65% for C60P40 and only about 40% for BCSAF.

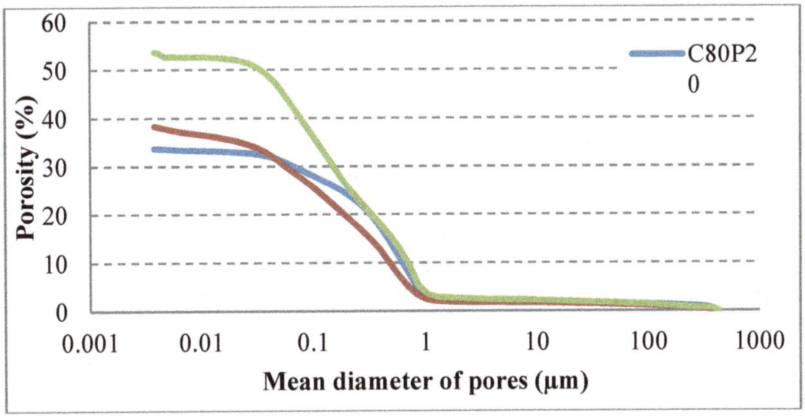

Figure 2: Porosity of the three hydrated binders by MIP.

Therefore, the growth of a big quantity of ettringite crystals into pores may reduce the quantity and size of pores. For the three hydrated binders, most pores range between 0.03–1 μm and they get less porosity from micro-pores with a diameter inferior to 0.01 μm or big pores with a diameter superior to 1 μm. Hydrated BCSAF is the only one with significant pore volume for pore diameter below 0.1 μm. These pores behave as water containers during carbonation.

3.2 Evolution of hydrated binder grains upon accelerated carbonation

At each set date of carbonation, the corresponding samples were analysed by TGA and XRD techniques. The error of thermal analysis is estimated about 1%. The TGA results are demonstrated in Fig. 3. The "turning points" of curves for C60P40 and BCSAF are at 2 h while at 6 h for C80P20. This trend may indicate that the carbonation of particles could be separated to two periods: before turning point, carbonation is dominated by chemical reaction at surface of grains; after a turning point, carbonation is a diffusion-controlled process.

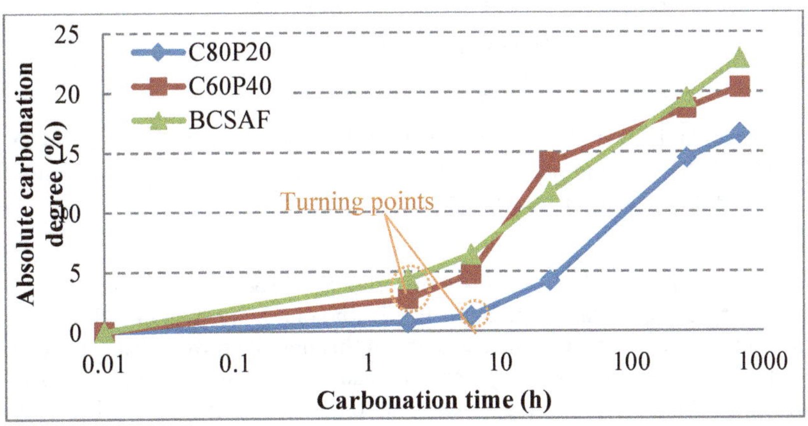

Figure 3: Carbonation degrees of three binders' grains (1–2mm) by TGA at different carbonation dates.

Among the three pastes, BCSAF seems to be the least resistant against carbonation while C80P20 is the most resistant. However, the difference between C60P40 and BCSAF is not high, which may be explained by their CO_2 binding capacities and porosities. Because of a higher porosity of BCSAF paste comparing to C60P40, the diffusion of CO_2 is supposed to be faster in BCSAF than in C60P40. However, this advantage may not be so significant due to that both of the grains have large pores and entry diameter (~1 μm) resulting in less influence led by CO_2 diffusion. Compared to BCSAF and C60P40, not only the carbonation rate but also the carbonation degree of C80P20 is less significant. This result comes out under both effects from relatively small porosity and CO_2 binding capacity. Typically, at 1 d, the absolute carbonation degree of C80P20, C60P40 and BCSAF are 4.2%, 14.1% and 11.6%, respectively reaches 16.8%, 45.5% and 35.2% of theoretical CO_2 binding capacities.

Similarly to TGA data, the change of ettringite content in C80P20 grains during first 6 h is low (Fig. 4). The little change of $CaCO_3$ quantity (V+A in Fig. 4) during first 6 h may suggest the formation of transition product containing CO_2. Then the carbonation of major ettringite before 11 d increases significantly the quantity of vaterite and aragonite, which corresponds to a large increase of carbonation degree in TGA curves from 6 h to 11 d. After 11 d, no ettringite exists yielding to no increase of gypsum content. The growth of $CaCO_3$ comes from the carbonation of a part of other hydrates in the grains. Here, the main $CaCO_3$ formed by carbonation of ettringite is equally confirmed as vaterite and aragonite since that hardly any calcite forms during whole process. Meanwhile amorphous aluminium hydroxyl is also detected during the carbonation of ettringite.

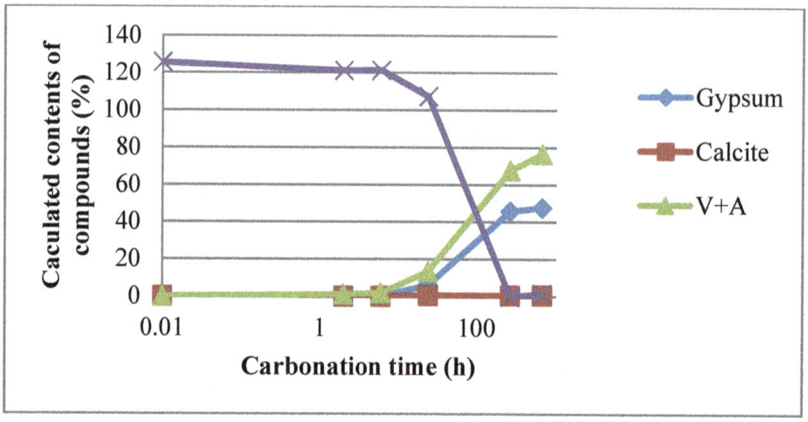

Figure 4: Mineralogical evolution and calculated contents of compounds by XRD for C80P20 1–2mm grains at different dates of carbonation. V+A presents the total content of vaterite and aragonite.

As for C60P40 (Fig. 5), the mechanism of carbonation seems to be different. First, a new hydrate containing CO_3^{2-} named calcium hemicarboaluminate hydrate (Hc) exists before carbonation experiment, eqn (6). The existence of this hydrate proves that some carbonation occurred during the preparation of grains. Monosulfoaluminate was equally found before experiments. At the very beginning of carbonation (before 6 h), a remarkable amount of ettringite in C60P40 is decomposed, firstly to monosulfoaluminate. On the one hand, this decomposition of ettringite may occur under CO_2 influence along with releasing of sulphates and calcium ions into pore solution. On the other hand, unreacted anhydrous phases can also react with ettringite to form monosulfoaluminate. Further, CO_3^{2-} produced by dissolved CO_2 in pore solution bonds calcium ions to form $CaCO_3$. After 6 h of carbonation, the pH of pore solution possibly reduces significantly which makes ettringite, monosulfate and Hc carbonated to $CaCO_3$ directly. At 24 h, more than half of ettringite is carbonated while no monosulfate remains. This result possibly indicates that monosulfate is more reactive to CO_2 than ettringite. The rate of consuming ettringite is about 3 times higher for C60P40 than C80P20 at 24 h given by XRD. After the complete depletion of ettringite at 11 d, the increased quantity of vaterite and aragonite comes from the carbonation of other hydrates in the C60P40 grains like strätlingite and katoite.

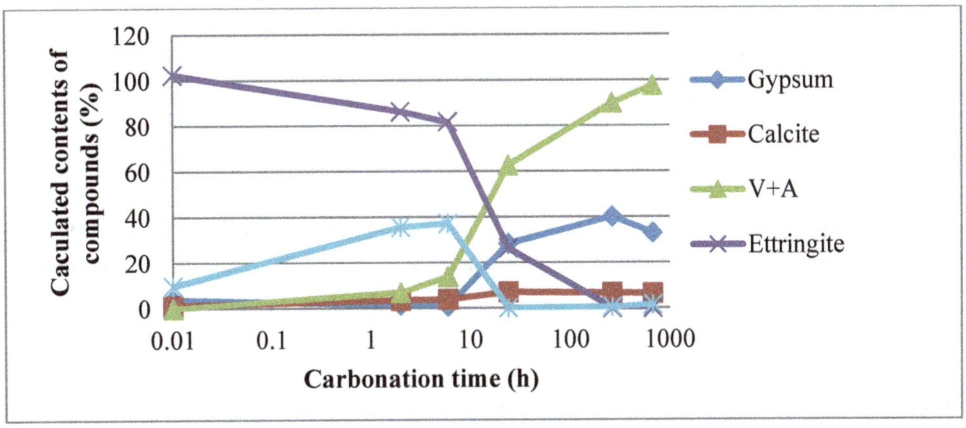

Figure 5: Mineralogical evolution and calculated content of compounds by XRD for C60P40 1 – 2mm grains at different carbonation periods. V+A: the total content of vaterite and aragonite.

$$3C_3A + 3CaSO_4 + 2Ca(OH)_2 + CO_2 + 55H_2O \rightarrow 3CaO\cdot Al_2O_3\cdot 3CaSO_4\cdot 32H_2O + 2\ 3CaO\cdot Al_2O_3\cdot 1/2CaCO_3\cdot 1/2Ca(OH)_2\cdot 11.5H_2O \quad (6)$$

Being consistent with TGA results, BCSAF demonstrates the highest $CaCO_3$ content in grains than other two other mixtures at 28 d in XRD patterns (Fig. 6). During first 2 h of carbonation, ettringite is scarcely carbonated. Therefore, the rise of $CaCO_3$ may be owing to the carbonation of amorphous hydrate CSH. Then ettringite becomes to be the major phase carbonated. Because of owning relative low quantity of ettringite in comparison with C80P20 and C60P40, almost all ettringite react at 24 h. Similarly to C80P20, once ettringite is totally consumed after 11 d, the increase of gypsum is stopped considering measurement error. After that, the increase of $CaCO_3$ is mainly related to the carbonation of other hydrates like strätlingite and katoite without producing any sulphate bearing minerals.

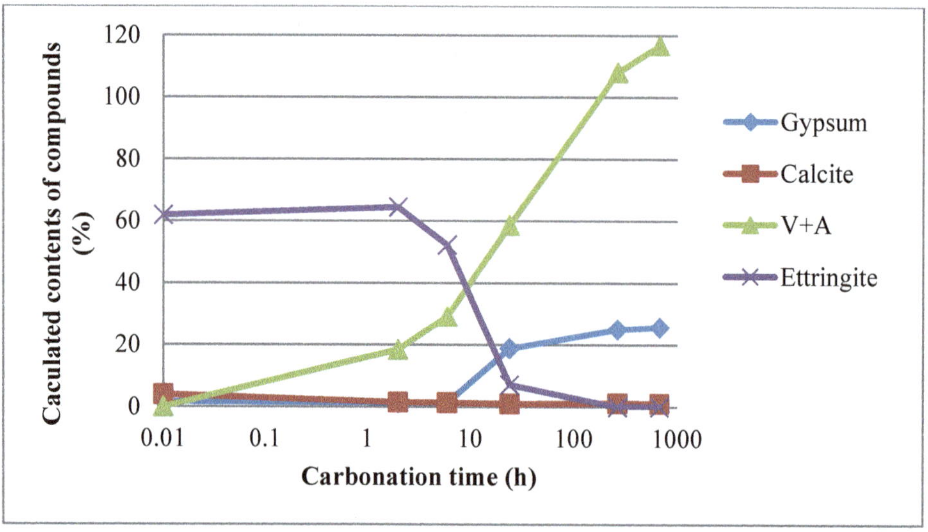

Figure 6: Mineralogical evolution and calculated contents of compounds by XRD for BCSAF 1–2mm grains at different dates of carbonation. V+A: the total content of vaterite and aragonite.

4 CONCLUSIONS

Accelerated carbonation tests at high relative humidity, carried out on 1–2mm grains of three materials containing different ettringite contents were investigated. The carbonation degree and evolution of carbonation products were demonstrated by analysing TGA and XRD data. At the same time, the changes of phase assemblage and their carbonation resistance were discussed. Main outcomes of this investigation are listed as follow:

- C80P20 has a better carbonation resistance than the other two ettringite enriched materials. Especially at 1 d, the absolute carbonation degree of C80P20 only reaches 16.8% of according theoretical CO2 binding capacity, while it is at least two times more for the other two binders.
- The carbonation products of ettringite at 90% RH are principally gypsum, amorphous aluminium hydroxyl, vaterite and aragonite as calcium carbonates that were found in XRD analysis.
- At the very beginning of accelerated carbonation for C60P40, the formation of monosulfoaluminate from ettringite has been observed. During the carbonation process, monosulfate shows more vulnerable to CO_2 than ettringite. Moreover, some hydrates like strätlingite, katoite and tobermorite, prove a better resistance against CO_2 than ettringite: they react with CO_2 after ettringite.
- The turning points of carbonation degree of three binders possibly suggest two steps of carbonation for the grains of three binders: a surface reaction stage and a CO_2 diffusion controlled process.

ACKNOWLEDGEMENTS

We express our appreciation to LafargeHolcim R&D and Association Nationale de la Recherche et de la Technologie (ANRT) of France for the funding of this research.

REFERENCES

[1] Chen, B., Kuznik, F., Horgnies, M., Johannes, K., Morin, M. & Gengembre E., Physicochemical properties of ettringite/meta-ettringite for thermal energy storage: Review. *Solar Energy Materials and Solar Cells*, **193**, pp. 320–334, 2019.

[2] Baquerizo, L.G., Matschei, T. & Scrivener, K.L., Impact of water activity on the stability of ettringite. *Cement and Concrete Research*, **79**, pp. 31–44, 2016.

[3] Ndiaye, K., Ginestet, S. & Cyr, M., Thermal energy storage based on cementitious materials: A review. *AIMS Energy*, **6**(1), pp. 97–120, 2018.

[4] Ndiaye, K., Cyr, M. & Ginestet, S., Durability and stability of an ettringite-based material for thermal energy storage at low temperature. *Cement and Concrete Research*, **99**, pp. 106–115, 2017.

[5] Kaufmann, J., Winnefeld, F. & Lothenbach, B., Stability of ettringite in CSA cement at elevated temperatures. *Advances in Cement Research*, **28**(4), pp. 251–261, 2016.

[6] Robl, T.L., Graham, U.M. & Taulbee, D.N., The effect of carbonation reactions on the long term stability of products made from dry FGD materials. *Preprints of Papers, American Chemical Society, Division of Fuel Chemistry*. United States, 1996. Available at: https://www.osti.gov/servlets/purl/430280.

[7] Pajares, I., Martínez-Ramírez, S. & Blanco-Varela, M.T., Evolution of ettringite in presence of carbonate, and silicate ions. *Cement and Concrete Composites*, **25**(8), pp. 861–865, 2003.

[8] Carmona-Quiroga, P.M. & Blanco-Varela, M.T., Ettringite decomposition in the presence of barium carbonate. *Cement and Concrete Research*, **52**, pp. 140–148, 2013.

[9] Grounds, T., Midgley, H.G. & Novell, D.V., Carbonation of ettringite by atmospheric carbon dioxide. *Thermochimica Acta*, **135**, pp. 347–352, 1988.

[10] Nishikawa, T., Suzuki, K., Ito, S., Sato, K. & Takebe, T., Decomposition of synthesized ettringite by carbonation. *Cement and Concrete Research*, **22**(1), pp. 6–14, 1992.

[11] Chen, X., Zou, R. & Chen, X., Kinetic study of ettringite carbonation reaction. *Cement and Concrete Research*, **24**(7), pp. 1383–1389, 1994.

[12] Zhou, Q. & Glasser, F.P., Kinetics and mechanism of the carbonation of ettringite. *Advances in Cement Research*, **12**(3), pp. 131–136, 2000.

SECTION 2
RECYCLED MATERIALS

USE OF RECYCLED AGGREGATES FROM DEMOLITION WASTES IN CONCRETE: ACOUSTIC PROPERTIES

CARLOS RODRÍGUEZ[1], ISIDRO SÁNCHEZ[2], SALVADOR MARTÍNEZ[1],
JESÚS CARBAJO[3], JAIME RAMIS[3] & IGNACIO GARCÍA-LEGAZ[1]
[1]Technological Centre of the Construction of Region of Murcia, Spain
[2]Department of Civil Engineering, University of Alicante, Spain
[3]Department of Applied Physics, University of Alicante, Spain

ABSTRACT

The construction industry is one of the most demanding in terms of natural resources, and of the most polluting in terms of emissions to the atmosphere. Concrete is one of the most used materials in engineering and is a paradigm of the consumption of rocks for aggregates, used in polluting industries such as cement production. In trying to reduce the consumption of natural resources, efforts are being made to use waste as recycled aggregates. This fact has evident economic and environmental advantages, but it has a drawback. The concrete prepared with recycled aggregates has lower mechanical strength than ordinary concrete. This decrease of resistance is mainly due to a higher porosity exhibited by the materials from recycled aggregates. But the increase of porosity can be a great advantage in terms of sound absorption. In this work several types of recycled aggregates have been tested, paying special attention to their acoustic behaviour and pore structure. The results show that concrete made with recycled waste are effective in absorbing sound. It might become an alternative to traditional concrete, since recycling samples behave similarly or better in terms of sound absorption, using recycled materials, and increasing their life-cycle. The measurement of acoustic properties, as well as density, has already been undertaken in previous works. This work presents an image analysis methodology that is completely novel, and that helps to understand the acoustic behaviour of concrete elements.

Keywords: recycled aggregates, pore structure, sound absorption.

1 INTRODUCTION

In recent history the development has mainly considered economic aspects. Nowadays the trends are changing towards a more sustainable development, in a circular economy model, with a reduction of natural resources, and decreasing the generated wastes. In this context, construction and demolition wastes have a great value, due to the huge volume generated, and that offered a great possibility of recycling. In this sense the European Parliament has stated the need of reusing these wastes to promote a more sustainable construction [1].

The Spanish standard for structural concrete promotes the use of these residues as aggregates, but for the case of structural concrete it is limited to a 20% (with no risk for the structural integrity), and to wastes containing only concrete [2]. The real fact is that most demolition wastes contain many other materials such as ceramics, asphalt, etc. Those wastes (recycled mixed aggregate, RMA) could not be used, according to the Spanish standard, even for non-structural concrete. This makes the recycling rates very low in Spain [3], [4]

However, many efforts are being made for producing non-structural elements of concrete, containing recycled aggregates, even at industrial scale [5]. In these works, the mechanical properties, some durability aspects are determined. It is a common fact that the use of recycled aggregates causes a decrease of the compressive strength depending on the type of recycled aggregate used. Some studies show that it is possible to use up to 50% of recycled concrete aggregates with no loss of mechanical properties [6], but samples prepared with RMA (a mix of construction and demolition wastes, with almost no classification) may cause an important decrease of the properties of concrete [7]. This fact is the reason why most of the elements prepared are non-structural elements [5], [8], [9]. The origin of the mechanical loss is mainly due to the greater porosity produced by the recycled aggregates, either because of the higher porosity of recycled aggregates [8], or because of the greater interfacial transition zone (ITZ) among paste and aggregate, due to the presence of undesired materials in the aggregates [10], [11]. In both cases the porosity of concrete increases, and it causes the loss of properties.

This increase in porosity can be an advantage in terms of sound absorption. To that end, different types of recycled materials (construction and demolition wastes) have been used to prepare concrete with recycled aggregates. They have been tested to establish acoustic properties, mainly, and to establish if the recycled samples have similar, or improved the acoustic behaviour of samples prepared with natural aggregates.

2 EXPERIMENTAL SETUP

In this section both the materials prepared, and the techniques used will be described.

2.1 Materials

Concrete samples were prepared using an ordinary Portland cement, CEM I 42.5R according to the standard [12], and the w:c ratio used was of 0.58. The dosage used for each element prepared follows the proportions given in Table 1.

Table 1: Concrete dosage used for every mix prepared.

Material	% in mass
Cement	7.4
Water	4.3
Limestone filler	0.8
Fine aggregate (0–4mm)	10.55
Coarse aggregate (2–6 mm)	76.95

Only the coarse fraction of the aggregates was replaced by recycled materials. The materials used to substitute the natural aggregates are recycled arlite, asphalt, ceramic, ethylene-vinyl acetate (EVA), hollow tiles, recycled concrete and rubber. The materials were crushed, some of them in a laboratory crusher, and the other provided by the suppliers of the wastes. The materials were selected to have dimensions in the indicated range, and to have a similar grain size distribution, compared with natural aggregates. The arlite is not a recycled aggregate, but since it is a very porous aggregate, it was included to grain size distributions of selected aggregates are shown in Fig. 1. The replacement of the coarse fraction was total, and all the aggregates were water saturated and surface dried prior to mixing, to ensure the hydration water for the cement. Samples were cast in molds with dimensions 50 mm height and 70 mm diameter and kept in humid chamber for 28 days. Only half of the samples were compacted with a manual procedure, compacting by layers using a steel bar, while the other half suffered no compaction. Non-compacted samples were only prepared with the natural aggregate, arlite, asphalt, ceramics, and hollow tiles wastes. The differences are evident and can be appreciated in Fig. 2. The sample that suffered no compaction presents a much more porous structure. Six samples of each type were prepared for testing.

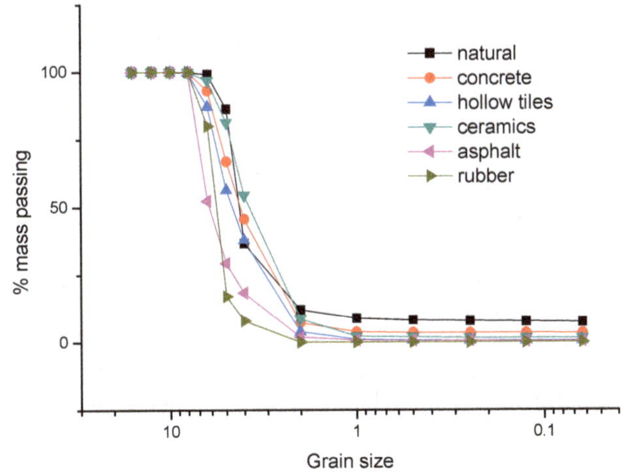

Figure 1: Grain size distribution of selected aggregates for the preparation of concrete.

Figure 2: Prepared samples using arlite as aggregate. The sample on the left was manually compacted, while the sample at the right suffered no compaction.

2.2 Testing procedures

The samples were tested to obtain the acoustic properties. Initially the bulk density and fraction of holes were determined using the hydrostatic balance method [13]. The sound absorption was determined using the impedance tube method [14] and the absorption coefficient was determined in the frequency range from 250 to 2,600 Hz. In order to try to understand the behaviour of the materials an image analysis has been done on samples.

2.2.1 Image analysis

Samples were coloured on their surface, by letting them lay on a coloured surface, to avoid paint penetration, and let them dry. The result is shown in Fig. 3(a). After that a 3 step methodology is followed to analyse the pore structure:

The process to make the analysis was in three steps:

Step 1: Capture of the photographs in a controlled situation of light, using the same focal length, and diaphragm opening.

Step 2: Treatment of the photographs to clean all the parts that do not belong to the sample.

Step 3: Use the program ImageJ to analyse the photos and get all the information about holes and surfaces. The program uses filters to separate the different parts, for this study we have decided to use three filters, one for the deep holes, one for the plane surface, and one for the rest of the surfaces. The parameters used for each one of the filters are reported in Table 2. The results of the application of the subsequent filters to the image are presented in Fig. 3(b)–(d).

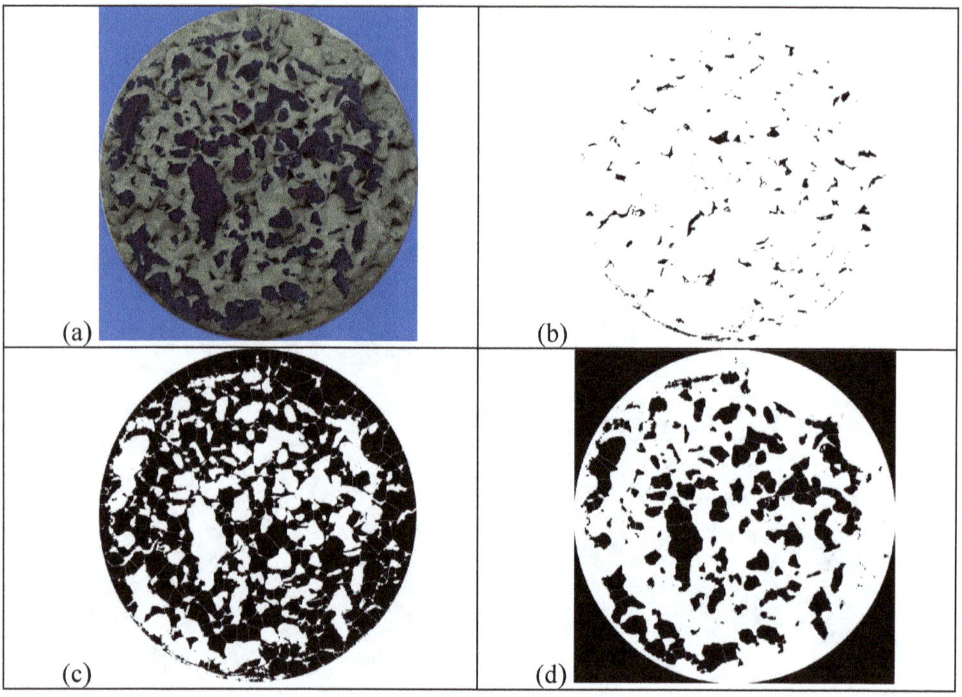

Figure 3: Example of the image analysis done on the coloured samples: (a) For the original picture taken of the coloured sample; (b) After applying filter 1; (c) After applying filter 2; and (d) After applying filter 3.

The first filter selects the deep holes that appear in the image. Before applying the filters, the noise, that appears as small black dots, was also removed. The filters were selected after several trials, the first one corresponded to the surface, that was coloured, and it was very easy to determine, in second place the darkest parts of the image, that correspond to the

Table 2: Parameters used for each filter used in the image analysis.

	Colour	Saturation	Brightness
Filter 1	0–116	0–255	0–70
Filter 2	0–116	0–255	70–255
Filter 3	116–255	0–255	0–255

deepest holes in the sample were identified, and the last filter studied the intermediate zone, that corresponds to pores, but that are not too deep. So, the three different parts in which the image is divided were analysed to obtain an idea of the pore network. The abovementioned program (ImageJ) draws the perimeter of the surfaces with the same colour and calculates the perimeter and the surface of each element, that is identified with a number.

3 RESULTS AND DISCUSSION

3.1 Density and porosity

As was explained, these parameters were determined according to the UNE-EN 12390-7 standard [13]. The average results are shown in Table 3.

Table 3: Parameters for each filter used in the image analysis.

Material	Bulk density (g/cm^3)	Porosity, %	Compacted
Reference	2.47	35.1	No
		31.77	Yes
Arlite	–	–	No
		–	Yes
Asphalt	2.40	37.36	No
		30.45	Yes
Ceramic	1.90	39.73	No
		32.88	Yes
EVA	–	–	Yes
Hollow tiles	2.46	32.37	No
		27.57	Yes
Recycled concrete	2.27	32.07	Yes
Rubber	–	32.2	Yes

As a first result it can be noticed that the density and pore volume fraction could not be measured due to the standardized method used, that implies immersion under water. The bulk density shows no variation from compacted to non-compacted samples, as it could be expected since all the samples came from the same mix. The reference concrete has high density, but the samples prepared with fractions of hollow tiles show a slightly higher density, possibly due to the compaction method usually applied to these elements (vibropressing) [5]. The samples prepared with ceramic aggregates exhibits the lowest porosity among those which allowed density measurement. This fact is quite common, due to the high porosity measured in ceramic aggregates in previous works [8], [15]. The elements prepared with arlite, EVA and rubber showed a density lower than water, and this did not allow the proper determination of the density.

Regarding the void fraction measured for each sample, the non-compacted samples exhibit a clearly higher pore fraction than the compacted ones for the same type of waste, as could be expected. Among the compacted samples the highest porosity is shown by the samples prepared with ceramic aggregates, in coincidence with the bulk density results. This fact confirms the theory of the influence of the highest porosity of ceramic aggregates on the properties of the concrete elements. Only compacted samples prepared with hollow tiles wastes and asphalt wastes have shown a slightly smaller porosity that the reference element, again possibly due to the compaction methods used for hollow tiles. The result for asphalt would need more in depth research, but it could be due to the adherence of the organic phase of the asphalt to the aggregate used. The non-compacted samples exhibit a quite different behaviour. Non-compacted samples prepared with ceramics and asphalt wastes are the only ones showing a higher porosity as compared to the non-compacted reference material. This could be possibly due to the highest porosity of ceramic samples, and also to the difficulty of paste flowing through the aggregates, due to their characteristics, that could increase the viscosity. Other option in the case of asphalt is the presence of a greater ITZ, that will cause a decrease of porosity when samples are not properly compacted [16]. The rest of the samples showed a lower porosity, possibly due to a better compaction, or a more compact ITZ due to the similarity of the cement mortar and the aggregates (in the case of hollow tiles and recycled concrete), or to the different granulometric distribution exhibited by rubber wastes that allow the mortar to flow through the grains of aggregate, as it has been already reported with the presence of coarse asphalt particles [17].

3.2 Acoustic properties

The plots for the sound absorption coefficient in the measured range can be seen in Fig. 4 for compacted samples and Fig. 5 for non-compacted samples.

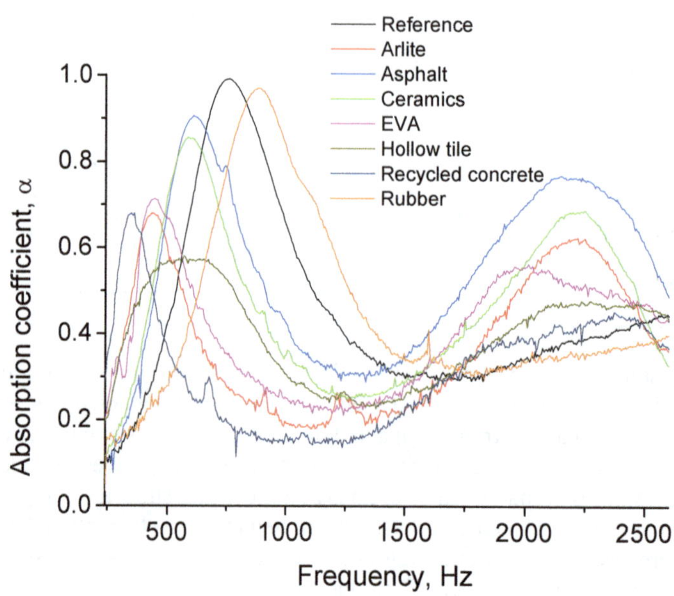

Figure 4: Absorption coefficient in the studied range of the compacted concrete samples prepared with each type of aggregates (including reference).

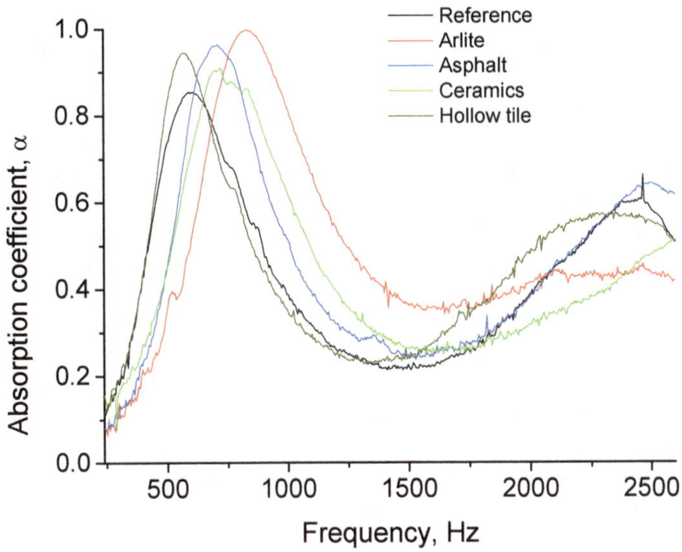

Figure 5: Absorption coefficient in the studied range of the non-compacted concrete samples prepared with each type of aggregates (including reference).

Two parameters will be considered to study the effectiveness of the sound attenuation on each set of samples. The maximum value of the absorption coefficient, α, and the percentage of frequencies for which this factor is higher than 0.5. This parameter would be used to study the amplitude in which the concrete could be used as an acoustic attenuator. Both parameters are shown in Table 4.

Table 4: Parameters obtained from the acoustic measurements on concrete samples.

Material	Max α	%α > 0.5	Compacted
Reference	0.85	36.43	No
	0.99	24.10	Yes
Arlite	0.997	26.92	No
	0.68	27.70	Yes
Asphalt	0.96	38.69	No
	0.91	55.63	Yes
Ceramic	0.91	26.70	No
	0.86	39.86	Yes
EVA	0.71	29.95	Yes
Hollow tiles	0.94	42.31	No
	0.58	16.21	Yes
Recycled concrete	0.68	6.30	Yes
Rubber	0.97	26.57	Yes

As it can be seen, there is a clear difference in the acoustic behaviour as a function of the type of aggregate used. Among the set of compacted materials, the one that shows a higher sound absorption is the reference material, but both rubber and asphalt present close values.

The worst peak values are presented by the hollow tiles, the recycled concrete and the arlite. The first one shows these results, that can be clearly related to the higher density and lower porosity. In the rest of materials there seems to be no clear correlation between density and maximum acoustic attenuation.

Some of the non-compacted samples give attenuation factors bigger than 0.9, and all the recycled materials under non-compaction conditions improve the maximum attenuation provided by the reference material. It is also clear that the tested materials improve the value of the maximum with respect to the compacted samples. From this point of view, it is possible to state that it is possible to use recycled aggregates for sound attenuation, and that it is better using non-compacted materials than using a manual compacting method.

Regarding the frequency range for which the attenuation factor is higher than 0.5, when samples are compacted almost every material (excepting recycled concrete and hollow tiles) have a bigger bandwidth with effective sound attenuation, so, they can be used for a wider range of acoustic applications. When samples are non-compacted only hollow tiles and asphalt present a bandwidth broader than the reference. The behaviour of this parameter is quite erratic and some more research is required, because it has nothing to do with the density, with the maximum peak of sound absorption and not even with the type of material.

3.3 Image analysis

The results of the proposed image analysis for this study are presented in Table 5. The percentage of area corresponding to each type of hole, or surface, and the average area of all the pores measured for each mask applied on the image are presented. It has to be pointed out that the more reliable data are those of the deep pores, for which the program makes no assumption or correction. The mean diameter calculated for both surface and the intermediate region can have small differences with the reality due to the assumptions made by the algorithm used by the program. The analysis has only been done for the compacted samples, as a first attempt to establish a correlation with this analysis, and the authors have decided to take the complete set of materials tested.

In a first approach, the image analysis data (Table 5) indicates that the mixes made from arlite or rubber exhibit a higher percentage of area of deep pores (22–32%) when compared to the others (3.2–8%), being also the samples that showed a higher sound absorption peak (see Figs 4 and 5). These results were expected given that samples having deeper pores should yield higher viscothermal losses as the acoustic wave propagates throughout them and thus an increased sound attenuation [18]. It should be pointed out that even though the porosity of the rubber mixes (unfortunately, the porosity of the arlite mixes was not possible to measure) is not as high as that of the others, its sound absorption performance clearly exceeds that of the remaining mixes. In this regard, it is worth remarking that apart from the porosity, the microstructure of a porous material plays a key role in these dissipation phenomena. Hence, the results confirm the need to measure not only the porosity but also other parameters (e.g. flow resistivity, tortuosity…) for a more complete characterization and a better understanding of the acoustic behaviour of these concrete-based materials [19], [20].

However, some conclusions can be drawn from the presented results. The results show that the samples with higher contribution of the deep holes (more accessible deep holes) correspond to the samples prepared with arlite, rubber and asphalt. If instead of doing a global study we focus on the samples with materials coming from the same family, we would obtain more useful information. Starting with the aggregates that have an organic origin, rubber and asphalt were the materials that produced a higher absorption coefficient after the reference

Table 5: Contribution of each pore type to the total surface.

Material	Type of pore	% area
Reference	Surface	46.6
	Intermediate	50.2
	Deep	4.1
Arlite	Surface	3.77
	Intermediate	63.66
	Deep	32.46
Asphalt	Surface	33.3
	Intermediate	58.2
	Deep	8.5
Ceramics	Surface	42.9
	Intermediate	50.8
	Deep	6.3
EVA	Surface	50
	Intermediate	46.8
	Deep	3.2
Hollow T.	Surface	39.8
	Intermediate	57
	Deep	3.2
Recycled Conc	Surface	32.9
	Intermediate	63
	Deep	4.1
Rubber	Surface	23.3
	Intermediate	55.4
	Deep	21.2

material (see Table 4), so it might indicate that the surface of deep holes has an influence on the sound attenuation, but there is also an effect of the type of material used. Looking at the materials prepared with cement-based materials wastes (recycled concrete, and hollow tiles) the same conclusion can be reached, the material with higher fraction of deep holes recycled concrete exhibits also a higher peak absorption of sound (0.68). The rest of the materials cannot be grouped in families, so we can say that for materials coming from the same type of waste, those that give a bigger volume of deep holes are the ones that are more effective for sound attenuation in terms of maximum absorption coefficient. This analysis has never been done, to the knowledge of the authors, and is quite innovative. Most of the papers dedicated to the study of the sound attenuation have tried different materials [21], [22] but no attention is usually paid to the relationship between pore or holes structure and acoustic properties.

Regarding the bandwidth for which the samples show an absorption coefficient higher than 0.5, it is clearly not related neither to the deep holes nor to the intermediate holes, but a relationship can be established with the surface (blue coloured, with no apparent pore). Taking a look at the aggregates from organic origin (asphalt, EVA and rubber) EVA samples have the highest closed surface, and rubber has the lowest surface and the have an absorption coefficient higher than 0.5 namely 55.63 for asphalt 29.95 for EVA and 26.57 for rubber, so there is no clear correlation with the surface either. More research is required for the study of the effect of using recycled aggregates on the bandwidth where there is a significant sound attenuation.

But, with the provided results it can be concluded that the maximum absorption is a function of both the type of material used and the fraction of deep pores in the materials used to this end.

4 CONCLUSIONS

After the data measured and the analysis done the following conclusions can be drawn:

- The use of recycled aggregates strongly affects the values of density and pore fraction of the concrete samples prepared. The use or not of a compacting method also affects the values of density and porosity.
- The use of recycled aggregates also affects the acoustic properties of concrete. The recommendation of using non-compacted materials could be established since they provide a better sound absorption than the compacted ones.
- In terms of sound absorption it is possible to use recycled aggregates with total satisfaction, especially when the materials are non-compacted.
- The fraction of deep holes for each family of materials used as recycled aggregate can be used to predict the sound absorption. This parameter will be a function of the family of materials and the fraction of deep holes.
- The study of the bandwidth with an attenuation higher than 0.5 should be more studied in depth, because, with the provided results, it seems not to correspond to any of the parameters studied.
- Recycling of materials as aggregates for concrete that will be used as a sound isolator is possible, and it can improve the acoustic behaviour of the concrete with natural aggregates. This will improve the recycling rates, that should be increased according to the legislation, and give more sustainable materials, even though more research is necessary to understand completely their behaviour.

ACKNOWLEDGEMENT

The authors of this study would like to thank D. Manuel Iniesta Castillo for collaboration in the testing of samples.

REFERENCES

[1] European Parliament, Directive 2008/98/EC of the European Parliament and of the Council of 19 November 2008 on waste and repealing certain Directives. *Off. J. Eur. Union*, pp. 312–330, 2008.

[2] P. Comission for Concrete, Instrucción de hormigón estructural EHE-08. Structural Concrete Instruction, EHE-08, 2008.

[3] Mália, M., de Brito, J., Pinheiro, M.D. & Bravo, M., Construction and demolition waste indicators. *Waste Manag. Res.*, **31**, pp. 241–55, 2013.

[4] Rodríguez, G., Medina, C., Alegre, F.J., Asensio, E. & Sánchez de Rojas, M.I., Assessment of construction and demolition waste plant management in Spain: In pursuit of sustainability and eco-efficiency. *J. Clean. Prod.*, **90**, pp. 16–24, 2015.

[5] Rodríguez, C. et al., The incorporation of construction and demolition wastes as recycled mixed aggregates in non-structural concrete precast pieces. *J. Clean. Prod.*, **127**, pp. 152–161, 2016.

[6] Yang, J., Du, Q. & Bao, Y., Concrete with recycled concrete aggregate and crushed clay bricks. *Constr. Build. Mater.*, **25**, pp. 1935–1945, 2011.

[7] Rodríguez, C., Sánchez, I., Miñano, I., Benito, F., Cabeza, M. & Parra, C., On the possibility of using recycled mixed aggregates and GICC thermal plant wastes in non-structural concrete elements. *Sustainability*, **11**, p. 633, 2019.

[8] Rodríguez, C., Miñano, I., Aguilar, M.Á., Ortega, J.M., Parra, C. & Sánchez, I., Properties of concrete paving blocks and hollow tiles with recycled aggregate from construction and demolition wastes. *Materials (Basel)*, **10**, 2017. DOI: 10.3390/ma10121374.

[9] Sousa, J.G.G., Bauer, E. & Sposto, R.M., Empleo de residuos de la construcción civil como áridos reciclados. Producción de bloques de hormigón. *Mater. Constr.*, pp. 59–70, 2003.

[10] Evangelista, L. & de Brito, J., Mechanical behaviour of concrete made with fine recycled concrete aggregates. *Cem. Concr. Compos.*, **29**, pp. 397–401, 2007.

[11] Brand, A.S. & Roesler, J.R., Bonding in cementitious materials with asphalt-coated particles: Part II – Cement-asphalt chemical interactions. *Constr. Build. Mater.*, **130**, pp. 182–192, 2017.

[12] AENOR, UNE-EN 197-1:2011. Composición, especificaciones y criterios de conformidad de los cementos comunes, 2000.

[13] AENOR, UNE-EN 12390-7:2009. Ensayos de hormigón endurecido. Parte 7: Densidad del hormigón endurecido (Testing hardened concrete. Part 7: Density of hardened concrete), 2009.

[14] ISO 10543-2:2002, Acoustics. Determination of sound absorption coefficient and impedance in impedance tubes. Part 2: Transfer function method, 2002.

[15] López Gayarre, F., Suárez González, J., Blanco Viñuela, R., López-Colina Pérez, C. & Serrano López, M.A., Use of recycled mixed aggregates in floor blocks manufacturing. *J. Clean. Prod.*, **167**, pp. 713–722, 2018.

[16] Brand, A.S. & Roesler, J.R., Bonding in cementitious materials with asphalt-coated particles: Part I – The interfacial transition zone. *Constr. Build. Mater.*, **130**, pp. 171–181, 2017.

[17] Evangelista, L. & de Brito, J., Mechanical behaviour of concrete made with fine recycled concrete aggregates. *Cem. Concr. Compos.*, **29**, pp. 397–401, 2007.

[18] Allard, N. & Atalla, J.F., *Propagation of Sound in Porous Media: Modelling Sound Absorbing Materials*. John Wiley & Sons: UK, 2009.

[19] Vašina, M., Hughes, D.C., Horoshenkov, K.V. & Lapčík, L., The acoustical properties of consolidated expanded clay granulates. *Appl. Acoust.*, **67**, pp. 787–796, 2006.

[20] Horoshenkov, K.V., Hughes, D.C. & Cwirzen, A., The sound speed and attenuation in loose and consolidated granular formulations of high alumina cements. *Appl. Acoust.*, **64**, pp. 197–212, 2003.

[21] Herrero, S., Mayor, P. & Hernández-Olivares, F., Influence of proportion and particle size gradation of rubber from end-of-life tires on mechanical, thermal and acoustic properties of plaster-rubber mortars. *Mater. Des.*, **47**, pp. 633–642, 2013.

[22] Ghizdăveţ, Z., Ştefan, B.M., Nastac, D., Vasile, O. & Bratu, M., Sound absorbing materials made by embedding crumb rubber waste in a concrete matrix. *Constr. Build. Mater.*, **124**, pp. 755–763, 2016.

PREPARATION AND INVESTIGATION OF GEOPOLYMERS GENERATED FROM CONSTRUCTION, DEMOLITION AND INDUSTRIAL WASTES

BELLA UDVARDI, KRISZTINA ROMÁN, EMESE KUROVICS, RÓBERT GÉBER & ISTVÁN KOCSERHA
Institute of Ceramics and Polymer Engineering, University of Miskolc, Hungary

ABSTRACT

Geopolymers are inorganic amorphous aluminosilicate materials, which can be produced by the alkali activation of materials with high aluminum and silicon content. In this research, we prepared geopolymer samples obtained from construction and demolition (C&D) waste materials (brick powder and concrete powder), as well as industrial waste material (aluminum dross), and we investigated them using different test methods. Density, particle size distribution and chemical elemental composition were determined on the dried powdered waste materials, and microstructural observations were also made using a scanning electron microscope (SEM) technique. To prepare the geopolymer sample mixtures of sodium-hydroxide and sodium silicate, we used an alkali activator solution at different concentrations (4 M, 8 M and 12 M). We used Fourier Transformation Infrared Spectroscopy (FTIR) to reveal new phases in geopolymer samples. Compression strength was determined after 7 days and after 28 days. In the cases when aluminum dross was used, a gas-forming reaction took place in the geopolymer pastes; therefore, foamed geopolymer was produced. According to these results, the used C&D and industrial wastes may be potential raw materials which could be used to prepare geopolymers.
Keywords: carbon dioxide, geopolymers, foamed geopolymer, construction waste, demolition waste, aluminum dross, compression strength, material microscopy.

1 INTRODUCTION

Geopolymers are amorphous materials with an inorganic polymer structure. They can be prepared in an alkaline medium (generally sodium hydroxide (NaOH) or potassium hydroxide (KOH)), by the reaction of clay minerals (aluminosilicate oxides) and alkali silicates. They have a polysyalate structure; therefore, geopolymers are so-called artificially produced rocks. Geopolymers are today's modern and environmentally friendly binder material for the construction and building industry, which can compete with traditional concretes. During the production of ordinary Portland cement (OPC) a large amount of carbon-dioxide (CO_2) at ~0.5–0.95 tons/1 ton are emitted into the atmosphere, promoting global warming [1], [2]. In contrast, the production of geopolymers produces significantly less CO_2 (~10–20% of the OPC emissions) [3], [4]. Some properties of these materials are: resistance against acids, heat and fire; good mechanical strength; and the ability to encapsulate toxic materials.

For the synthesis of geopolymers, metakaolin and silicate-containing sources and industrial wastes can be used as the raw materials, such as: power plant waste (e.g. fly ash), metallurgical slag (e.g. ground granular blast furnace slag), waste incineration by-products (e.g. incinerator bottom ash), red mud, waste glass, or construction and demolition (C&D) waste materials. Researchers are continuously looking for other possible materials which can be used for geopolymer production. The following section summarizes the actual trends and research results.

In the work of Nazari and Sanjayan [5], the synthesis possibility of geopolymeric paste and concrete through alkali activation of aluminum and grey cast iron slags were studied. Mixtures of slags were used as the aluminosilicate source; while waterglass, sodium

WIT Transactions on Engineering Sciences, Vol 124, © 2019 WIT Press
www.witpress.com, ISSN 1743-3533 (on-line)
doi:10.2495/MC190051

hydroxide, polycarboxylate superplasticizer and natural sand were used to prepare geopolymer samples.

Compressive strength and workability were studied. The results indicated the possibility of geopolymers being produced by use of proper ratios of aluminum slag and grey cast iron slag. Islam et al. [6] prepared lightweight geopolymer mortar using palm oil fuel ash (POFA), ground granulated blast furnace slag (GGBFS) and low calcium fly-ash (FA) with manufactured sand (M-sand). During their research, compressive strength and density changes were studied. The authors found that the compressive strength of geopolymer samples increase up to 70% with the increase of ground granulated blast furnace slag. The highest compressive strength (60 MPa) was achieved with the mixture of 30% palm oil fuel ash and 70% ground granulated blast furnace slag. Puligilla and Mondal [7] used Class F fly ash and GGBFS to produce geopolymer. In their research, the hardening rate was tested and microstructural changes were observed. They found that calcium strongly affects the early and late age properties of a hardened geopolymer. They also concluded that calcium enhances geopolymer gel formation. Nath and Kumar [8] prepared geopolymer samples with the use of GGBFS and granulated corex slag (GCS) with Class F fly ash. Compressive strength, isothermal conduction calorimetry, Fourier transformation infrared spectroscopy and microstructural observation were done on these samples. They studied the effect of slag addition on the reaction kinetics. The authors found that compressive strength is increased with an increase in slag amounts. They also concluded that GCS can be a potential substitute material for GGBFS. Other researchers [9]–[12] also dealt with the utilization of different slags (primary lead slag, zync slag, alkaline activated ground blast furnace slag, lead smelting slag) in order to prepare geopolymers.

Tuyan et al. [13] produced waste clay brick powder-based geopolymer composites and the effect of the alkali solution concentration on the curing conditions and compressive strength were investigated. The morphology and structure of the composites were characterized by use of X-ray powder diffraction (XRD) analysis, thermogravimetric (TG) analysis, Fourier transform infrared spectroscopy (FTIR) analysis and scanning electron microscopy (SEM). The results of their examination show that the maximum compression strength was 36.2 MPa, which was achieved at 90°C, 40% relative humidity and 5 days of curing. Muhammad et al. [14] made a review that was about the suitability of clay application as a geopolymer material in brick production. According to their findings in the literature, it had been established that clay has high SiO_2 and Al_2O_3 content; therefore, clay-based geopolymer has a good potential for geopolymer brick production. Brînduş-Simuţ et al. [15] investigated the effect of particle size of the waste brick powder on the properties of geopolymers. Five different fractions of waste bricks were used in their research, and the effect of this particle size on the rheological properties (yield stress and viscosity) were then revealed. The flexural and compressive strength of the hardened products were determined after 7, 14 and 90 days. The microstructure and pore structure of the geopolymer samples was examined by porosimetry and by SEM. The authors found that the flexural strength of geopolymer samples increased at 20°C with the increase of the particle size of waste brick powder. Compressive strength, depending on the particle size, was continuously increased until 28 days. Ren and Zhang [16] made a complete investigation in the field of recycling and utilization of waste concrete (with coarse and fines particles) for making new geopolymer concrete products. Compressive strength and initial setting time were determined. According to the results, the authors found that waste concrete is suitable for recycling and to produce new geopolymer.

In order to prepare geopolymers that are even more environmentally friendly, it is preferable to use industrial by-products and waste materials as the solid raw materials. The amount of environmental pollutants is growing rapidly, and these materials are mainly deposited in landfills, but it is not a definitive solution. This problem may be solved by the utilization of these wastes in geopolymers. The most frequently used secondary materials for geopolymer production are fly ash and slags [17]–[21].

As the above research results also confirm, utilization of different waste materials is economically and environmentally important. Therefore, the aim of this research is to prepare geopolymeric materials for different applications, using waste materials from different sources (C&D wastes: brick and concrete powder; and industrial waste: aluminum dross).

2 MATERIALS AND METHODS

2.1 Sample preparation

During this research, the following materials were used to prepare geopolymer samples: aluminum dross (D), brick powder (B), concrete powder (C) and sand (S), as well as alkali activator (NaOH solution) with waterglass (Na_2SiO_3) and water. Also aluminum dross, a by-product of the aluminum casting process. B and C are originated from C&D wastes. In addition, high purity natural sand was also used. A NaOH solution was prepared from NaOH flakes and distilled water, in order to achieve different concentrations (4 M, 8 M and 12 M).

The required particle size of powders was produced by means of a hammer, then by planetary ball mill. Samples were then dried to a weight constancy and closed air-proof for further examination. Fig. 1 illustrates the raw materials used for sample preparation. The geopolymer samples were prepared by the following procedure, according to Karyawan et al. [22] (Fig. 2).

Figure 1: Materials used for geopolymer preparation.

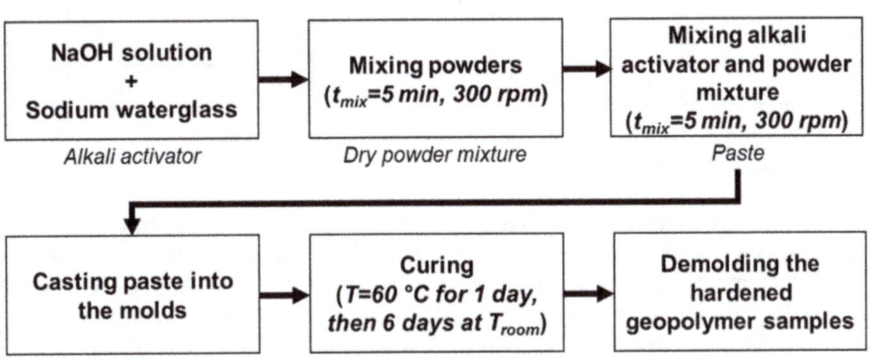

Figure 2: Sample preparation steps for geopolymer samples.

Geopolymer samples were produced using different proportions of powdered waste materials and different concentrations of alkali activator solution according to Table 1. Samples were prepared also standalone and in mixtures. A total of 24 mixtures were made. After mixing the powder and liquid components, geopolymer pastes were cast into the molds. The samples were then put in a sealed container and were cured at 60°C for 24 hours, in a confined space.

Table 1: Composition of geopolymer samples made with 4 M, 8M and 12 M NaOH.

Sample ID	Concentration of NaOH (M)	Alkali activator (m/m%)		Water (m/m%)	Sand (m/m%)	Aluminum dross (m/m%)	Brick powder (m/m%)	Concrete powder (m/m%)
		NaOH	Na_2SiO_3					
D	4	10	20	0	0	70	0	0
B		10	20	10		0	60	0
C		10	20	0	0	0	0	70
DB		10	20	10	0	30	30	0
DSB		10	20	0	8.33	30.83	30.83	0
DSBC		10	20	0	8.33	41.66	10	10
D	8	10	20	0	0	70	0	0
B		10	20	0	0	0	70	0
C		10	20	0	0	0	0	70
DB		10	20	10	0	35	35	0
DSB		10	20	0	8.33	30.83	30.83	0
DSBC		10	20	0	8.33	41.66	10	10
D	12	10	20	0	0	70	0	0
B		10	20	10	0	0	60	0
C		10	20	0	0	0	0	70
DB		10	20	0	0	35	35	0
DSB		10	20	0	8.33	30.83	30.83	0
DSBC		10	20	0	8.33	41.66	10	10

2.2 Experimental methods

For the morphological tests, a Carl Zeiss EVO MA10 scanning electron microscope (SEM) was used. Further observation was done by taking high resolution electron micrographs. Chemical elemental analysis (Energy Dispersive X-Ray Spectroscope (EDX)) was also made on the powders, during microstructural examination. We determined chemical composition by X-ray fluorescence spectrometry (XRF) on a Rigaku Super Mini 200 tabletop wavelength dispersive spectrometer.

The particle size distribution (PSD) of the ground materials was measured by a Malvern Mastersizer X particle size analyzer. To study the effect of geopolymerization on the geopolymers and to reveal chemical bonds, we used the Fourier transformation infrared spectroscopy method (FTIR) with a Bruker Tensor 27 instrument. Compressive strength of the hardened geopolymer samples was determined by the Instron 5566 Universal Material Tester instrument. We also determined the density of our geopolymer samples.

3 RESULTS AND DISCUSSION

The chemical element composition of powders as determined by EDX are summarized in Table 2.

Table 2: Chemical element composition of raw materials.

Chemical element	Sand (wt%)	Aluminum dross (wt%)	Brick powder (wt%)	Concrete powder (wt%)
C	34.13	24.32	25.94	15.76
O	35.39	20.58	34.94	37.40
Na	0.09	7.60	0.35	0.48
Mg	0.25	2.32	0.75	0.95
Al	1.33	22.47	8.92	2.39
Si	25.68	0.79	21.30	23.73
S	0	0	0	0.65
K	0.65	4.02	1.67	0.98
Ca	2.34	1.82	0.53	15.90
Fe	0	0	5.31	1.74
Cl	0.13	16.08	0	0
Ti	0	0	0.29	0
TOTAL	99.99	100	100	99.98

Note: Al: Aluminum; C: carbon; Ca: calcium; Cl: chloride; Fe: iron; K: potassium; Mg: magnesium; Na: sodium; O: oxygen; S: sulfur; Si: silica; Ti: titanium.

Our results show that sand contains a high amount of Si, which indicates the presence of SiO_2. Besides, a low amount of Al was also detected, which refers to the presence of alumina (Al_2O_3). A significant amount of Al, Na, K and Cl were detected in aluminum dross. These elements indicate the presence of alumina and the slag-forming agents NaCl and KCl. Brick powder contains a high amount of Si and a significant amount of Fe and Al. Si refers to SiO_2, while having Al and Fe shows the presence of alumina and Fe_2O_3. High amount of Si and Ca were detected in concrete powder. This suggests the presence of SiO_2 and CaO, which are the main oxides of cement minerals.

During our research work, scanning electron micrographs were taken to observe the microstructure and the surface features of powders. Fig. 3 shows that sand and Al dross contain coarser particles than brick and concrete powders; however, these powders contain high amount of fine particles, as well. Smaller particles were stuck on the surface of bigger grains. In the case of aluminum dross, the brick powder and concrete powder there are particles that are glued together.

Table 3 shows the chemical composition of powders, as determined by XRF. The XRF results show that sand contains a high amount of SiO_2. The loss of ignition (LOI) is due to the evaporation of absorbed water. The LOI values of brick powder and concrete powder are linked to the amorphous materials. Opposite to this, the LOI of aluminum dross is significantly higher. The reason for this is the evaporation of slag-forming agents.

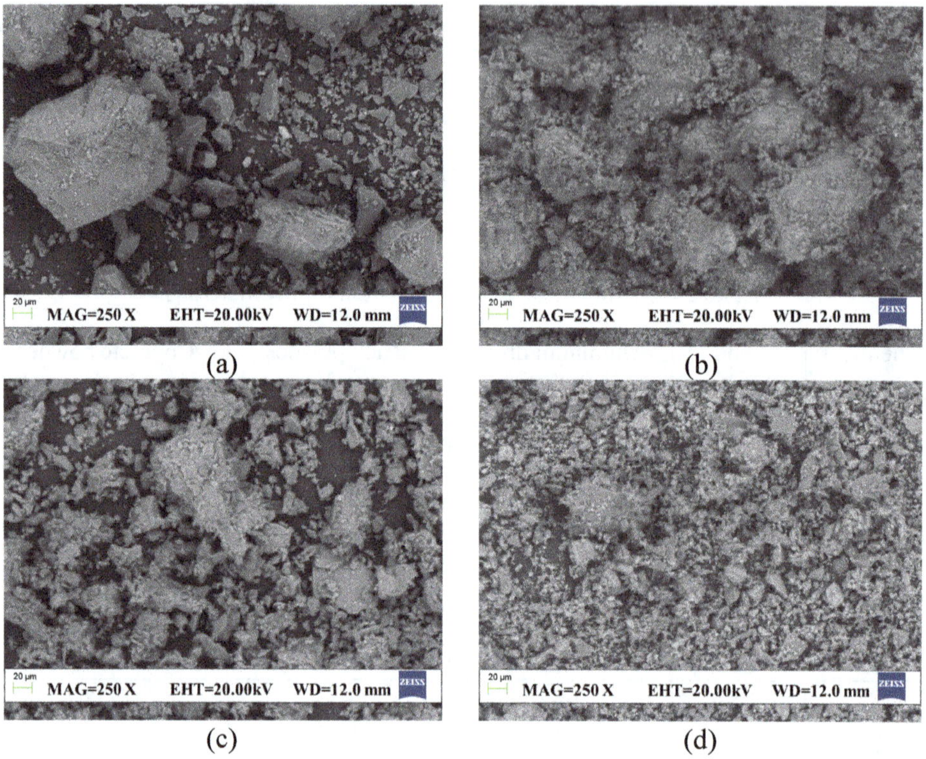

Figure 3: SEM micrographs of powders. (a) Sand; (b) Aluminum dross; (c) Brick powder; and (d) Concrete powder.

Table 3: Chemical composition of powders.

Compounds	Sand (wt%)	Aluminum dross (wt%)	Brick powder (wt%)	Concrete powder (wt%)
SiO_2	83.55	18.14	60.97	67.4
Al_2O_3	2.46	35.93	19.49	4.2
MgO	0.26	3.81	1.65	1.73
CaO	2.33	1.91	0.72	14.6
Na_2O	0.09	10.57	0.79	0.46
K_2O	0.52	3.36	2.26	0.84
Fe_2O_3	1.48	0.32	7.03	1.14
MnO	0.046	0.104	0.105	0.147
TiO_2	3.340	0.207	1.230	0.155
P_2O_5	0.014	0.050	0.090	0.062
TOTAL	94.09	74.40	94.34	90.73
LOI	5,91	25,60	5,66	9,27

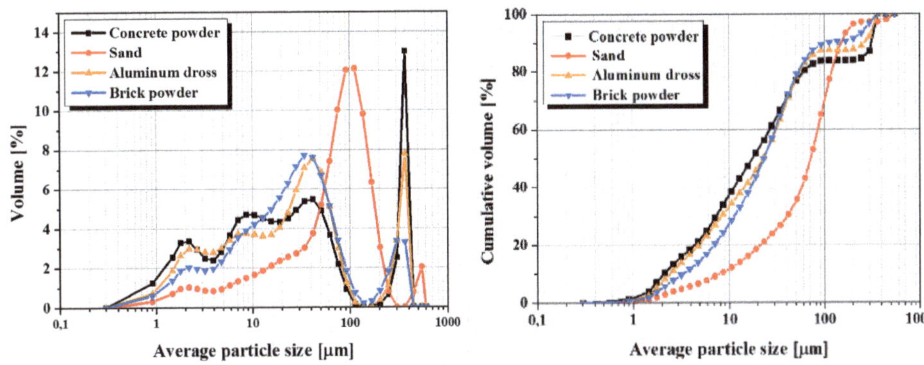

Figure 4: Particle size distribution of powder materials.

Fig. 4 shows the results of volume distribution and cumulative volume.

It can be concluded that all of the powders are polydispersed materials. Concrete powder has four modes, three of these are below 100 µm. This powder has a narrow peak of around 350 µm. Aluminum dross has a similar distribution, while brick powder has a significant volume, around 40 microns. Sand has a low volume in fine particles, while it has a wide peak that can be seen around 100 microns. Average particle sizes (d_{50}) are the following:$d_{50\,C}$: 20 µm < $d_{50\,D}$: 26 µm < $d_{50\,B}$: 26 µm < $d_{50\,S}$: 79 µm. Concrete powder, aluminum dross and brick powder have a significant volume of fine particles below 10 microns.

Particle density of the powders studied were determined by the pycnometer method, according to standard (EN 1097-7) [23]. Our results show that concrete powder has the lowest density (ρ_C=1.24 g/cm^3); while the other three materials have the following particle density: ρ_D=2.90 g/cm^3, ρ_B=2.78 g/cm^3 and ρ_S=2.62 g/cm^3.

Fig. 5 shows a summarized example of the FTIR spectras from Table 4, which contains all the characteristic compounds or ions of geopolymer samples, as well.

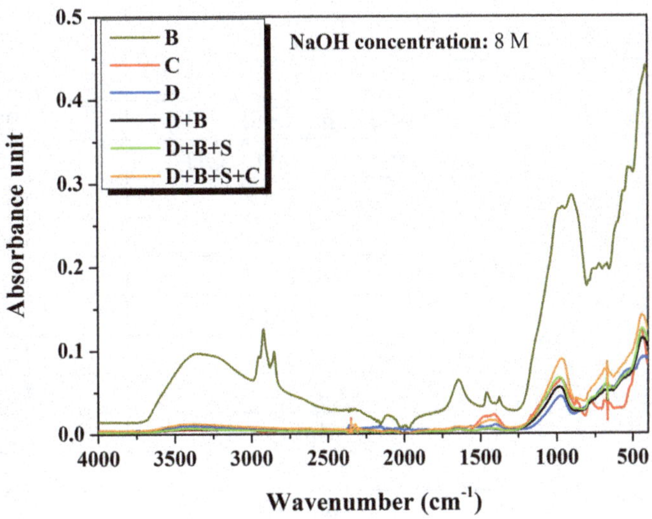

Figure 5: FTIR results of geopolymer samples (NaOH concentration: 8M).

Table 4: Compounds' characteristic absorption.

Concentration of NaOH (M)	Characteristic compounds or ion	Wavenumber (cm^{-1})
4	OH^-	3368.79
	OH^-	3238.18
	organic compound	2922.79
	NH_3	2325.40
	$Ca_2MgSi_2O_7$	1643.13
	MgO	1385.94
	Si-O-Si	1004.09–1001.20
	$MgCO_3$	775.88
	$Si_2O_3^{2-}$	692.63
	SiO_2	442.64–439.29
8	OH^-	3333.20
	organic compound (CH_8)	2922.02
	organic compound (CH_2)	2852.93
	NH_3	2349.33–2322.86
	$Ca_2MgSi_2O_7$	1643.55
	CO_3^{2-}	1461.21–1376.98
	Si-O	990.02–906.87
	$MgCO_3$	795.13–770.22
	O-Si-O	671.25–664.55
	Ca=O	615.15
	MgO	563.51–532.81
	SiO_2	451.65–445.54
12	OH^-	3200.80
	organic compound (CH_8)	2921.90
	organic compound (CH_2)	2852.52
	$Ca_2MgSi_2O_7$	1683.82
	Crystal water	1662.19
	$MgCO_3$	1653.44
	$Ca_2MgSi_2O_7$	1646.22
	NH_3	1635.95
	CO_3^{2-}	1457.17
	MgO	1375.92
	Si-O	989.02
	$MgCO_3$	774.05
	SiO_2	444.00

Typical bands of multi-atom inorganic ions are generally much more intense and wider than those of organic compounds. The frequency of the vibrations corresponding to the different bonds depends on the strength of the bond and the weight of the atoms involved in the bond. The most intense changes in material structure were observed in the case of 12 M NaOH concentration.

Table 5 contains the results of compression strength values and density values of the geopolymer samples, depending on the age and concentration of NaOH used to make them. Note that in some cases, samples were broken during demolding; therefore, the determination of strength and density were not possible.

Table 5: Density of geopolymer samples.

Sample ID	Concentration of NaOH (M)	Compression strength (MPa)		Density of sample (g/cm^3)
		7 days	28 days	
B	4	–	–	–
D	4	0.62	0.82	1.16
C	4	3.19	4.69	1.42
DB	4	–	–	–
DSB	4	0.58	0.50	1.38
DSBC	4	1.36	1.52	1,51
B	8	4.27	9.85	1.76
D	8	1.77	1.86	0.93
C	8	3.38	17.13	1.96
DB	8	0.07	0.87	0.89
DSB	8	0.54	1.07	0.91
DSBC	8	0.90	0.87	0.87
B	12	–	–	–
D	12	1.92	1.81	1.06
C	12	9.04	11.72	1.43
DB	12	–	–	–
DSB	12	0.73	0.97	0.99
DSBC	12	0.89	1.15	1.02

The results show the compression strengths and density values of the geopolymer samples tested. Note that density values of these geopolymers were determined according to the dimensions of sample particles. It can be concluded that using only brick powder or the mixture of brick powder and aluminum dross geopolymers could not be formed, in the case of both 4 M and 12 M NaOH concentration. Only 8 M NaOH was appropriate to form a geopolymer structure.

Using aluminum dross, geopolymer pastes were foamed according to the reaction of the metallic aluminum and water. During this reaction hydrogen gas was formed, resulting in a foaming process. Consequently, geopolymers containing aluminum dross have a lower density and low mechanical strength. Geopolymer structures were formed in all concentrations, when concrete powder was used. The highest compressive strength (17.13 MPa) was achieved with 8 M NaOH concentration and the use of only concrete powder.

4 CONCLUSIONS

In this research, the possibility of utilization of different C&D and industrial wastes was studied for their ability to prepare useful geopolymer materials. Complex analyses were made on these powdered materials, and then their geopolymer samples were made using alkali activators. The most important properties, compressive strength and density, were determined. It can be concluded that not only the type and amount of waste materials affect the forming of a geopolymer structure, but the concentration of alkali activator is important. According to our results, concrete powder showed the highest compressive strength after 28 days; therefore, these wastes may be a potential raw material for making geopolymers. Brick powder may also be suitable as a raw material; however, in this case the use of a proper concentration of alkali activator is required. Using aluminum dross, a low density foamed geopolymer was prepared, which may be useful for thermal insulating.

ACKNOWLEDGEMENTS

The described article was carried out as part of the "Sustainable Raw Material Management Thematic Network – RING 2017" EFOP-3.6.2-16-2017-00010 project, performed within the framework of the Széchenyi2020 Program. The realization of this project is supported by the European Union, and was co-financed by the European Social Fund.

REFERENCES

[1] Onutai, S., Development of geopolymer materials sourced with fly ash and industrial waste. PhD thesis, 2016.

[2] Worrell, E., Price, L., Martin, N., Hendriks, C. & Ozawa Meida, L., Carbon dioxide emissions from the global cement industry. *Annual Review of Environment and Resources*, **26**, pp. 303–329, 2001. DOI: 10.1146/annurev.energy.26.1.303.

[3] Davidovits, J., *Geopolymer Chemistry and Application*, Institut Geopolimére: France, pp. 283–286, 2011.

[4] Mucsi, G., Lakatos, J., Molnár, Z. & Szabó, R., Development of geopolymer using industrial waste materials. *The 9th International Conference on Environmental Engineering*, Vilnius Gediminas Technical University Press, **39**, pp. 1–8, 2014. DOI: 10.3846/enviro.2014.039.

[5] Nazari, A. & Sanjayan, J.G., Synthesis of geopolymer from industrial wastes. *Journal of Cleaner Production*, **99**, pp. 297–304, 2015. DOI: 10.1016/j.jclepro.2015.03.003.

[6] Islam, A., Alengaram, U.J., Jumaat, M.Z. & Bashar, I.I., The development of compressive strength of ground granulated blast furnace slag-palm oil fuel ash-fly ash based geopolymer mortar. *Materials and Design*, **56**, 833–841, 2014. DOI: 10.1016/j.matdes.2013.11.080.

[7] Puligilla, S. & Mondal, P., Role of slag in microstructural development and hardening of fly ash-slag geopolymer. *Cement and Concrete Research*, **43**, pp. 70–80, 2013. DOI: 10.1016/j.cemconres.2012.10.004.

[8] Nath, S.K. & Kumar, S., Influence of iron making slags on strength and microstructure of fly ash geopolymer. *Construction and Building Materials*, **38**, pp. 924–930, 2013. DOI: 10.1016/j.conbuildmat.2012.09.070.

[9] Onisei, S. et al., Synthesis of inorganic polymers using fly ash and primary lead slag. *Journal of Hazardous Materials*, **205**, pp. 101–110, 2012. DOI: 10.1016/j.jhazmat.2011.12.039.

[10] Alex, T.C. et al., Utilization of zinc slag through geopolymerization: Influence of milling atmosphere. *International Journal of Mineral Processing*, **123**, pp. 102–107, 2013. DOI: 10.1016/j.minpro.2013.06.001.

[11] Yusuf, M.O., Megat Johari, M.A., Ahmad, Z.A. & Maslehuddin, M., Evolution of alkaline activated ground blast furnace slag–ultrafine palm oil fuel ash-based concrete. *Materials and Design*, **55**, pp. 387–393, 2014. DOI: 10.1016/j.matdes.2013.09.047.

[12] Ogundiran, M.B., Nugteren, H.W. & Witkamp, G.J., Immobilisation of lead smelting slag within spent aluminate–fly ash based geopolymers. *Journal of Hazardous Materials*, **248–249**, pp. 29–36, 2012. DOI: 10.1016/j.jhazmat.2012.12.040.

[13] Tuyan, M., Andiç-Çakir, Ö. & Ramyar, K., Effect of alkali activator concentration and curing condition on strength and microstructure of waste clay brick powder-based geopolymer. *Composites Part B: Engineering*, **135**, pp. 242–252, 2018. DOI: 10.1016/j.compositesb.2017.10.013.

[14] Muhammad, F., Mustafa, A.B., Kamarudin, H., Binhussain, M., Ruzaidi, C.M. & Izzat, A.M., Application of clay-based geopolymer in brick production: A review. *Advanced Materials Research*, **626**, pp. 878–882, 2013. DOI: 10.4028/www.scientific.net/AMR.626.878.

[15] Brînduş-Simuţ, J., Vyšvařil, M., Bayer, P., Keppert, M. & Rovnaníková, P., Effect of particle size of waste brick powder on the properties of alkaline activated materials. *International Conference Building Materials, Product and Technologies*, **379**, pp. 1–8, 2018. DOI: 10.1088/1757-899X/379/1/012019.

[16] Ren, X. & Zhang, L., The complete recycling of waste concrete to produce geopolymer concrete. *Civil Engineering and Engineering Mechanics*, **272**, pp. 103–111, 2016. DOI: 10.1061/9780784480151.011.

[17] Janosevic, N., Djoric-Veljkovic, S., Toplicic-Curcic, G. & Karamarkovic, G., Properties of geopolymers. *Architecture and Civil Engineering*, **1**(16), pp. 45–56, 2018. DOI: 10.2298/FUACE161226004J.

[18] Kumar, S. et al., Mechanical activation of fly ash and its influence on micro and nano-structural behaviour of resulting geopolymers. *Advanced Powder Technology*, **28**, pp. 805–813, 2017. DOI: 10.1016/j.apt.2016.11.027.

[19] Daval, S. & Soundarapandian, N., Effect of fly-ash based geopolymer coated aggregate on bituminous mixtures. *Journal of the Croatian Association of Civil Engineers*, **70**, pp. 187–199, 2018. DOI: 10.14256/JCE.1678.2016.

[20] Dassekpo, J. et al., The effects of the sequential addition of synthesis parameters on the performance of alkali activated fly ash mortar. *Results in Physics*, **7**, pp. 1506–1512, 2017. DOI: 10.1016/j.rinp.2017.04.019.

[21] Rafeet, A. et al., Guidelines for mix proportioning of fly ash/GGBS based alkali activated concretes. *Construction and Building Materials*, **147**, pp. 130–142, 2017. DOI: 10.1016/j.conbuildmat.2017.04.036.

[22] Karyawan, I., Ahyudanari, E. & Ekaputri, J., Potential use of fly ash base-geopolymer as aggregate substitution in asphalt concrete mixtures. *International Journal of Engineering and Technology*, **5**(9), pp. 3744–3752, 2017. DOI: 10.21817/ijet/2017/v9i5/170905005.

[23] EN 1097-7, Tests for mechanical and physical properties of aggregates. Part 7: Determination of the particle density of filler. Pycnometer method, 2000.

SECTION 3
METALLIC MATERIALS

EFFECTS OF MAGNESIUM CONCENTRATION AND LAYER THICKNESS ON THE ADHESION OF PHYSICAL VAPOR DEPOSITED ZNMG-ZN BI-LAYER COATINGS

SOHEIL SABOONI[1], EMAD GALINMOGHADDAM[1], RUUD JOHANNES WESTERWAAL[2], EDZO ZOESTBERGEN[2] & YUTAO PEI[1]

[1]Department of Advanced Production Engineering, University of Groningen, The Netherlands

[2]Tata Steel Nederland Technology, The Netherlands

ABSTRACT

For many years, zinc coatings have been regarded as one of the most effective anti-corrosion protective coatings for steel. Recently, it was shown that the addition of even small amounts of magnesium (Mg) to a zinc (Zn) coating can noticeably increase its corrosion protection performance; however, it has also been observed that there is poor adhesion of zinc-magnesium (ZnMg) coatings to advanced high-strength steels. The addition of a more ductile Zn interlayer between the steel substrate and the ZnMg coating is a solution to improve the adhesion problem. In the present study, a series of ZnMg-Zn bi-layered coatings with different Mg concentrations (up to 14.1 wt.% Mg), and also different thicknesses of the Zn and ZnMg layers, were prepared by a thermal evaporation process (physical vapor deposition (PVD)) in order to investigate the adhesion performance and interfacial adhesion strength. The adhesion performance of these coatings was qualified by the BMW crash adhesion test (BMW AA-M223), while the interfacial adhesion strength at the ZnMg/Zn interface was quantified by the scratch test. It was found that the interfacial adhesion strength decreases gradually with an increase in the Mg content of the top layer. The novel finding is that the interfacial adhesion strength at the ZnMg/Zn interface is independent of the thickness of the Zn interlayer; however, the adhesion performance of a ZnMg-Zn bi-layered coating during a bending test is a complex function of different parameters, such as the thickness of the Zn and ZnMg layers, the interfacial adhesion strength and the interfacial defects density.

Keywords: adhesion strength, bi-layered coating, crash adhesion test, magnesium, physical vapor deposition, scratch test, zinc.

1 INTRODUCTION

For many years, zinc (Zn) coatings have been known as one of the most effective corrosion protective coatings for steel [1]. Recently, it was shown that the addition of alloying elements such as nickel (Ni) [2], chromium (Cr) [3], aluminum (Al) [4] and magnesium (Mg) [5] to the pure Zn increases its corrosion protection performance, considerably. The advantage of Mg over other elements can be described by its effectiveness even in a low alloying concentration [6]. The higher corrosion resistance of ZnMg coatings, as compared to the pure Zn, is related to the formation of a dense protective layer called "simonkolleite" by the addition of magnesium, which acts as a barrier to the corrosive media [7]. Different methods are commonly used to deposit pure or alloyed Zn coatings on steel substrates, such as hot-dip galvanizing (HDG) and electrodeposition [8]; however, some drawbacks occur, such as the inability of producing highly alloyed ZnMg coatings, depositing multilayered coatings, hydrogen embrittlement (especially for advanced high-strength steels), high temperature impact on the steel, high costs and environmental impact that can limit their applicability. Physical vapor deposition (PVD) can be considered as a favorable technique to replace the conventional methods, as it can be performed at much lower substrate temperatures (~250–300°C) and fulfills strict environmental regulations [9].

www.witpress.com, ISSN 1743-3533 (on-line)
doi:10.2495/MC190061

Although the addition of Mg to the Zn is beneficial for corrosion resistance, it reduces the adhesion of the coating to the steel substrate [10], [11]; however the effect of the Mg content of the ZnMg layer and also the relation between the thickness of the Zn interlayer and of the ZnMg top layer on the adhesion of ZnMg-Zn coatings are not yet fully understood. To fill this gap, a series of ZnMg-Zn bi-layered coatings with different Mg concentrations and also different thicknesses of the Zn and ZnMg layers were deposited on steel substrates, and then each was evaluated by both bending and scratch tests.

2 EXPERIMENTS

ZnMg-Zn bi-layered coatings (with different Mg concentrations, 0 to 14.1 wt.% Mg) and also different thicknesses of the Zn and ZnMg layers, were deposited on a low-carbon steel substrate (commonly called black plate steel) and a DP800 type of steel, using a thermal evaporation process. The chemical composition of the steel substrates is shown in Table 1.

Table 1: Chemical composition (wt.%) of DP800 and black plate steels.

Substrate	C	Si	Mn	P	S	Ni	Cr	Cu	Fe
DP800	0.153	0.386	1.487	0.013	0.007	0.018	0.02	0.015	Bal.
Black plate	0.04–0.08	0.03	0.18–0.35	0.02	0.03	0.08	0.08	0.08	Bal.

The vacuum chamber of the PVD machine is equipped with two crucibles containing pure Zn and a ZnMg alloy, respectively. The surface of the steel strip is pretreated by a plasma magnetron-based sputter unit, to remove any surface oxides. The evaporators use an induction coil system to melt and thermally evaporate the Zn-ZnMg source material. The metal vapor passes through a vapor distribution box and is deposited on the surface of the running steel strip. The ZnMg top layers are labelled as ZnMgX, with X indicating the Mg content in weight percent.

Grazing angle X-ray diffraction was used to analyze the phases present in the ZnMg layer. The incidence angle was set at 2° for all of the coatings, to ensure that the diffracted X-ray is only generated from the top layer, in order to discard the effect of the Zn interlayer. The cross-sections of the samples were mounted in an epoxy resin, ground by SiC paper, and finally polished up to 1 μm diamond particles in an ethanol lubricant. The microstructure of these coatings was evaluated using scanning electron microscope (SEM), a Philips XL30 ESEM.

A CSM Revetest scratch tester was used to quantify the adhesion strength of the ZnMg-Zn bi-layered coatings at the ZnMg/Zn interface. In this test, a diamond stylus (Rockwell C indenter with the tip radius of 200 μm) is moved along the coating with a progressively increasing load. The load at which the first visible failure (observing the pure Zn layer underneath) is recognized, is considered to be the critical load (L_C).

The maximum load and scratch length were considered as 20 N and 10 mm, respectively, for all bi-layered coatings. A minimum of 5 scratches were made and the average L_C was reported for each sample. BMW crash adhesion tests (BMW AA-M223) were also carried out, to study the adhesion performance of the coatings during bending load. This method is a standard adhesion test method, broadly used in industry to qualify the adhesion of galvanized coatings. Initially, a line of an adhesive (Betamate 1496V DOW Automotive Systems) at least 150 mm in length, 4–5 mm thick and at least 10 mm in width was applied to the surface of the coating. The samples were kept at 175°C for half an hour (30 minutes) to cure the adhesive. After cooling, the samples were quickly bent over a 90° angle. To pass the test, the adhesively bonded joint should fail in the adhesive, not in the coating/substrate interface. The BMW adhesion test was repeated three times for each sample. More information about this test can be found elsewhere [12].

3 RESULTS AND DISCUSSION

Fig. 1 shows the phase fraction of ZnMg coatings versus the Mg concentration, as determined by the grazing angle XRD. When the Mg content is low (<1.5 wt.% Mg), the top layer mostly consists of pure Zn. With an increase in the Mg content, the phase fraction of Zn decreases gradually, while the fraction of Mg_2Zn_{11} increases. The coating is composed of a mixture of Mg_2Zn_{11} and $MgZn_2$ at Mg concentrations that are higher than 6 wt.%. The phase fraction of $MgZn_2$ increases with further increases in Mg; and the ZnMg top layer coating is almost fully covered by this phase at ~14 wt.% Mg. Fig. 2 shows the cross-sectional SEM micrographs of the ZnMg-Zn bi-layer coatings, with different Mg concentrations. The thicknesses of the Zn interlayer and the ZnMg top layer for each coating are presented in Table 2. Microstructural studies also confirm the results as found by the XRD measurements.

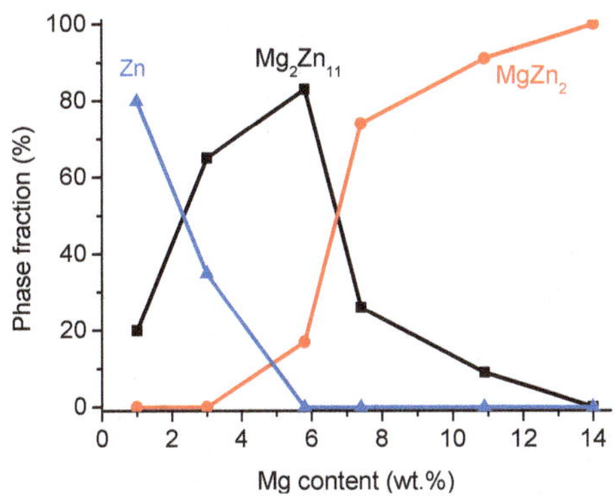

Figure 1: Phase fraction of ZnMg versus Mg content in the range of 1.5–14.1 wt.% Mg.

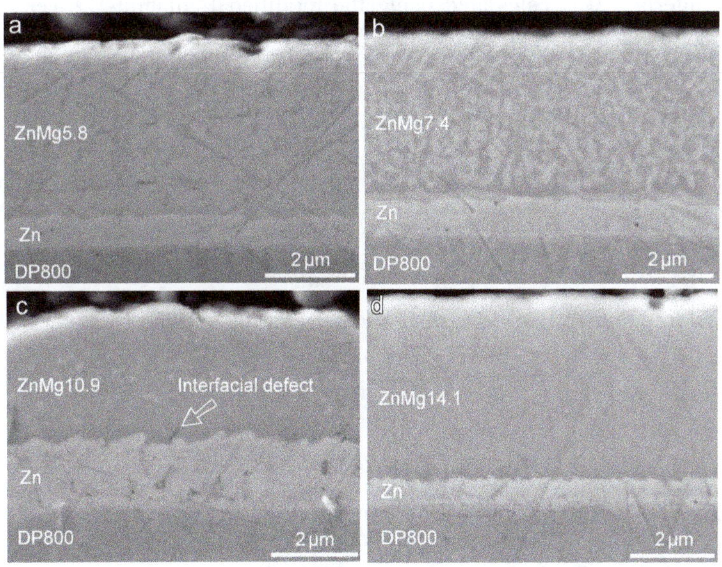

Figure 2: SEM micrographs showing cross sections of ZnMg-Zn bi-layered coatings with a ZnMg top layer, at the different Mg concentrations (wt.%): (a) 5.8; (b) 7.4; (c) 10.9; (d) 14.1.

Table 2: Adhesion strength of pure Zn and ZnMg-Zn bi-layer coatings containing different Mg contents measured using scratch test.

Coating	Thickness of Zn layer (μm)	Thickness of ZnMg layer (μm)	Critical load Lc (N)	Residual depth at Lc (μm)	Weight factor (ω)	Adhesion strength (MPa)	BMW adhesion test
Pure zinc	4.9	0	38.5 ± 1.5	-	-	171	pass
ZnMg5.8-Zn	0.7	3.9	18.1 ± 0.6	4.3	0.16	129	pass
ZnMg7.4-Zn	0.9	3.7	13.2 ± 0.7	3.3	0.27	103	pass
ZnMg10.9-Zn	1.6	3.1	9.3 ± 0.6	2.0	0.80	54	not pass
ZnMg14.1-Zn	0.6	4.4	8 ± 0.6	1.9	0.31	78	pass

Scratch tests were carried out to study the effect of the Mg concentration on the interfacial adhesion strength between the ZnMg and the Zn layer. Table 2 summarizes the critical load of delamination for each coating. It is worth mentioning that the overall thickness of all the coatings is in the same range, 4.6–5 μm. The critical load of the delamination for the pure Zn coating (38.5 N) is higher than that of all other ZnMg-Zn coatings. L_C decreases with increasing Mg concentration, reaching 18.1 N and 13.2 N, at 5.8 wt.% and 7.4 wt.% Mg, respectively. Further reduction of L_C is observed at even higher Mg content. The BMW crash adhesion test revealed that all of the coatings could pass the test, except the ZnMg10.9-Zn coating; however, the L_C of this coating is still higher than that of ZnMg14.1-Zn. Therefore, it can be concluded that L_C is not a suitable criterion to compare the adhesion of different ZnMg-Zn bi-layered coatings with different Mg concentrations and/or different layer thicknesses.

To find a proper way to quantify adhesion, the interfacial adhesion strength at the ZnMg/Zn interface was calculated using the modified Benjamin-Weaver model [13], as follows:

$$F = \frac{K\,a\,H}{\sqrt{R^2 - a^2}}, \tag{1}$$

where F is the adhesion strength (in MPa); K is constant; R is the radius of the indenter tip; H is considered as the hardness of the substrate; and a is the radius of the contact circle at L_C, as shown in eqn 2.

$$a = \left(\frac{L_C}{\pi H}\right)^{0.5}. \tag{2}$$

To consider the effect of both steel and zinc as the substrate for the ZnMg top layer, composite hardness was calculated, using a defined weight factor, as follows:

$$\omega = \frac{\text{Thickness of zinc interlayer}}{\text{Residual depth at } L_C}, \tag{3}$$

$$H_{composite} = \omega\, H_{Zn} + (1 - \omega)\, H_{steel}. \tag{4}$$

The adhesion strength of the PVD pure Zn on the steel substrate is 171 MPa, which is very comparable with the 180 MPa reported earlier for the adhesion of hot-dipped pure Zn onto steel [14]. The adhesion strength at the ZnMg/Zn interface decreases gradually with an increase in the Mg concentration, and reaches up to 78 MPa at 14.1 wt.% Mg.

The results of the interfacial adhesion strength obtained by the Scratch test follow the same trend; with the results of the BMW adhesion test and the ZnMg10.9-Zn coating failing during the bending, which also showed the lowest adhesion strength among samples. Such low adhesion strength is related to the interfacial defects, which are present at the interface (Fig. 2(c)). It is also worth mentioning that the modified model can nicely reproduce the results of the BMW adhesion test for the ZnMg14.1-Zn coating, in contrast to only the L_C method. The ZnMg14.1-Zn coating, which pass the BMW adhesion test, shows a higher interfacial adhesion strength when compared to the ZnMg10.9-Zn coating.

The coatings that contain 5–7 wt.% Mg are apparently good candidates for practical applications which require simultaneously good adhesion and corrosion performance of the ZnMg coatings. Therefore, to study the effect of the thickness of the Zn interlayer on adhesion, we prepared a series of coatings with the same Mg concentration (~6.5 wt.% Mg) in the ZnMg top layer and different thicknesses of the Zn interlayer (0.2, 0.7 and 1.3 μm) (Fig. 3).

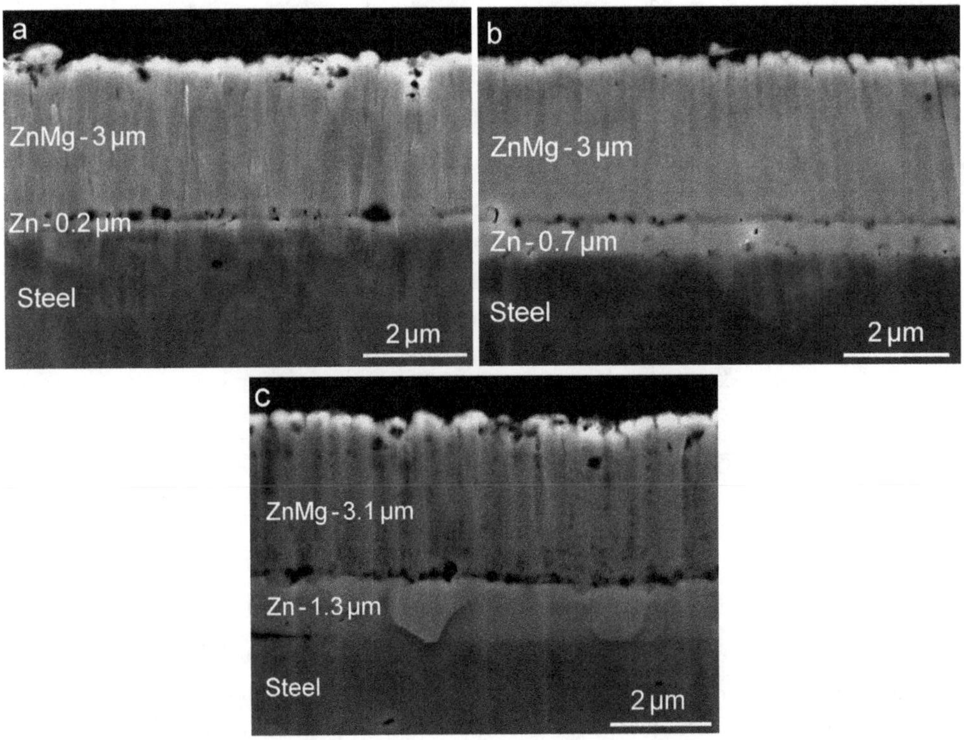

Figure 3: SEM micrographs of ZnMg6.5-Zn coatings with different thickness of the Zn interlayer: (a) 0.2 μm; (b) 0.7 μm; (c) 1.3 μm.

Fig. 4 shows the effect of the thickness of the Zn interlayer on the L_C (Fig. 4(a)) and the interfacial adhesion strength (Fig. 4(b)) of the ZnMg-Zn bi-layer coatings. The L_C increases with increases in the thickness of Zn from 11.6 to 13.25 N; however, the adhesion strength at the ZnMg/Zn interface is independent of the thickness, and is around 110 MPa for the ZnMg6.5-Zn coatings. It indicates that the Mg concentration plays the most dominant role in the interfacial adhesion strength, rather than the thickness. It should be noted that although the interfacial adhesion strength of all these coatings is the same, the coating with the lowest thickness of Zn failed in the BMW adhesion test, while the others passed. It can be concluded that the adhesion performance of a substrate/coating system during bending does not only depend on the interfacial adhesion strength, but also the thickness of the Zn interlayer.

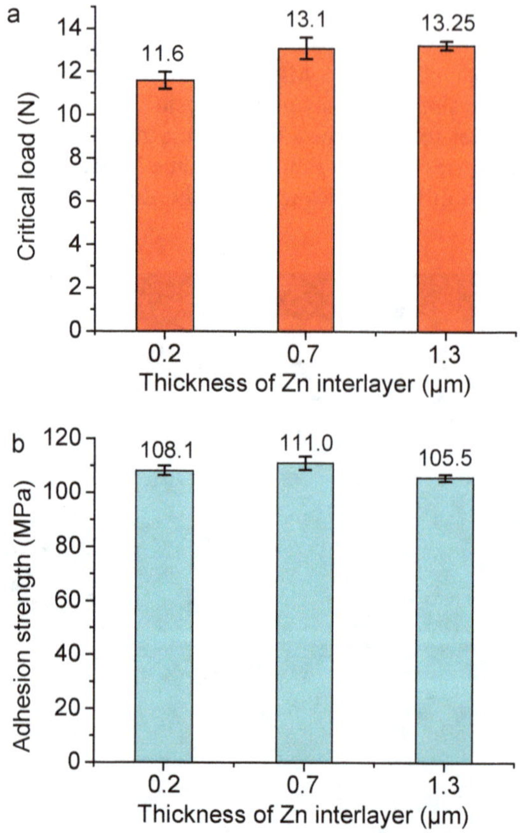

Figure 4: (a) Critical load of delamination and (b) interfacial adhesion strength of ZnMg6.5-Zn coatings, versus the thickness of the Zn interlayer.

It was shown previously that ZnMg-Zn bi-layered coatings with a sufficient thickness of the Zn interlayer and no interfacial defects between the Zn and ZnMg layer can pass the BMW adhesion test; however, it was also observed that the thickness of the ZnMg top layer can influence the adhesion performance of the coatings. The coatings fail in the BMW adhesion test when the thickness of ZnMg increases from a threshold level. Fig. 5(a) shows the cross-section SEM micrograph of a ZnMg14-Zn coating, with 6.8 μm thick top layer, which failed the BMW adhesion test. The SEM micrographs of the exposed side of the coating that remained on the substrate after the failure are shown in Fig. 5(b)–(d). The results of EDS analysis of different areas of the failed surface are shown in Table 3. Spot 1 consisted of 95.4 wt.% Zn and 4.6 wt.% Mg, while Spot 2 consisted of 84.5 wt.% Zn and 15.3 wt.% Mg. It should be mentioned that no sign of Fe is detected on the failed surface. These results clearly indicate that the ZnMg-Zn coatings with a relatively thick top layer experience failure in the BMW adhesion test, as cohesive brittle failure inside the ZnMg layer instead of adhesive failure at the interface.

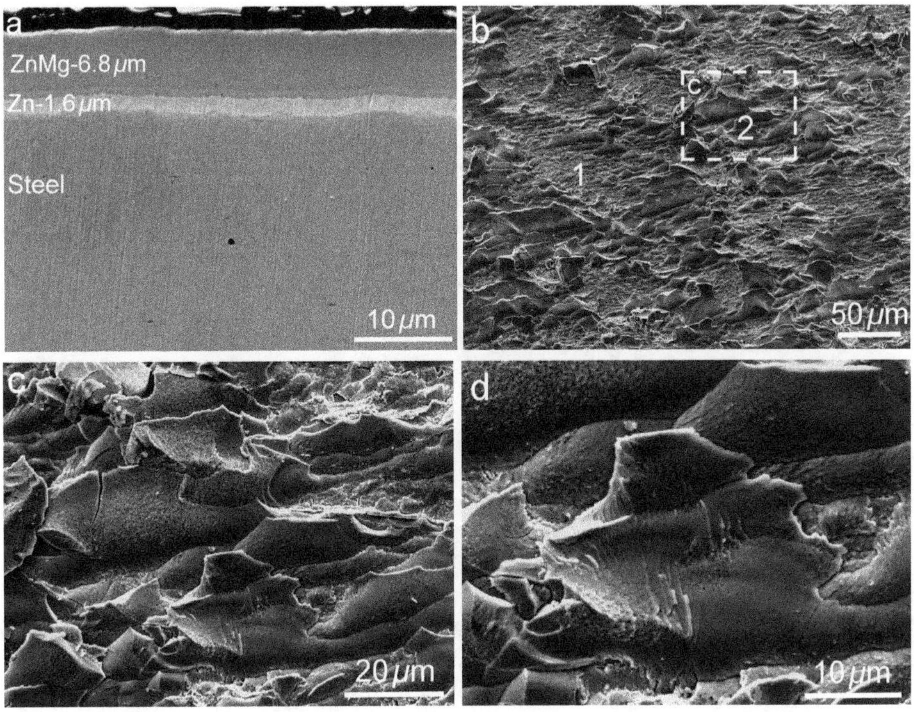

Figure 5: (a) SEM micrograph showing the cross section of ZnMg14-Zn coating containing 6.8 μm thick ZnMg top layer; (b)–(d) SEM micrographs of the exposed side of the coating which remains on the substrate after failure in the BMW crash adhesion test.

Table 3: Chemical composition of different points in Figure 5(b).

Location	Zn content (wt.%)	Mg content (wt.%)	Fe content (wt.%)
Spot 1	95.4	4.6	0
Spot 2	84.7	15.3	0

As a summary, it can be concluded that the adhesion performance of a ZnMg-Zn bi-layered coating during a bending test is a complex function of different parameters, such as the thickness of both the Zn and ZnMg layer, interfacial adhesion strength, and the interfacial defect density; as is schematically drawn in Fig. 6.

4 CONCLUSIONS

The effects of the Mg concentration and the thickness of Zn and ZnMg layers were studied on the adhesion of ZnMg-Zn bi-layered coatings, by the scratch and BMW crash adhesion tests. The novel findings are summarized as follows:

1. The critical load of the delamination is not a proper criterion through which to compare the adhesion of the ZnMg-Zn bi-layered coatings with different composition and/or thickness ratios.

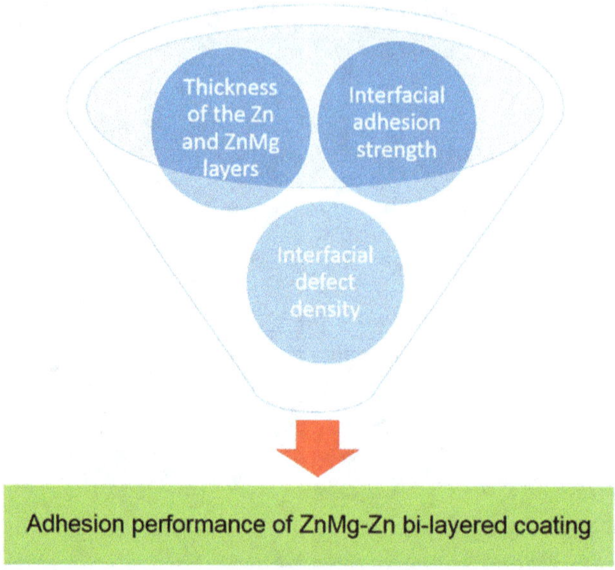

Figure 6: Parameters influencing the adhesion performance of ZnMg-Zn bi-layered coatings in the BMW crash adhesion test.

2. The Benjamin-Weaver model needs to be modified to be able to calculate the adhesion strength at the ZnMg/Zn interface by taking into account the L_C, thickness of the Zn interlayer, and the hardness of both substrate and interlayer.
3. The same trend is observed between the interfacial adhesion strength calculated by the scratch test and the BMW adhesion test, which is currently being used in industry for qualification of the adhesion.
4. The adhesion strength at the ZnMg/Zn interface depends on the Mg concentration of the top layer, but not on the thickness of the Zn interlayer. The interfacial adhesion strength decreases gradually, with increasing Mg concentrations.
5. The adhesion performance of a ZnMg-Zn bi-layered coating during a bending test is a complex function of different parameters, such as the thickness of the Zn and ZnMg layers, interfacial adhesion strength, and the defect density at the interface.

ACKNOWLEDGEMENTS

This research was carried out under project number S22.3.13513a within the framework of the Partnership Program of the Materials Innovation Institute M2i (www.m2i.nl) and the Technology Foundation TTW (www.stw.nl), which is part of the Netherlands Organization for Scientific Research (www.nwo.nl).

REFERENCES

[1] Asgari, H., Toroghinejad, M.R. & Golozar, M.A., On texture, corrosion resistance and morphology of hot-dip galvanized zinc coatings. *Applied Surface Science*, **253**, pp. 6769–6777, 2007.

[2] Rashmi, S., Elias, L. & Hedge, A.C., Multilayered Zn-Ni alloy coatings for better corrosion protection of mild steel. *Engineering Science Technology, an International Journal*, **20**, pp. 1227–1232, 2017.

[3] Chakarova, V., Boiadjieva-Scherzer, T., Kovacheva, D., Kronberger, H. & Monev, M., Corrosion behaviour of electrodeposited Zn-Cr alloy coatings. *Corrosion Science*, **140**, pp. 73–78, 2018.

[4] LeBozec, N., Thierry, D., Persson, D. & Stoulil, J., Atmospheric corrosion of zinc-aluminum alloyed coated steel in depleted carbon dioxide environments. *Journal of the Electrochemical Society*, **165**, C343–C353, 2018.

[5] Dong, A. et al., Effect of Mg on the microstructure and corrosion resistance of the continuously hot-dip galvanizing Zn-Mg coating. *Materials*, **10**, p. 980, 2017.

[6] Oh, M.S., Kim, S.H., Kim, J.S., Lee, J.W., Shon, J.H. & Jin, Y.S., Surface and cut-edge corrosion behavior of Zn-Mg-Al alloy-coated steel sheets as a function of the alloy coating microstructure. *Metals and Materials International*, **22**, pp. 26–33, 2016.

[7] Pohl, K., Surface Chemistry and Corrosion Studies of Zn–Al and Zn–Mg–Al Alloy Coatings. PhD thesis, University of Paderborn, Germany, 2013.

[8] Marder, A.R., The metallurgy of zinc-coated steel. *Progress in Materials Science*, **45**, pp. 191–271, 2000.

[9] Navinsek, B., Panjan, P. & Milosev, I., PVD coatings as an environmentally clean alternative to electroplating and electroless processes. *Surface and Coatings Technology*, **116–119**, pp. 476–487, 1999.

[10] La, J.H., Bae, K.T., Lee, S.Y. & Nam, K.H., A hybrid test method to evaluate the adhesion characteristics of soft coatings on steel substrates: Application to Zn-Mg coated steel. *Surface and Coatings Technology*, **307**, pp. 1100–1106, 2016.

[11] Jung, W.S., Lee, C.W., Kim, T.Y. & De Cooman, B.C., Mg content dependence of EML-PVD Zn-Mg coating adhesion on steel strip. *Metallurgical and Materials Transactions A*, **47**, pp. 4594–4605, 2016.

[12] Zoestbergen, E., Van de Langkruis, J., Maalman, T.F.J. & Batyrev, E., Influence of diffusion on the coating adhesion of zinc-magnesium thin films onto steel. *Surface and Coatings Technology*, **309**, pp. 904–910, 2017.

[13] Sabooni, S., Galinmoghaddam, E., Ahmadi, M., Westerwaal, R.J., Van de Langkruis, J., Zoestbergen, E., De Hosson J.Th.M. & Pei, Y.T., Microstructure and adhesion strength quantification of PVD bi-layered ZnMg-Zn coatings on DP800 steel. *Surface and Coatings Technology*, **359**, 227–238, 2019.

[14] Song, G.M., Sloof, W.G., Pei, Y.T. & De Hosson, J.T.M., Interface fracture behavior of zinc coatings on steel: Experiments and finite element calculations. *Surface Coating Technology*, **201**, pp. 4311–4316, 2006.

CHARACTERIZATION AND MODELING OF ADVANCED OPEN-POROUS COPPER HEAT EXCHANGER MATERIALS

JAKUB SKIBINSKI
Faculty of Materials Science and Engineering, Warsaw University of Technology, Poland

ABSTRACT

Open-porous copper foams applied as heat exchanger materials are addressed within the study. Commercially available materials are characterized with the use of microcomputed tomography. The results of materials characterization act as assumptions for the development of a numerical design procedure, with the ability to control the selection of porous structure parameters, including constrictivity and pore size distribution. An adaptation of Laguerre-Voronoi tessellations is performed allowing for exact representation of complex porous structures. The design algorithm has been developed in such a way as to enable determination of the influence of pore size distribution on utility properties such as temperature transfer efficiency.

Keywords: copper foams, open-porous materials, Laguerre-Voronoi tessellations.

1 INTRODUCTION

Open-porous copper foams are high performance materials with high thermal conductivity and high specific strength. These features make such materials attractive for application in cross-flow heat exchangers, catalytic converters, electrodes of high-temperature fuel cells or solar collectors [1], [2], where high performance materials with thermal conductivity being an essential factor are required.

Structural parameters of porous materials, such as porosity, mean pore size, pore size distribution or strut shape can be controlled during the manufacturing process by adjustment of foaming powders or foaming gas. The microstructure of open-porous foams is complex and irregular, which rules out the possibility to apply analytical approaches for studying such materials. Rapid increase in computational power in recent years, together with the development of more efficient algorithms made modeling of open-porous structure possible.

Various models have been proposed in the literature in order to properly describe the complexity of the structure, representing crucial features of real porous structures. Models representing the structure by ordered set based on regular polyhedrons (usually tetracaidecahedrons) are frequently used in studies on the properties of open porous materials [3]–[5]. It turns out that – despite the random structure – a set of macroscopic properties of porous structures can be modeled with sufficient accuracy with the use of regular geometry. Experimental observations, however, clearly indicate the distribution of pore size [6] and sometimes anisotropy in real materials, which makes such simplified models insufficiently accurate. In order to increase consistency of models with real materials other approaches for simulating properties of porous materials have been developed, namely tessellation methods and methods based on computed tomography. In tessellation methods the volume of representative cell of the material is divided to smaller polyhedral volumes taking into account predefined size and distribution. Different variants of Voronoi tessellations are successfully applied to represent the random nature of foam materials [7], [8], most commonly Laguerre-Voronoi tessellations (LVT) [9]–[11]. This algorithm, applied also by the authors of this study, enables one to develop a model of foam materials with controlled pore size distribution. 2D/3D reconstruction methods [12]–[14] have also drawn the attention

WIT Transactions on Engineering Sciences, Vol 124, © 2019 WIT Press
www.witpress.com, ISSN 1743-3533 (on-line)
doi:10.2495/MC190071

of scientist in recent years, where model of geometry is developed on the basis of reconstruction and processing of tomographic image and subsequently used in dedicated software for simulations.

Quantitative relationships between microstructural parameters and porous materials properties is the main goal of most of the studies applying models of porous structures found in literature [15]–[17]. Heat transfer in foam materials is also a subject described in scientific research. A wide literature review in this area has been performed by Zhao [18]. Thermal conductivity of porous materials has been reported to be strongly influenced by porosity. The influence of mean pore size (expressed in pores per inch – ppi) on heat transfer properties of porous materials is a subject undertaken in research performed by Bağcı and Dukhan [19]. Also the influence of strut shape on thermal properties of porous materials has been studied [20]. Our previous study exhibited the influence of pore size distribution on the permeability of open-porous foams [21].

The aim of this study is to develop numerical model properly representing commercial copper foams. Commercial materials are characterized with the use of micro-computed tomography, serving as reference structures for the development of numerical design procedure. The developed model is based on Laguerre-Voronoi tessellations. The purpose of the model is to enable numerical modeling of heat and mass transfer processes for further studies of microstructure influence on utility parameters of open-porous metallic foams. A set of numerical simulations has been performed employing the finite volume method, in order to verify applicability of the developed model.

2 MATERIALS AND METHODS

2.1 Commercial copper foams tomography

Commercial copper foams for heat transfer applications were used as reference for the design of virtual models. Foams with pore size of 10 ppi have been studied with use of SkyScan X-Radia XCT-400 in order to determine their structural parameters. Porosity, mean pore diameter and mean strut diameter have been determined for validation of the virtual models generated by the design procedure. Three-dimensional image acquisition has been performed under acceleration voltage of 150kV and current of 50A, which assured a voxel resolution of 41.49 microns. A more detailed description of the applied methodology can be found in Regulski et al. [22].

2.2 Numerical model of open-porous foams

Numerical models of complex porous geometries were designed by using of Laguerre-Voronoi tessellations algorithm. LVT enables generation open-cell porous structures with predefined porosity, mean pore size and pore size variation. More details about the modeling procedure can be found in Wejrzanowski et al. [11]. Adaptation of the described procedure included mainly the adaptation of cell strut shape. In our previous studies we used cylindrical strut shape, since the model was based on alumina foams applied as filters for molten metals. Strut shape of copper foams studied within the present work clearly indicates struts which are narrow between the triple points and thicker at the connections. Proposed model strut shape is shown in Fig. 1.

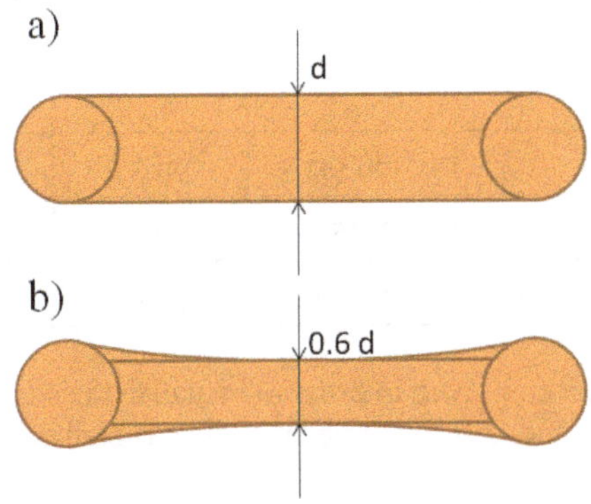

Figure 1: Cross-section of a single strut: (a) Standard; (b) Modified LVT procedure.

2.3 Temperature transfer efficiency calculations

Temperature transfer efficiency of the studied materials has been determined with use of finite element method (FEM) implemented in ANSYS Fluent software. Numerical simulations of fluid flowing through the heated copper foam with constant temperature in order to determine temperature transfer efficiency of the material. Studied material was placed in rectangular calculation channel, where fluid flow was applied. Navier-Stokes equations for Newtonian fluid at steady state have been solved with assumption of constant temperature of 352.5 K on the walls of the foam structure. Temperature value has been chosen based on working temperature of cross-flow heat exchanger with copper foam as a heating material. Temperature transfer coefficient μT was be calculated:

$$\mu T = \frac{(T_3 - T_1)}{(T_2 - T_1)},$$

where T_1 is the inlet temperature, T_2 is the struts temperature and T_3 is the outlet temperature. Performed mesh convergence study indicated a mesh size of approximately 7 million elements to be optimal for valid representation of the microstructure of the samples. The aim of initial calculations performed within the presented study was to verify the utility of the developed modelling method for further studies of the influence of structural parameters on utility properties of the materials.

3 RESULTS AND DISCUSSION

The goal of the performed studies was to characterize open-porous copper foam structures, indicating crucial structural parameters that would serve as guidelines for development of numerical model. The developed model, properly representing complexity of the foam structure, will allow for further studying the influence of chosen structural parameters on utility properties of open-cell metallic foams.

3.1 Characterization of copper foams

The images obtained with computer tomography has been analysed with CTAn software in order to determine crucial structural parameters of studied foams. Parameters obtained are listed in Table 1.

Table 1: Microstructural parameters of 10 ppi commercial copper foams.

Structure	10 ppi No 1	10 ppi No 2	10 ppi No 3	Standard deviation
Porosity (%)	90.98	90.37	91.19	+/- 0.82
Mean pore diameter (mm)	3.753	3.836	3.917	+/- 0.164
Mean strut diameter (mm)	1.144	1.192	1.137	+/- 0.055

Avizo software has been employed in order to visualize the studied sample and to perform assessment of the strut shape. An example of reconstructed 3D copper foam structure is shown in Fig. 2.

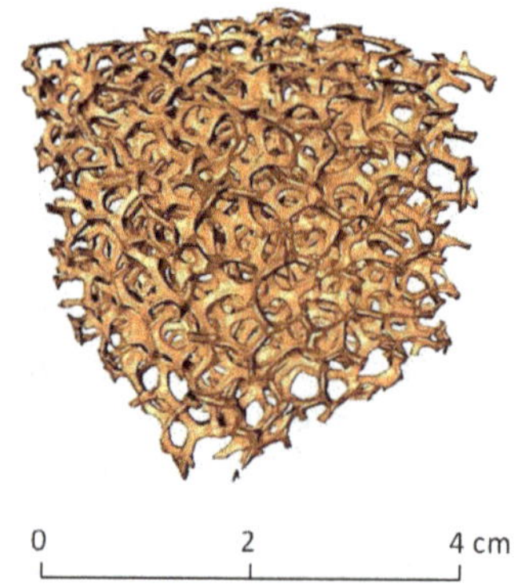

Figure 2: Reconstruction of tomography results of 10 ppi open-porous copper foam.

3.2 Numerical model of open-porous foams

A set of 11 structures with coefficient of pore volume variation CV(V) ranging from 0.45 to 2.0 was developed. Coefficient of pore volume variation was set using a selected range of sphere diameters in the first step of LVT procedure described above. The value of CV(V) for each structure was calculated as CV(V) for LVT cells. Each structure was limited by a bounding box of cubic shape and consisted of 200 pores. According to the literature data [12], [23] this is expected to be sufficiently high number of pores to study their effect on the properties of the structures in question. Mean strut diameter value was 1.0 mm in accord with data obtained from materials characterisation. This resulted in porosity of ~91%, which is in accord with porosity of commercial materials (Table 2).

An example of the designed structure is shown in Fig. 2.

3.3 Temperature transfer efficiency calculations

Temperature transfer efficiency of the designed models has been determined with use of ANSYS Fluent software. The inlet temperature of the fluid was 300K (room temperature), the temperature of the struts was chosen in accord with working temperature of commercial materials, 352.5K, and the inlet velocity was 5 cm/s. Calculated values have been plotted vs.

Table 2: Microstructural parameters designed model foams.

Structure	Porosity (%)	Mean strut diameter (mm)	Coefficient of pore size variation (-)
Model 1	91.07	1.00	0.478
Model 2	91.08	1.00	0.484
Model 3	91.07	1.00	0.459
Model 4	91.06	1.00	0.587
Model 5	91.31	1.00	0.859
Model 6	91.41	1.00	0.640
Model 7	91.25	1.00	1.400
Model 8	91.21	1.00	1.223
Model 9	91.32	1.00	1.391
Model 10	91.58	1.00	1.433
Model 11	91.61	1.00	1.740

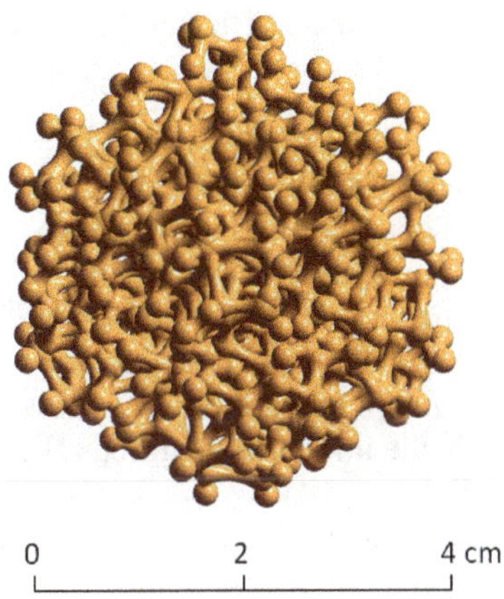

Figure 3: Numerical model of open-porous foam obtained with developed LVT design procedure.

coefficient of pore size variation CV(V) in Fig. 3 in order to determine the influence of pore size distribution (dimensionless coefficient of pore size variation) on temperature transfer efficiency of the materials.

The results clearly indicate the influence of pore size variation on the heat transfer properties of metallic foams. With increasing value of pore size variation coefficient, the temperature transfer efficiency of the material drops. This phenomenon can be explained with higher permeability of the materials with higher inhomogeneity of the pore size. Structures with homogeneous pore size distribution have lower permeability [21], hence they retain flowing fluid easier, facilitating the heat exchange. The effect can also be assigned to constrictivity, a factor determining narrow connections between the neighbouring pores, however this is just an assumption, and separate research needs to be conducted in order to confirm such influence on heat exchange phenomena.

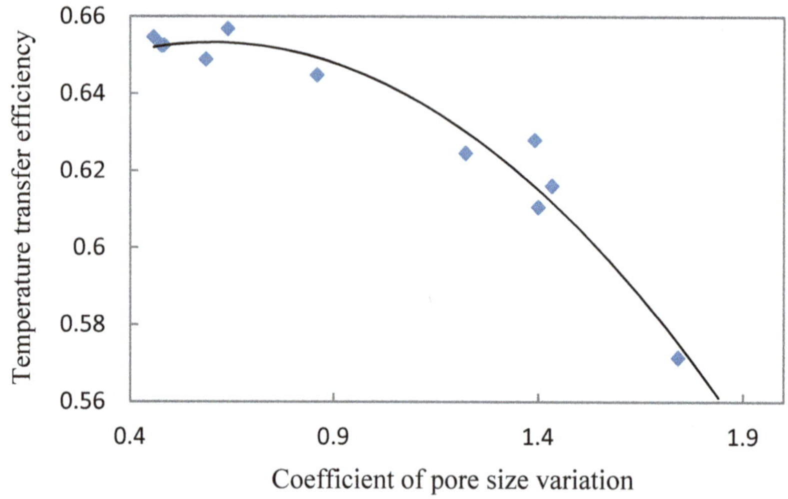

Figure 4: Calculated temperature transfer efficiency of the designed open-cell foams.

4 CONCLUSIONS

Commercial open-porous copper foams were characterized with use of computer tomography. Numerical model of open-porous metallic foams, employing LVT tessellations, has been developed based on the structural parameters of characterised materials. The developed model has been applied to develop a representative set of 11 structures with various pore size distribution. Calculations of temperature transfer efficiency has been performed with use of finite volume method in order to assess the ability of the developed materials design method to study the influence of structural parameters on utility properties of open-porous ceramic materials.

The results obtained indicate a strong influence of pore size distribution on heat transfer properties of open-cell copper foams. Lower temperature transfer efficiency values have been obtained for higher values of coefficient of pore size variation. Further research is necessary in order to determine the influence of structural parameters on heat transfer properties of foams. More structures need to be designed and tested including different inlet velocities, range of material porosities. Nevertheless, the results obtained confirmed the utility of the developed design procedure for application in further research of the influence of structural parameters on utility properties of open-porous metallic foams.

ACKNOWLEDGEMENTS

This work was financially supported by the Polish National Science Centre in the framework of the SONATINA research project (Grant no. 2018/28/C/ST8/00311).

REFERENCES

[1] Han, X.-H., Wang, Q., Park, Y.-G. T'Joen, C., Sommers, A. & Jacobi, A., A review of metal foam and metal matrix composites for heat exchangers and heat sinks. *Heat Transf. Eng.*, **33**, pp. 991–1009, 2012. DOI: 10.1080/01457632.2012.659613.

[2] Banhart, J., Manufacture, characterisation and application of cellular metals and metal foams. *Prog. Mater. Sci.*, **46**, pp. 559–632, 2001. DOI: 10.1016/S0079-6425(00)00002-5.

[3] Liu, R. & Antoniou, A., A relationship between the geometrical structure of a nanoporous metal foam and its modulus. *Acta. Mater.*, **61**, pp. 2390–2402, 2013. DOI: 10.1016/j.actamat.2013.01.011.

[4] Pia, G. & Delogu, F. On the elastic deformation behavior of nanoporous metal foams. *Scr. Mater.*, **69**, pp. 781–784, 2013. DOI: 10.1016/j.scriptamat.2013.08.027.

[5] Ko, W.L., Deformations of foamed elastomers. *J. Cell. Plast.*, **1**, pp. 45–50, 1965. DOI: 10.1177/0021955X6500100107.

[6] Lautensack, C. & Sych, T., 3D Image analysis of open foams using random tessellations. *Image Anal. Stereol.*, **25** pp. 87–93, 2006. DOI: 10.5566/ias.v25.

[7] Watson, D., Spatial tessellations: concepts and applications of Voronoi diagrams, *Comput. Geosci.*, **19**, pp. 1209–1210, 1993. DOI: 10.1016/0098-3004(93)90024-Y.

[8] Ferenc, J.S. & Néda, Z., On the size distribution of Poisson Voronoi cells. *Physica. A.*, **385**, pp. 518–526, 2007. DOI: 10.1016/j.physa.2007.07.063.

[9] Redenbach, C., Microstructure models for cellular materials. *Comput. Mater. Sci.*, **44**, pp. 1397–1407, 2009. DOI: 10.1016/j.commatsci.2008.09.018.

[10] Wejrzanowski, T., Haj Ibrahim, S., Skibiński, J., Cwieka, K. & Kurzydlowski, K.J., Appropriate models for simulating open-porous material. *Image. Anal. Stereol.*, **36**, pp. 107–112, 2017.

[11] Wejrzanowski, T., Skibinski, J., Szumbarski, J. & Kurzydlowski, K.J., Structure of foams modeled by Laguerre-Voronoi tessellations. *Comput. Mater. Sci.*, **67**, 2013. DOI: 10.1016/j.commatsci.2012.08.046.

[12] Veyhl, C., Belova, I.V., Murch, G.E. & Fiedler, T., Finite element analysis of the mechanical properties of cellular aluminium based on micro-computed tomography. *Mater. Sci. Eng. A.*, **528**, pp. 4550–4555, 2011. DOI: 10.1016/j.msea.2011.02.031.

[13] Magnico, P., Analysis of permeability and effective viscosity by CFD on isotropic and anisotropic metallic foams. *Chem. Eng. Sci.*, **64**, pp. 3564–3575, 2009. DOI: 10.1016/j.ces.2009.04.036.

[14] Westhoff, D., Skibinski, J., Šedivý, O., Wysocki, B., Wejrzanowski, T. & Schmidt, V., Investigation of the relationship between morphology and permeability for open-cell foams using virtual materials testing. *Mater. Des.*, **147**, pp. 1–10, 2018.

[15] Haj Ibrahim, S., Skibiński, J., Oliver, G.J. & Wejrzanowski, T., Microstructure effect on the permeability of the tape-cast open-porous materials. *Mater. Des.*, **167**, 2019.

[16] Bhattacharya, A., Calmidi, V.V. & Mahajan, R.L., Thermophysical properties of high porosity metal foams. *Int. J. Heat Mass Transf.*, **45**, pp. 1017–1031, 2002. DOI: 10.1016/S0017-9310(01)00220-4.

[17] Wejrzanowski, T. et al., Multi-modal porous microstructure for high temperature fuel cell application. *J. Power Sources*, **373**, pp. 85–94, 2018.

[18] Zhao, C.Y., Review on thermal transport in high porosity cellular metal foams with open cells. *Int. J. Heat Mass Transf.*, **55**, pp. 3618–3632, 2012. DOI: 10.1016/j.ijheatmasstransfer.2012.03.017.

[19] Bağcı, Ö. & Dukhan, N., Experimental hydrodynamics of high-porosity metal foam: Effect of pore density. *Int. J. Heat Mass Transf.*, **103**, pp. 879–885, 2016. DOI: 10.1016/j.ijheatmasstransfer.2016.07.097.

[20] Ambrosio, G., Bianco, N., Chiu, W.K.S., Iasiello, M., Naso, V. & Oliviero, M., The effect of open-cell metal foams strut shape on convection heat transfer and pressure drop. *Appl. Therm. Eng.*, **103**, pp. 333–343, 2016. DOI: 10.1016/j.applthermaleng.2016.04.085.

[21] Skibinski, J., Cwieka, K., Kowalkowski, T., Wysocki, B., Wejrzanowski, T. & Kurzydlowski, K.J., The influence of pore size variation on the pressure drop in open-cell foams. *Mater. Des.*, **87**, pp. 650–655, 2015. DOI: 10.1016/j.matdes.2015.08.079.
[22] Regulski, W. et al., Pressure drop in flow across ceramic foams-A numerical and experimental study. *Chem. Eng. Sci.*, **137**, 2015.
[23] Roberts, P. & Garboczi, E.J., Elastic properties of model porous ceramics. *J. Am. Ceram. Soc.*, **83** pp. 3041–3048, 2000. DOI: 10.1111/j.1151-2916.2000.tb01680.x, 10.1016/j.ces.2015.06.043.

DEVELOPMENT OF MATERIAL OPTIMIZATION TECHNOLOGY FOR SMALL STRUCTURES

IKUO TANABE & PAULO DA SILVA
Department of Mechanical Engineering, Nagaoka University of Technology, Japan

ABSTRACT

Many innovative products with multi-functions and of excellent quality are being manufactured. Although new ideas, concepts and innovations emerge, products with higher properties remain the market's favourite. Therefore, in this research, the material optimization technology for a small structure was developed and evaluated. Some small structures were desired for miniaturizing, weight and energy saving in the products. The concept of "material optimization" is a new and general idea. This material is of the composite type. The material optimization technology regarding Young's modulus, density, coefficient of linear expansion, specific heat and thermal conductivity were the main consideration of this research. The calculation models for the properties were discussed first and presented. The manufacturing method for a small structure with the desired properties was then developed. The small structure with the material optimization was manufactured and evaluated. It is concluded from the research that the technology was very effective and useful for the development of a new composite material with several hybrid properties.
Keywords: small structure, material optimization, new material, hybrid property, composite.

1 INTRODUCTION

In the 21st century, the demand for products with high quality [1], hybrid properties [2], multi-function [3] and considerable environment-friendliness [4] has been rapidly increasing. Here, several optimization technologies are currently being used to address these issues. Particularly, topology optimization technology is considered as useful in the manufacturing field due to the high quality, high reliability and safety that it offers. However, it has been observed that there is a lack of proper material optimization techniques in the technology development process. There are hundreds of materials used in the industrial field yet, surprisingly, the study on material property optimization for innovative development is very minimal.

Thus, previous research [5], through a developed software, defined a material optimization technology for innovation. The technology relied on a software that creates new materials with hybrid properties, a hybrid materials manufacturing method, and an algorithm for material optimization. The aforementioned software can effectively calculate up to five desired material properties that are requested (i.e. Young's modulus, density, coefficient of linear expansion, specific heat and thermal conductivity). In addition, the algorithm for material optimization was defined through FEM inverse analysis. The material optimization technology was then evaluated with a simple experiment. In this regard, it was thought that material optimization technologies were highly relevant to explore due to the design and manufacturing implications of having a customized material.

In the present research, the material optimization technology for small structures was developed and evaluated. A new manufacturing process was developed for a small structure with a new composite. The previous software which could effectively calculate up to five desired material properties that are requested at once (i.e. Young's modulus, density, coefficient of linear expansion, specific heat and thermal conductivity) was evaluated for a small structure in several experiments. A material optimization algorithm using FEM inverse analysis and the previous software [5] was considered.

WIT Transactions on Engineering Sciences, Vol 124, © 2019 WIT Press
www.witpress.com, ISSN 1743-3533 (on-line)
doi:10.2495/MC190081

2 PROPERTIES CALCULATION MODELS FOR NEW COMPOSITE MATERIALS

In this section, a previously developed software and the several equations are quoted [5]–[8]. There are density, thermal conductivity, specific heat, coefficient of linear expansion and Young's modulus among properties considered in the calculation models. Their calculation accuracies are shown in [9].

3 PROPOSED MANUFACTURING PROCESS FOR SMALL STRUCTURES WITH A NEW COMPOSITE

In a similar way, the defined proposed manufacturing method for the designed materials using the aforementioned equations in a software (which defined two scenarios (1) Selecting materials and calculating appropiate ratios to achieve intended effects; (2) Calculating the properties generated with these ratios and materials) and its method that was used in the previous research are applied in this study. Consequently, a review of the process was deemed appropiate. For the aforementioned software, a solid body uniformity was assumed for each calculation model but conventional manufacturing of composite or composites (mixing) yields an heterogeneous mix which as a result would be considered as a material defect [9], [10]. Thus, the proposed new manufacturing process was meant to achieve high quality by avoiding this issue. The new manufacturing procedure is shown in Fig. 1. Here, a bonding material (e.g. epoxy resin), a fine aggregate, a middle aggregate and a coarse aggregate were firstly put into an under small mould and filled up to about 6 vl.%, 10 vl.%, 24 vl.% and 60 vl.% respectively as revealed in Fig. 1(a). The bond and the aggregates in the mould were mannually mixed. Then they were continuously mixed by an ultrasonic scaler with 2–3 μm amplitude and 28–30 kHz vibration frequency in Fig. 1(b) (see Fig. 2(a)).

This process was performed for removing the unnecessary air in the mixing composite, and was continued for a few minutes till all air bublles were fully transfered to the surface of the composite. Lastly, an upper small mould was put on the upper surface of the composite, and the upper mould was knocked by the ultrasonic scaler in Fig. 1(c) (see Fig. 2(b)). The upper mould was gradually sank till the composite reached the most high density condition. In the final condition, the coarse aggregate becomes about 60 vl.% in the mould, the available air given space for the middle aggregate was about 40 vl.% relative to the coarse aggregate. Thus, given that the middle aggregate filled up about 60 vl.% of the total space, 40 vl.% would mean that the aggregate filled about 24 vl.% (=0.4×0.6). Consequently, the coarse and

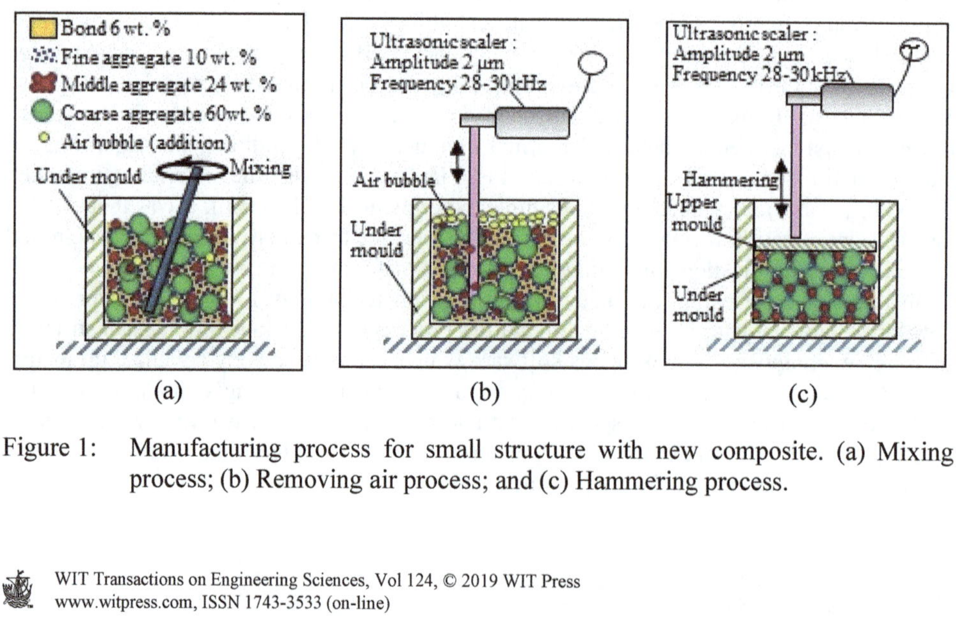

(a) (b) (c)

Figure 1: Manufacturing process for small structure with new composite. (a) Mixing process; (b) Removing air process; and (c) Hammering process.

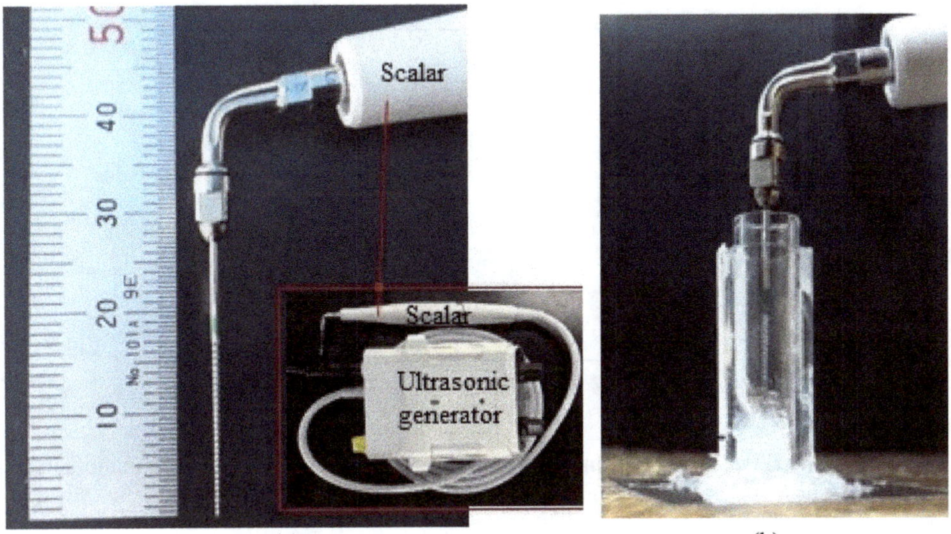

(a) (b)

Figure 2: Schematic view for small structure with new composite in manufacturing process. (a) Ultrasonic scalar for removing air and hammering; and (b) Schematic view for removing air.

the middle aggregates uniformly filled up 84 vl.% (=60 vl.%+24 vl.%). In addition, fine aggregate was also poured into the mix to fill the gaps between the middle aggregates. It was thought that this aggregate would have filled up to 10 vl.% (=(100-60-24)×0.6). As a result, all aggregates uniformly filled around 94 vl.% (=60 vl.%+24 vl.%+10 vl.%). It must be considered that space was meant to be filled by the aggregates in a rhombic display and, thus, the mixture becomes a mortar structure. Bonding materials (e.g. epoxy resin) were meant to be poured from upper surface over all aggregates. Upon solidification of the bonding materials, the generated composite material was removed from the mould. Here, the component distribution was similar to that of the calculation model. For these, the size of the middle aggregate was restricted to be less than one seventh of the coarse aggregate. Meanwhile, the relationship between the fine and the middle aggregates is also the same. Therefore when the mould volume V_m is decided, the volumes and the weights regarding the coarse, the middle and the fine aggregates and the bond can very easily be calculated, and the good timing for agreement between the under and upper moulds can also be estimated. This timing is the limit of the upper mould sinking (Fig. 1(c)) and the highest density condition.

Working times for manufacturing the small sturcture with the new composite are shown in Fig. 3. The small structure was the right cylinder with 5 mm diameter and 10 mm length. It consists of coase aggregate (Glass: 0.7 mm diameter and 60 wt.%), middle aggregate (Glass: 0.08 mm diameter and 24 wt.%), fine aggregate (Carbon: 0.005 mm diameter and 10 wt.%) and bond (Epoxy resin: 6 wt.%) (see Table 1, Sample No. 1). The working times for the three processes; mixing, air removing and hammering are 0.5 min, 2.1 min and 1.3 min respectively. Working times for the samples No. 2 and No. 3 in Table 1 were similar to the sample No. 1. The working times is very effective in the industrial field.

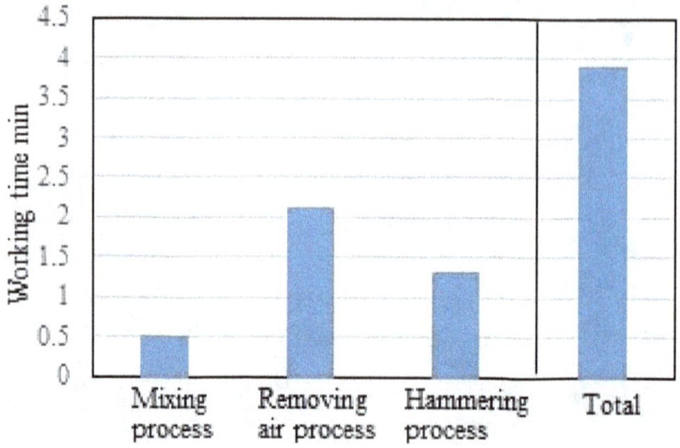

Figure 3: Working times for small structures with new composite in the manufacturing process.

Table 1: The specification of ultrasonic scaler, used materials and its properties for the test.

Specification of the used devices for the small structure with the new composite			
Ultrasonic scaler	Varios 370 (NSK-Nakanishi JAPAN)	Power	11 W
		Amplitude	2–3 μm
		Vibration frequency	28–32 kHz
		Size	W80×D115×H32 mm

The used material, wt.%, size for test pieces						
Sample No.	Coarse aggregate	Middle aggregate	Fine aggregate	Bond	Calculated properties	
No. 1	Glass 0.7 mm 60 vl.%	Glass 0.08 mm 24 vl.%	Carbon 0.005 mm 10 vl.%	Epoxy resin 6 vl.%	Young's modulus GPa	41
					Density kg/m^3	2210
					Thermal conductivity W/(mK)	0.61
					Coefficient of linear expansion K^{-1}	10×10^{-6}
					Specific heat kJ/(kg・K)	0.84
No. 2	Alumina 0.7 mm 60 vl.%	Alumina 0.08 mm 24 vl.%	Silicon nitride 0.01 mm 10 vl.%	Epoxy resin 6 vl.%	Young's modulus GPa	136
					Density kg/m^3	3540
					Thermal conductivity W/(mK)	20
					Coefficient of linear expansion K^{-1}	10×10^{-6}
					Specific heat kJ/(kg・K)	0.98
No. 3	Steel 0.7 mm 60 vl.%	Alumina 0.08 mm 24 vl.%	Carbon 0.005 mm 10 vl.%	Epoxy resin 6 vl.%	Young's modulus GPa	79
					Density kg/m^3	5770
					Thermal conductivity W/(mK)	5.9
					Coefficient of linear expansion K^{-1}	12.1×10^{-6}
					Specific heat kJ/(kg・K)	0.58

Properties of the used materials for the small structure with the new composite					
Material	Young's modulus GPa	Density kg/m^3	Thermal conductivity W/(mK)	Coefficient of linear expansion 1/K	Specific heat kJ/(kg・K)
Steel	210	7800	30	12×10^{-6}	0.46
Glass	69	2460	0.75	9.0×10^{-6}	0.84
Alumina	340	3920	34.7	10×10^{-6}	1.1
Silicon nitride	300	3200	17	3×10^{-6}	0.7
Carbon	8.9	740	0.12	2.1×10^{-6}	1.3
Epoxy resin	2.6	1200	0.4	96×10^{-6}	0.6

Specification of the ultrasonic scaler used in the manufacturing process, the materials for the test pieces used in Section 4 and each material property for the all new composites are shown in Table 1. The ultrasonic scaler used in the manufacturing process is a device for a dental clinic. Three small structures with new composite; samples No. 1, No. 2 and No. 3 were manufactured for evaluation purposes in Section 4. At that time, several properties for samples No. 1, No. 2 and No. 3 were also calculated by the calculation models in Section 2. All samples are manufactured by the manufacturing process described in Section 3.

In this research, the smallest size for the small structure (right cylinder) was 5 mm diameter because of the suitable sizes for coarse, middle and fine aggregates. It is no problem for the manufacturing process (Section 3) to produce a required small size for the structure.

4 EVALUATION OF DEVELOPED CALCULATION MODELS AND MANUFACTURING PROCESS

The calculation models for Young's modulus, density, coefficient of linear expansion, thermal conductivity and specific heat in Section 2 were evaluated. As both the density and the specific heat of the composite can easily and exactly be calculated by the composite ratio of the used materials, the evaluation of these properties was not performed in this section.

4.1 Evaluation for Young's modulus

A schematic view of the experimental set-up regarding Young's modulus for the new composites is shown in Fig. 4(a). Samples No. 1, No. 2 and No. 3 in Table 1 were used for this Young's modulus evaluation. Test piece sizes are 5 mm diameter and 20 mm height. (1) The test piece is put up in the guide pipe with 6 mm inner diameter. (2) 100 N of compression force is loaded on the test piece. (3) The displacement ℓ of the top of the test piece is measured by a CCD camera. (4) The Young's modulus E can be calculated by knowing the cross section area A and height h of the test piece, the displacement ℓ and the compression load P ($E=P \div A \div \ell \times h$). The results for evaluation of the Young's modulus and calculated value E_{cal} and the experimental value E_{ex} are shown in Fig. 4(b). When the experimental value E_{ex}

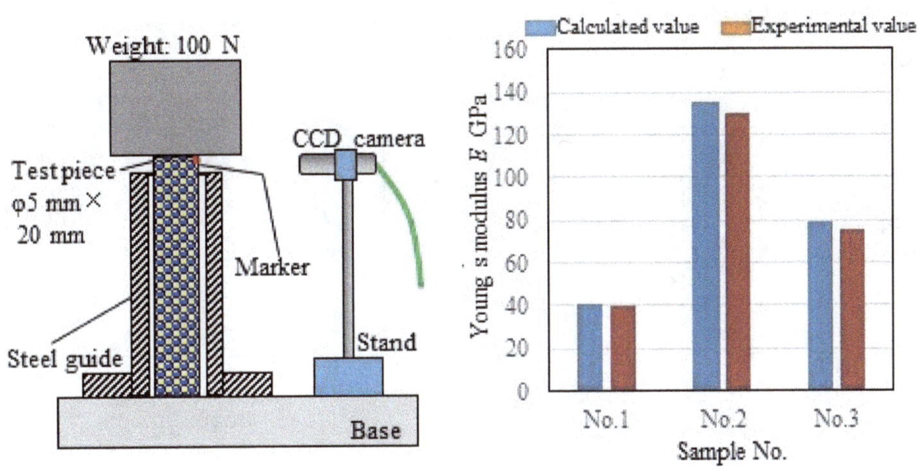

(a) (b)

Figure 4: Evaluation of Young's modulus. (a) Schematic view of experimental set-up; and (b) Experiment results of Young's modulus.

is compared with the calculated value E_{cal}, the experimental error for Young's modulus is around -3.6% to -4.8% (=($E_{ex} - E_{cal}$)$\div E_{cal} \times$100). The reasons were that the new composite was not perfectly in the highest density condition and the residual air in the new composite had very small influence on the Young's modulus.

4.2 Evaluation for coefficient of linear expansion

A schematic view of the experimental set-up regarding the coefficient of linear expansion for the new composites is shown in Fig. 5(a). Samples No. 1, No. 2 and No. 3 in Table 1 were used for this coefficient of linear expansion evaluation. Test piece sizes are 5 mm in diameter and 20 mm in height h. (1) The side of the test piece is insulated for heat transfer by styrene foam with 10 mm thickness. (2) The test piece with styrene foam is inserted and kept for very long time (about 24 hours) in the homoisothermal vessel at 100°C. (3) Then the test piece is put on an air conditioning room at 20°C±1°C the displacement ℓ of the top of the test piece is measured by a CCD camera till the test piece temperature is 20°C. (4) The coefficient of linear expansion α can calculated by the heaght h and temperature diffrence ΔT of the test piece and the displacement ℓ ($\alpha = \ell \div \Delta T \div h$). Thermal behaviour of the test piece in the initial stage which is from the removal of test piece in the homoisothermal vessel to start the measurement was complemented by using the relationship between the test piece temperature and the test piece displacement. It was confirmed that the coefficient of linear expansion for steel could be measured by using this method. The result for evaluation of the coefficient of linear expansion is shown in Fig. 5(b), which also reveals the calculated value α_{cal} and the experimental value α_{ex}. When the experimental value α_{ex} is compared with the calculated value α_c, the experimental error for coefficient of linear expansion is around +1.8%–+1.2% (=($\alpha_{ex} - \alpha_{cal}$)$\div \alpha_{cal} \times$100). The reasons are that the new composite is not perfectly in the highest density condition and the residual air in the new composite has small influence on the coefficient of linear expansion because of small Young's modulus.

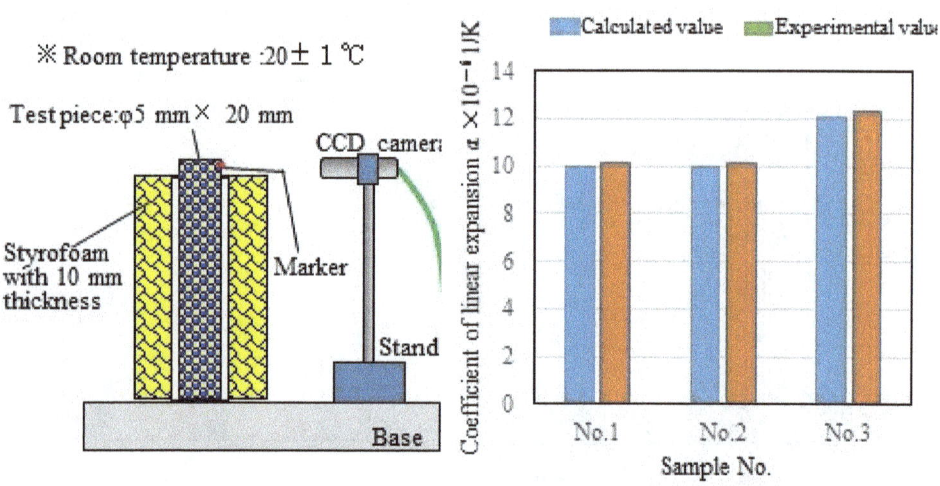

(a) (b)

Figure 5: Evaluation of coefficient of linear expansion. (a) Schematic view of experimental set-up; and (b) Experiment results of Young's modulus.

4.3 Evaluation for thermal conductivity

The schematic view of thermal conductivity experimental set-up for the new composites is shown in Fig. 6(a). Sample No. 1, Sample No. 2 and Sample No. 3 in Table 1 were used for the evaluation. Test piece sizes are 5 mm in diameter and 10 mm in height *h.* (1) The two test pieces are put up on both sides of the ceramics heater (5 W, 5 mm diameter and 1.5 mm thickness). (2) The side of the test piece is insulated for heat transfer by styrene foam with 10 mm thickness. (3) The top T_T and the bottom T_B temperatures of the test piece at the steady state condition are measured by the thermo-couples with T type in the air conditioning room at 20°C. (4) The thermal conductivity λ can be calculated by knowing the top T_T and the bottom T_B temperatures of the test piece at steady state, cross section area A and height h of the test piece, the displacement ℓ and the heat generation W compression load L ($\lambda = W \times (T_T - T_B) \div 2 \div A \div h$). The result from the evaluation of the thermal conductivity is shown in Fig. 6(b). The calculated value λ_{cal} and the experimental value λ_{ex} are shown in this figure. When the experimental value λ_{ex} is compared with the calculated value λ_{cal}, the experimental error for thermal conductivity is around -2.9% to -3.8% ($=(\lambda_{ex} - \lambda_{cal}) \div \lambda_{cal} \times 100$). The reasons are that the new composite is not perfectly in the highest density condition and the residual air in the new composite has small influence on the thermal conductivity.

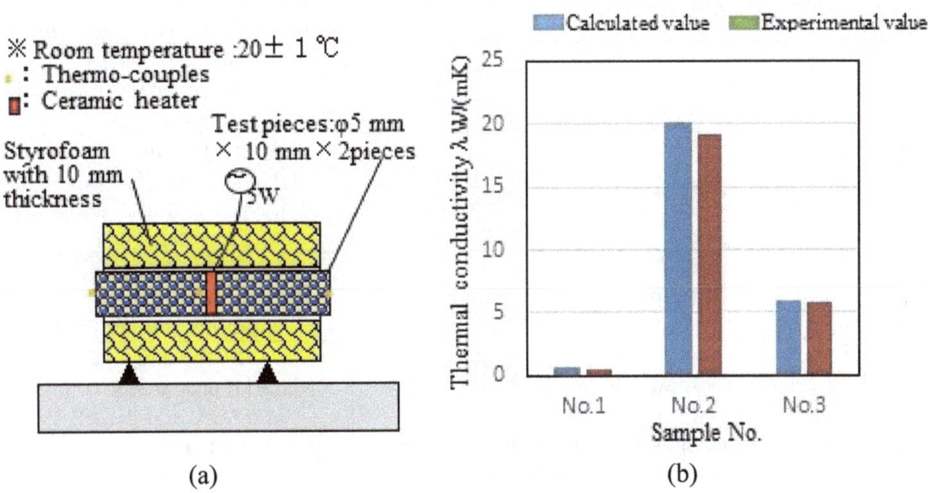

Figure 6: Evaluation of thermal conductivity. (a) Schematic view of experimental set-up; and (b) Experiment results of Young's modulus.

The calculation accuracy was evaluated in the section. The calculations for coefficient of linear expansion was accurate to around +1.8%–+1.2% but the Young's modulus and thermal conductivity were around -2.9% to -4.8%. This was thought to be because of air formation in the composite material and the FEM simulations were modified accordingly; as a result, the software calculation accuracy for the Young's modulus and thermal conductivity reached an error down to a ±2% each. The developed calculation models and manufacturing process were evaluated in several cases with the generation of new composites with hybrid properties.

5 MATERIAL OPTIMIZATION ALGORITHM USING FEM INVERSE ANALYSIS

Material optimization software using inverse analysis has already being developed in our research [9]. And in this section, the proposed software was used for the small structure. An algorithm of the material optimization using inverse analysis of the FEM was developed as depicted. At first, an innovative product with the most suitable function was decided by using a marketing data. The most suitable properties for the innovative product were calculated by both the inverse analysis of the FEM simulation and the method of successive substitution in the personal computer. The FEM simulations features are static analysis, thermal analysis and vibration analysis which are outlined in Table 2. It also shows the relationship between the different simulation features and the necessary properties for each analysis. The necessary materials and the composite ratio for the new materials with the most suitable properties were finally calculated by our software. Then the new composite material with the most suitable property was manufactured by the procedure described in Section 3.

Table 2: Various properties for the input data in certain FEM simulations.

Types of FEM simulation		Boundary conditions	Input data for FEM
Static analysis		Fix condition	Shape and size Material properties (Young's modulus and Poisson's ratio)
Thermal analysis	Steady state	Heat transfer coefficient	Shape and size Material properties (Thermal conductivity)
	Non-steady state	Heat transfer coefficient	Shape and size Material properties(Density, specific heat and thermal conductivity)
Vibration analysis	Free vibration	Fix condition	Shape and size Material properties(Young's modulus and density)
	Forced vibration	Fix condition and damping ratio	Shape and size Material properties(Young's modulus and density)

Note: At thermal deformation analysis, both Static and Thermal analysis are used, and coefficient of linear expansion is also used for input data. Poisson's ratio is always 0.3.

6 CONCLUSIONS

It is concluded from the results that:

1. The expanded proposed software was able to calculate Young's modulus, density, coefficient of linear expansion, specific heat and thermal conductivity for several material samples as previously proposed. Through the selection of adequate materials and the calculation of their ratios, it was possible to achieve specific desired properties in a new composite material.
2. The calculation of resultant properties was carried out given that the component materials and their ratios in a composite material are known. A small structure with material optimization was manufactured and evaluated. It is concluded from the results that the technology was very effective and useful for the development of a new composite material with several hybrid properties.

REFERENCES

[1] Moriwaki, T., Shamoto, E. & Tokunaga, T., Thermal deformation of an ultra-precision machine tool due to environmental temperature change. *Transactions of the Japan Society of Mechanical Engineers, Transaction of JSME, Series C*, **63**(615), pp. 4025–4030, 1997. (In Japanese.)

[2] Nakajima, H., Lotus-type porous metals. *Bulletin of The Iron and Steel Institute of Japan*, **6**(9), pp. 701–707, 2001. (In Japanese.)

[3] Kobayashi, T., Matsubayashi, T. & Shibata, K., On the improvement of the damping capacity of a steel structure by overlaying with fiber-reinforced plastics. *Transaction of JSME, Series C*, **58**(554), pp. 3096–3101, 1992. (In Japanese.)

[4] Okada, M., Hosokawa, A., Asakawa, N., Fujita, Y. & Ueda, T., Influence of minimum quantity lubrication on tool temperature in end milling of difficult-to-cut materials having low thermal conductivity. *Transaction of JSME, Series C*, **78**(792), pp. 3093–3103, 2012. (In Japanese.)

[5] Tanabe, I., Development of software for the creation of new material with hybrid properties. *WIT Transactions on Engineering Sciences*, vol. 116, WIT Press: Southampton and Boston, pp. 139–147, 2017.

[6] Russell, H.W., Principles of heat flow in porous insulators. *J. Am. Ceram. Society*, **18**, pp. 1–5, 1956.

[7] Tanabe, I., Development of technology for creating composite materials of machine tool. *Int. J. of Automation Technology*, **9**(6), pp. 714–719, 2015.

[8] Tanabe, I. & Silva, P.D., Development of material optimization technology for innovation. *WIT Transactions on the Built Environment*, vol. 175, WIT Press: Southampton and Boston, pp. 101–109, 2018.

[9] Tanabe, I., Takiguchi, S. & Iyama, T., Development of software for creating new materials with hybrid properties. *Transaction of JSME, Series C*, **78**(786), pp. 595–604, 2012. (In Japanese.)

[10] Tanabe, I., Takada, K. & Nakamura, A., Thermal and mechanical characteristics of epoxy resin composite used in machine tool structures. *Transaction of JSME, Series C*, **56**(525), pp. 1314–1321, 1990. (In Japanese.)

SECTION 4
MECHANICAL PROPERTIES

FLEXURAL PROPERTIES OF FLAX SLIVER REINFORCED GREEN COMPOSITE BY MOLDING PRESSURE AND CHITOSAN FIBER ADDITION

HIDEAKI KATOGI[1] & KENICHI TAKEMURA[2]
[1]Department of Human Environmental Sciences, Jissen Women's University, Japan
[2]Department of Mechanical Engineering, Kanagawa University, Japan

ABSTRACT

This study investigated the flexural properties of unidirectional flax sliver reinforced polylactic acid by molding pressure and chitosan fiber addition. As a molding condition, the range of molding pressures was 1 MPa to 5 MPa. Molding was done using vacuum compression molding. The amount of added chitosan fiber was 5 wt.%. The chitosan fiber was applied to the flax sliver surface. After chitosan fiber addition, the flax sliver surface was observed using scanning electron microscopy (SEM). Static three-point flexural tests of the green composite were conducted at room temperature. As a result, the flexural strength and modulus of green composite increased with increased molding pressure. The flexural strength and the modulus of chitosan fiber with added green composite greatly decreased compared with those of green composite with no added chitosan fiber when the molding pressure was 5 MPa. SEM observations revealed many chitosan fibers on the flax sliver surface. The flexural properties of green composites were affected because matrix penetration among the flax fibers readily occurred at molding pressures. However, the flexural properties of the green composite with added chitosan fiber at a molding pressure of 5 MPa was lower than that of the green composite with no added chitosan fiber at a molding pressure of 5 MPa because matrix penetration between flax fibers was probably prevented by the chitosan fiber addition.

Keywords: chitosan fiber, green composite, flax sliver, flexural property, molding pressure, polylactic acid.

1 INTRODUCTION

The mechanical properties of green composites have been examined for wider application [1]–[11]. The tensile properties of green composites were improved using flax slivers [12]. A green composite was used to form body frames of microelectric vehicles in 2017 [13]. However, interfacial adhesion between natural fibers and the biodegradable polymer was not good. Therefore, the effects of nanofiber addition on the mechanical properties of green composites have been studied for improvement.

Takemura et al. [14] reported the impact properties of a green composite with added cellulose nanofiber using plain woven jute fiber after water immersion. The impact strength of the green composite incorporating plain woven jute fiber was increased by cellulose nanofiber addition to jute fiber surface before and after water immersion. Takagi et al. [15] reported stretching strain effects on the tensile properties of 10 wt.% cellulose nanofiber added green composite using gel-polyvinyl alcohol. The tensile strength and Young's modulus of a green composite with added cellulose nanofiber increased with increased stretching strain by up to 40% by fiber orientation control of cellulose nanofiber.

Recently, chitosan fibers found in crab and shrimp shells have been specifically examined as nanofibers for use in the next generation of materials development. The chemical structure of chitosan fiber resembles that of cellulose. However, the effects of molding pressure and chitosan fiber addition on the flexural properties of green composites remain unclear: a fact that limits the wider application of these materials.

www.witpress.com, ISSN 1743-3533 (on-line)
doi:10.2495/MC190091

This study investigated the effects of molding pressure and chitosan fiber addition on the flexural properties of a unidirectional flax sliver reinforced green composite.

2 SPECIMENS

2.1 Materials

The matrix was a PLA resin sheet (Ecodear 250-1B01CA; Toray Co., Ltd.). The reinforcement was done using flax slivers (Shigaasa kogyo Co., Ltd.). The added fiber was chitosan fiber (Bfo-08002; Sugino Machine Limited), 5 wt.%.

2.2 Chitosan fiber addition method

The flax slivers were aligned in a longitudinal orientation. The chitosan fiber was applied to the flax sliver surface on the front and back side. Subsequently, the chitosan fiber with the added flax slivers was dried at room temperature for 24 h for fabrication of the preform.

2.3 Molding method and specimen size

For the fabrication of the green composite, chitosan fiber and PLA sheets, with no added flax slivers, were alternately laminated in the mold. Then, control of flax sliver direction was done using aluminum spacers at both sides on the mold. In addition, the unidirectional preform and PLA sheet were laminated. Vacuum compression molding was used for green composites with and without chitosan fibers and unidirectional flax slivers. The molding pressures were 1.0–5.0 MPa. The molding time was 10 min. The molding temperature was 190°C. Thereafter, the mold was cooled to room temperature. Fiber volume fractions of green composites with and without added chitosan fiber were 55%. Their specimen size for static flexural tests was 80 mm long, 15 mm wide, and 2.0 mm thick.

3 TESTING METHODS

3.1 Surface observations of green composites with and without added chitosan fibers

The surfaces of green composites with and without added chitosan fibers were observed using an optical microscope (VHX-5000; Keyence Co.). From photographs taken at three areas, we measured the fiber orientation angle on the specimen surface. Fig. 1 presents a schematic drawing of fiber orientation angle measurement.

3.2 Static three-point flexural test

Static three-point flexural tests of green composites with and without added chitosan fibers were conducted based on Japanese Industrial Standard (JIS) K 7017. The gauge length was 60 mm. The environmental temperature was room temperature. The crosshead speed was 1.0 mm/min. The number of specimens was five.

4 RESULTS AND DISCUSSION

4.1 Surface observations of flax sliver and green composite

Fig. 2 portrays the surface of flax sliver after chitosan fiber addition. The surface of chitosan fiber with added flax slivers became rough. Some chitosan fibers were found between flax fibers. The flax sliver surface in some areas was covered by added chitosan fibers.

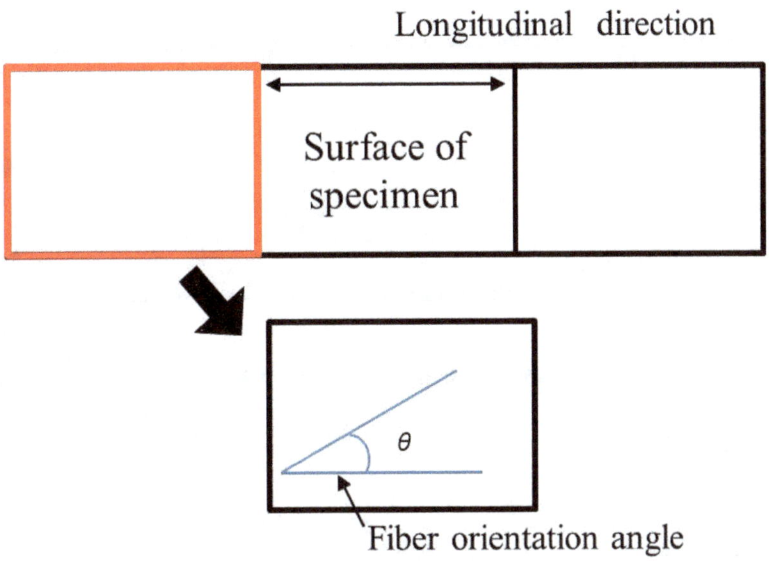

Figure 1: Schematic drawing of fiber orientation angle measurement.

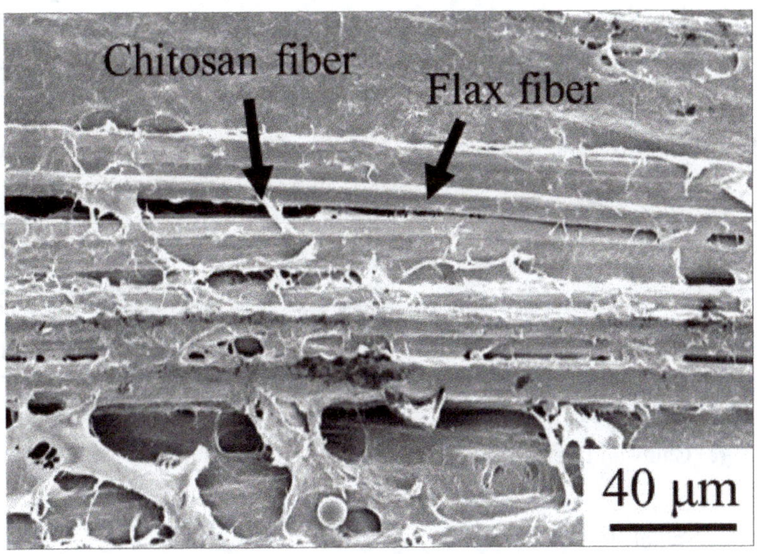

Figure 2: Surface of flax sliver after chitosan fiber addition.

Fig. 3 shows the effects of molding pressure and chitosan fiber on the surface of flax sliver reinforced green composites. Fig. 4 shows the fiber orientation angle on the surface of the flax sliver reinforced green composite. Scattering of fiber orientation angles of green composites under all conditions became small. Fiber orientation angles of green composites were very similar under almost all conditions. Therefore, the flexural properties of the green composite used for this study were unaffected by the fiber orientation angle.

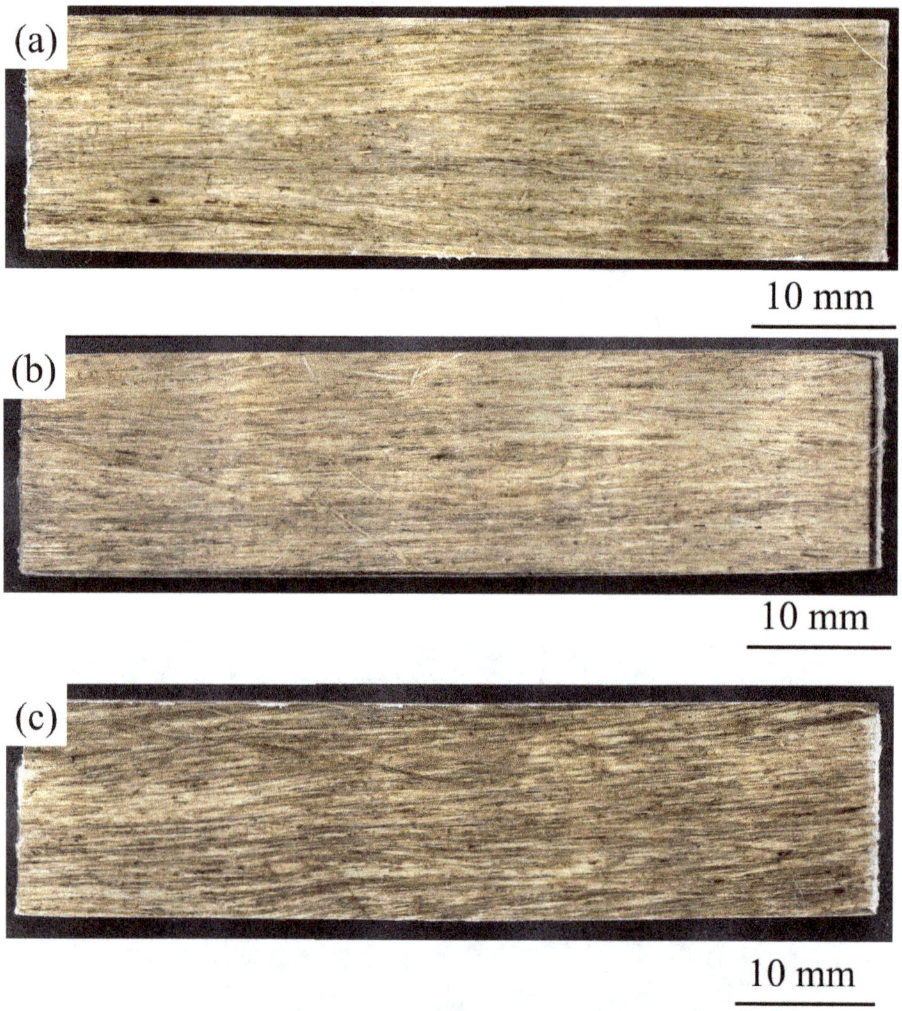

Figure 3: Specimens. (a) Molding pressure 1 MPa; (b) Molding pressure 5 MPa; and (c) Chitosan fiber addition and molding pressure 5 MPa.

4.2 Static three-point flexural test

Figs 5 and 6 show the effects of molding pressure and chitosan fiber addition on the flexural properties of the flax sliver reinforced green composite. The flexural strength of green composite increased with increased molding pressure. The flexural strength of the green composite molded at 5 MPa was much greater than that of green composite molded at 3 MPa. The flexural modulus of the green composite slightly increased with increased molding pressure up to 3 MPa. Subsequently, the flexural modulus of green composite formed at 5 MPa was much higher than that of the green composite molded at 3 MPa. However, the flexural strength and modulus of the green composite molded at 5 MPa were decreased 31% and 30%, respectively, by chitosan fiber addition.

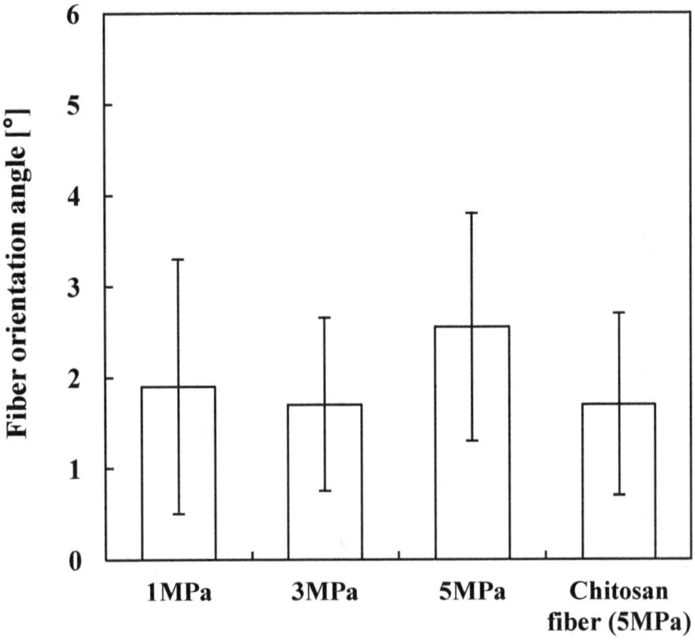

Figure 4: Fiber orientation angle of green composite.

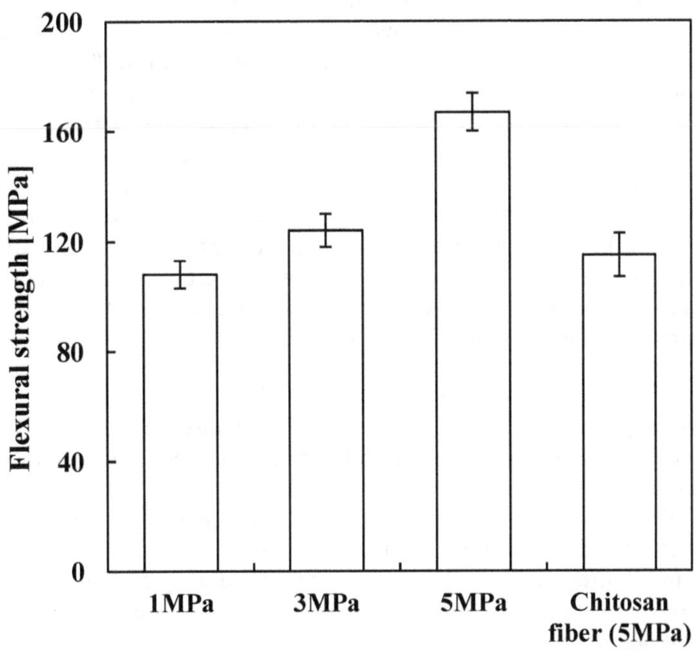

Figure 5: Effects of molding pressure and chitosan fiber addition on flexural strength of green composite.

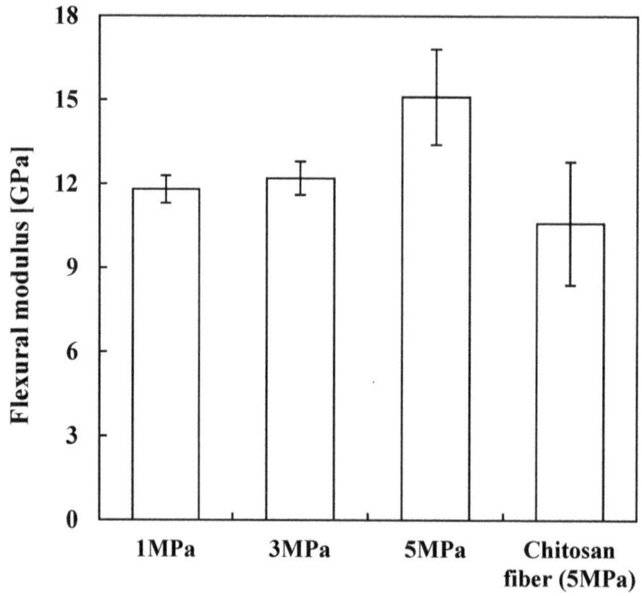

Figure 6: Effects of molding pressure and chitosan fiber addition on flexural modulus of green composite.

The results described above indicate that the fiber orientation angles of green composites were small under all conditions. Moreover, the flax sliver surfaces in some areas were covered by added chitosan fibers (Fig. 2) because chitosan has film-forming ability. When thermogravimetric analysis of chitosan was conducted, the initial temperature of chitosan was about 230°C [16]. The matrix was unable to penetrate among flax fibers because of chitosan fibers' film-forming ability and high temperature resistance. Subsequently, many voids formed between flax fibers during molding. Therefore, the flexural property of green composites was unaffected by the fiber orientation angle. The flexural properties of green composites were affected mainly by matrix penetration into flax slivers readily occurring due to molding pressure. By contrast, the flexural properties of green composites with added chitosan fiber molded at 5 MPa decreased because chitosan fiber addition probably prevented matrix penetration among flax fibers.

5 CONCLUSIONS

This study investigated the flexural properties of a unidirectional flax sliver reinforced green composite by molding pressure and chitosan fiber addition. The results show that the following conclusions can be drawn.

The flexural strength and modulus of the green composite increased with increased molding pressure. The flexural strength and modulus of the green composite was decreased considerably by chitosan fiber addition when the molding pressure was 5 MPa. From surface and SEM observations, fiber orientation angles on surfaces of green composites under all conditions were small. Many chitosan fibers were found on the flax fiber surface. Therefore, the flexural properties of green composites were affected because matrix penetration among the flax fibers occurred easily with adequate molding pressure. However, the flexural properties of green composite formed with added chitosan fiber at 5 MPa decreased probably because chitosan fiber addition prevented matrix penetration among the flax fibers.

REFERENCES

[1] Katogi, H., Shimamura, Y., Tohgo, K. & Fujii, T., Fatigue behavior of unidirectional jute spun yarn reinforced PLA. *Advanced Composite Materials*, **21**, pp. 1–10, 2012.
[2] Kobayashi, S. & Takeda, K., Transverse properties of hemp/PLA composite fabricated with micro-braiding technique. *Advanced Composite Materials*, **24**, pp. 509–518, 2015.
[3] Takagi, H., Nakagaito, A.N., Nishimura, K. & Matsui, T., Mechanical characterisation of nanocellulose composites after structural modification. *WIT Transactions on the Built Environment*, vol. 166, WIT Press: Southampton and Boston, pp. 335–341, 2016.
[4] Tanaka, K., Shiga, T. & Katayama, T., Fabrication of hydroxyapatite/PLA composite nanofiber by electrospinning. *WIT Transactions on the Built Environment*, vol. 166, WIT Press: Southampton and Boston, pp. 371–379, 2016.
[5] Sukmawan, R., Takagi, H. & Nakagaito, A.N., Strength evaluation of cross-ply green composite laminates reinforced by bamboo fiber. *Composites Part B: Engineering*, **84**, pp. 9–16, 2016.
[6] Wei, L. & McDonald, A.G., A review on grafting of biofibers for biocomposites. *Materials*, **9**, pp. 303–326, 2016.
[7] Katogi, H., Takemura, K. & Akiyama, M., Residual tensile property of pain woven jute fiber/poly(lactic acid) green composites during thermal cycling. *Materials*, **9**, pp. 573–583, 2016.
[8] Morreale, M., Mistretta, M.C. & Fiore, V., Creep behavior of poly(lactic acid) based biocomposites. *Materials*, **10**, pp. 395–407, 2017.
[9] Ullah, S.A.M.M., Shahinur, S. & Haniu, H., On the mechanical properties and uncertainties of jute yarn. *Materials*, **10**, pp. 450–464, 2017.
[10] Katogi, H. & Takemura, K., Creep rupture of water-absorbed green composite. *WIT Transactions on Engineering Sciences*, vol. 116, WIT Press: Southampton and Boston, pp. 281–288, 2017.
[11] Katogi, H., Shimamura, Y., Tohgo, K., Fujii, T. & Takemura, K., Effect of matrix ductility on fatigue strength of unidirectional jute spun yarns impregnated with biodegradable plastics. *Advanced Composite Materials*, **27**, pp. 235–247, 2018.
[12] Ren, B., Mizue, T., Goda, K. & Noda, J., Effects of fluctuation of fibre orientation on tensile properties of flax sliver-reinforced green composites. *Composite Structures*, **94**, pp. 3457–3464, 2012.
[13] JEC Group, www.jeccomposites.com/knowledge/international-composites-news/lina-bio-composite-based-car-pla-honeycomb-sandwich-design.
[14] Takemura, K., Takai, S. & Katogi, H., Effect of microfibrillated cellulose addition and water absorption on mechanical properties of jute/PLA composites. *WIT Transactions on the Built Environment*, vol. 124, WIT Press: Southampton and Boston, pp. 387–394, 2012.
[15] Takagi, H., Nakagaito, A.N. & Sakaguchi, Y., Structural modification of cellulose nanocomposites by stretching. *WIT Transactions on Engineering Sciences*, vol. 116, WIT Press: Southampton and Boston, pp. 251–256, 2017.
[16] Tsurugai, K. & Hiraide, T., Preparation and antibacterial activities of N-Trimethylammonium salts of chitosan. *Sen'i Gakkaishi*, **50**, pp. 215–220, 1994. (In Japanese.)

INFLUENCE OF SILANE TREATMENT ON WATER ABSORPTION AND MECHANICAL PROPERTIES OF PLA/SHORT BAMBOO FIBER-REINFORCED GREEN COMPOSITES

SHOMA MARUYAMA, HITOSHI TAKAGI & ANTONIO N. NAKAGAITO
Tokushima University, Japan

ABSTRACT

Green composites (GCs) consist of biodegradable resin and natural fibers. Due to the hydrophilicity of constituents, however, GCs are usually susceptible to water absorption. When GCs absorb water, the interface between the resin and fibers peels off, resulting in the shrinkage of the fibers after drying. In this study, we investigated the influence of silane treatment on water absorption and mechanical properties of polylactic acid/short bamboo fiber (PLA/SBF) composites. The silane treatment reduced water absorption and improved the mechanical strength of PLA/SBF composites, namely the tensile strength and Young's modulus were increased by 72% and 30% respectively, compared with those of untreated PLA/SBF composites. The silane treatment also increased interfacial adhesion of resin/fiber resulting in improved water absorption rate. Thus, it is concluded that the silane treatment is effective for improving the mechanical and water-proofing performances of PLA/SBF composites.

Keywords: bamboo fiber, green composites, polylactic acid, silane treatment.

1 INTRODUCTION

In recent years, while industries have developed based on petroleum-based chemistry, many environmental problems such as air pollution and global warming are matters of concern. Therefore, researchers' special attention is being paid to the development of biodegradable composite materials called green composites (GCs) as alternative materials to petroleum-derived plastics and composites [1]–[4]. The GC is a composite material in which biodegradable resin is reinforced with natural fibers, and it is made from various renewable resources as it is derived from natural plants. Hence, the emission of carbon dioxide generated when incinerating GCs is unchanged due to the concept of carbon neutral.

However, as a disadvantage of GCs, it is pointed out that their mechanical properties deteriorate in underwater environment [5]. The GCs swell when GCs absorb water, and after subsequent drying, the fiber shrinkage occurs, and the interfaces between the resin and the fiber are often peeled off. Thus, it is considered that the mechanical strength of GCs decreases because natural fibers do not fulfill the reinforcing function. It is an urgent matter to improve the interfacial adhesion between resin and fibers during water absorption.

An effective method for improving the interfacial adhesion between resin and fibers of GCs is silane coupling treatment. The silane treatment generally refers to a surface treatment using a silane coupling agent as a bonding aid to chemically bond inorganic substances to organic substances. It has been confirmed that when silane treatment is applied to GCs made with polylactic acid (PLA) and wheat husks, the mechanical strength of the composites has been improved and their water absorption has been reduced [6]. It is speculated from these results that by applying silane treatment to GCs, the interfacial adhesion strength between resin and fibers can be improved in other GCs, leading to improvement of mechanical strength and suppression of water absorption.

In this study, therefore, we fabricated GCs using PLA which is one of the widely used biodegradable resins and high strength bamboo fiber among natural fibers. The effects of

WIT Transactions on Engineering Sciences, Vol 124, © 2019 WIT Press
www.witpress.com, ISSN 1743-3533 (on-line)
doi:10.2495/MC190101

silane treatment on the mechanical properties and water absorption of the PLA/bamboo fiber-reinforced GCs were investigated.

2 EXPERIMENTAL METHODS

2.1 Fabrication method of composite samples

In this research, PLA/short bamboo fiber (SBF) composites in which SBFs are randomly oriented in the PLA matrix were prepared by using a hot-pressing method. First, long bamboo fibers of 300–450 mm extracted by steam-explosion treatment [7] of raw bamboo were cut into short fibers of about 30 mm. These short fibers were subjected to alkali treatment with an aqueous NaOH solution (5 wt.% NaOH) for 30 min and blender treatment was then carried out for 20 min to prepare SBFs.

Silane treatment was performed on SBFs. In this study, two kinds of silane agents, 3-aminopropyltrimethoxysilane (APMS) and 3-glycidyloxypropyltrimethoxysilane (GPMS), were used. A silane coupling agent in an amount of 1 wt.% based on the fiber weight was added dropwise in a solution prepared by mixing methanol and distilled water at a ratio of 1:1. A silane solution was prepared by adjusting the pH of the solution to 4.0 by adding acetic acid. After that, the SBFs were immersed in the silane solution for 1 h. The treated SBFs were dried at room temperature for 2 days and further dried with a constant temperature drier at 50°C for 24 h.

After drying, the silane-treated SBFs and dispersion-type PLA resin (PL-2000, Miyoshi Oil and Fat Co., Ltd., Japan) were mixed to be 40wt.% fiber loading and then treated for 4 h in a vacuum defoaming stirrer. The mixed sample was poured into a rectangular container and then dried at 50°C for 2 days to prepare composite preform sheets. Thereafter, the dried preforms were cut into strips having a size of 10 × 100 mm with a table-top circular saw and dried at 105°C for 1 h, then hot pressed at 190°C, 5 MPa, 5 min to prepare PLA/SBF composite samples. A detailed preparation method is also mentioned elsewhere [8].

2.2 Mechanical and fractographical characterization of composite samples

All the tensile tests were conducted at a tensile speed of 1.0 mm/min on an Instron material testing machine (Model 5567, USA). The shape of the specimen was 10 mm wide and 100 mm long, and the gauge length was 30 mm. Both the tensile strength and Young's modulus were evaluated from the stress-strain curves obtained.

The fractographical morphology of the PLA/SBF composites and their interfaces between PLA and bamboo fibers was examined by a scanning electron microscope (SEM: S-4700, Hitachi Ltd., Japan). All samples were sputter-coated with gold/platinum prior to the SEM observation.

2.3 Water absorption and drying test of composite samples

The water absorption rate of the PLA/SBF composites was evaluated according to JIS K 7209 (2000) standard. Five PLA/SBF composite samples were dried at 105°C for 1 h, then soaked in distilled water at room temperature for 5, 25, 100, and 250 h. After immersion in water, the samples were taken out, their surface water was entirely wiped off with a cloth, and the weight of the wet state sample was measured. Thereafter, drying again at 105°C for 1 h, and then weight gain rate, M_g and weight loss rate, M_l of the sample were calculated from the following equations:

$$M_g = \frac{W_w - W_d}{W_0} \times 100, \tag{1}$$

and

$$M_l = \frac{W_0 - W_d}{W_0} \times 100, \tag{2}$$

where W_0 is the weight of the dry sample before water absorption, W_w is the weight of the wet sample, and W_d is the weight of the re-dried sample after water absorption.

The tensile properties of the PLA/SBF composites after water absorption were evaluated by carrying out the tensile tests in the manner as described in Section 2.2. In addition, the fracture surface of the test specimen was also examined by SEM.

3 RESULTS AND DISCUSSION

3.1 Effect of silane treatment on mechanical properties of PLA/SBF composites

Figs 1 and 2 show the mechanical properties of neat PLA, PLA/SBF composites, alkali-treated PLA/SBF composites, and silane-treated PLA/SBF composites (APMS and GPMS). The PLA/SBF composites without any surface treatment have higher strength and higher Young's modulus than neat PLA. The strength and Young's modulus of the alkali-treated PLA/SBF composites are higher than those of untreated PLA/SBF composites. The highest mechanical properties are obtained in the silane-treated PLA/SBF composites. The tensile strength of the silane-treated sample was 80.3 MPa and Young's modulus was 5.2 GPa. Compared with the untreated sample, the tensile strength and Young's modulus are increased by 72% and 30%, respectively. The effect of these surface treatment, the interfacial bonding was increased and the stress transmission ability between the fiber and matrix was improved, which resulted in an improvement in mechanical properties.

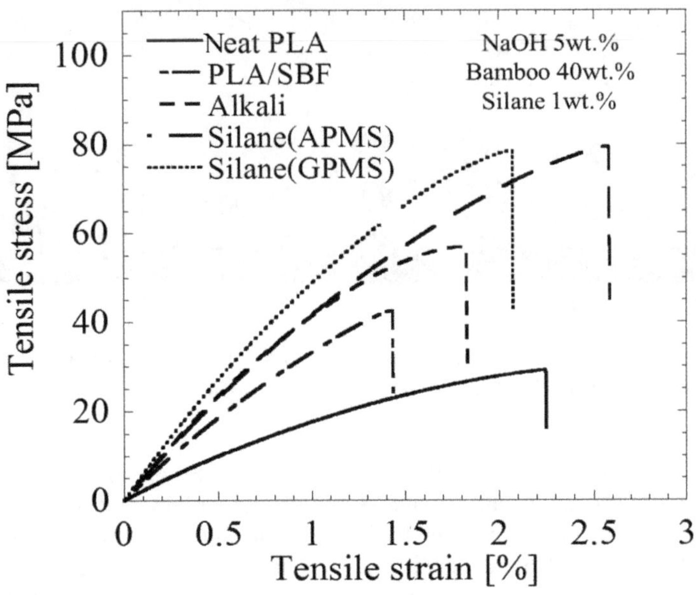

Figure 1: Typical stress-strain curves of surface-treated and untreated PLA/SBF composites.

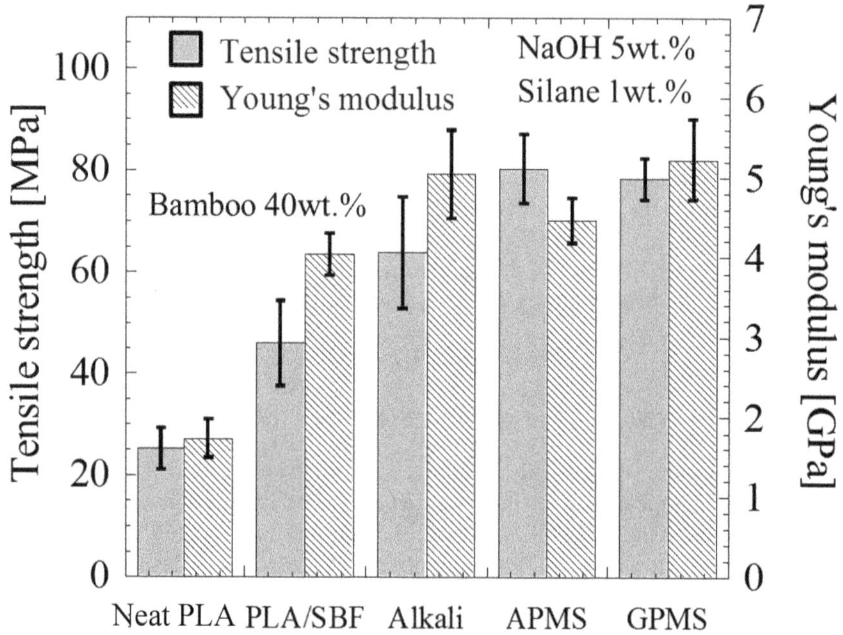

Figure 2: Effect of surface treatment on tensile properties of PLA/SBF composites.

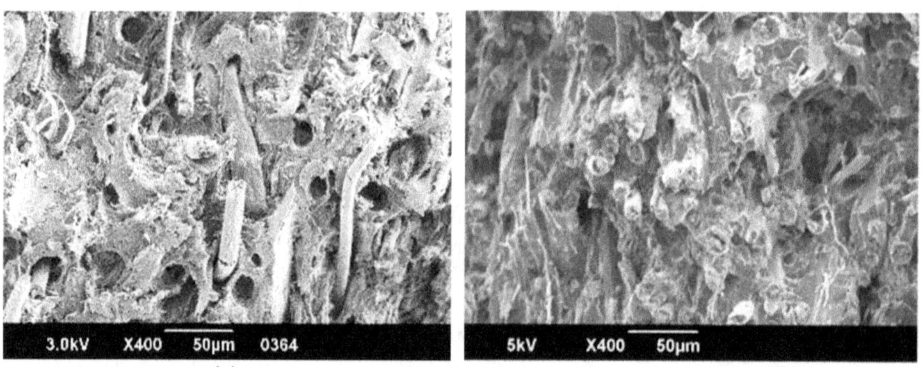

(a) (b)

Figure 3: SEM images of fracture surface. (a) Untreated PLA/SBF composites; and (b) Silane (GPMS)-treated PLA/SBF composites.

The SEM photomicrographs of the untreated PLA/SBF composites and silane-treated PLA/SBF composites are indicated in Fig. 3, respectively. Many pull-out cavities are found on the fracture surface of the untreated PLA/SBF composites. However, the fracture surface of the silane-treated PLA/SBF composites has few traces of pull-out and most of the bamboo fibers are covered with matrix PLA resin. It is suggested from these results that the silane treatment improves the interfacial adhesion between the resin and the fibers and that the mechanical properties of the PLA/SBF composites are improved. A similar increase in mechanical properties related to the alkali and silane treatment was reported elsewhere [9].

3.2 Effect of silane treatment on absorption behavior of PLA/SBF composites

Figs 4 and 5 show the results on how the water absorptivity and mechanical properties of the silane (APMS)-treated PLA/SBF composites are affected when the composites are immersed in water for up to 250 h.

The water absorption rate and the weight loss rate of each PLA/SBF composite sample after immersion in water for 250 h are shown in Fig. 4. The water absorption rate of the silane-treated PLA/SBF composites shows 17.6%. It decreased by about 4.8 points compared with that of alkali-treated composites. As for the weight loss rate, it is about 0.3 points larger than that of neat PLA, however, it is seen that it decreases compared with other composites containing bamboo fibers. From these results, it is suggested that the water absorption was improved because water penetration in the silane-treated PLA/SBF composites was prevented by improving the interfacial adhesion between the PLA matrix and bamboo fibers.

Fig. 5 shows the variation in tensile strength after immersion in water. It was found that after water immersion for 250 h, the tensile strength of all PLA/SBF composites decreased. In the untreated PLA/SBF composites, the tensile strength decreased by 55%, however, in the silane (APMS)-treated composites the reduction rate of the tensile strength was 45%, thus the reduction rate of about 10 points could be improved. It is considered from this result that the decrease in tensile strength could be suppressed because the interfacial adhesion between resin matrix and fiber was improved by silane treatment.

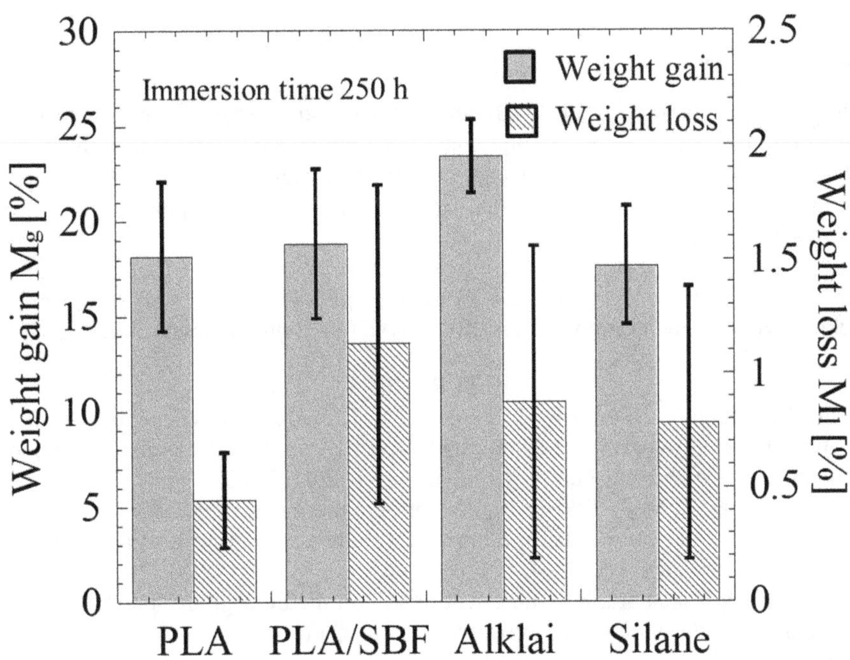

Figure 4: Comparison of weight gain and weight loss of treated and untreated PLA/SBF composites after immersion of 250 h.

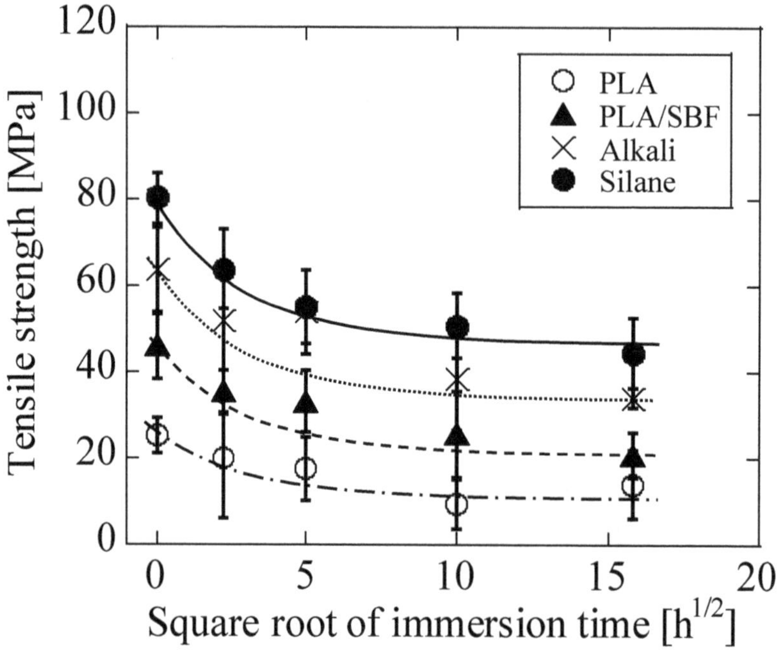

Figure 5: The relationship between tensile strength of PLA/SBF composites and immersion time.

4 CONCLUSIONS

In this study, we fabricated PLA/SBF composites and examined the effects of silane treatment on mechanical properties and water absorption of the composites. In conclusion, the present results demonstrate that silane treatment is useful as a method for improving the resin/ fiber interfacial adhesion, the water absorption, and mechanical properties of the PLA/SBF composites.

ACKNOWLEDGEMENT

We acknowledge financial support from JSPS, KAKENHI project number 16H01790.

REFERENCES

[1] Liu, K., Takagi, H., Osugi R. & Yang, Z.M., Effect of physicochemical structure of natural fiber on transverse thermal conductivity of unidirectional abaca/bamboo fiber composites. *Composites Part A*, **43**(8), pp. 1234–1241, 2012.

[2] Liu, K., Yang, Z. M. & Takagi, H., Anisotropic thermal conductivity of unidirectional natural abaca fiber composites as a function of lumen and cell wall structure. *Composite Structures*, **108**, 987–991, 2014.

[3] Katogi, H. & Takemura, K., The effect of soil on the biodegradable behavior of green composites under various environmental (ambient) temperature. *WIT Transactions on Engineering Sciences*, vol. 90, pp. 47–55, 2015.

[4] Sukmawan, R., Takagi, H. & Nakagaito, A.N., Strength evaluation of cross-ply green composite laminates reinforced by bamboo fiber. *Composites Part B*, **84**, pp. 9–12, 2015.

[5] Takemura, K., Takai, S. & Katogi, H., Effects of microfibrillated cellulose addition and water absorption on mechanical properties of jute/PLA composites. *WIT Transactions on The Built Environment*, vol. 124, pp. 387–394, 2012.
[6] Asumani, O.M.L., Reid, R.G. & Paskaramoorthy, R., The effects of alkali-silane treatment on the tensile and flexural properties of short fibre non-woven kenaf reinforced polypropylene composites. *Composites: Part A*, **43**(9), pp. 1431–1440, 2012.
[7] Takagi, H., Kako, S., Kusano K. & Ousaka, A., Thermal conductivity of PLA-bamboo fiber composites. *Advanced Composite Materials*, **16**(4), pp. 377–384, 2012.
[8] Maruyama, S., Takagi, H., Nakamura, Y., Nakagaito, A.N. & Sasaki, C., Influence of alkali treatment on mechanical properties of poly lactic acid bamboo fiber green composites. *Advanced Materials Research*, **1110**, pp. 56–59, 2015.
[9] Cai, M., Takagi, H., Nakagaito, A.N., Li, Y. & Waterhouse, G.I.N., Effect of alkali treatment on interfacial bonding in abaca fiber-reinforced composites. *Composites Part A*, **90**, pp. 589–597, 2016.

DETERMINING FRACTURE TOUGHNESS OF WOOD FLOUR/HEMP-REINFORCED PVC POLYMER COMPOSITES

KRISZTINA ROMÁN, BELLA UDVARDI, EMESE KUROVICS,
TAMÁS J. SZABÓ & KÁLMÁN MAROSSY
Institute of Ceramics and Polymer Engineering, University of Miskolc, Hungary

ABSTRACT

Rigid PVC wood flour composites were prepared with various volume fractions of wood flour to investigate the effect of the structure properties on the fractural and tensile behavior. During the mixture preparation the first question is the compatibility of the wood flour additive with the rigid PVC and the safe concentration range of the additive. The second question is the fractural behavior of the composites. Accordingly, the 300–400 micrometer particle size beech wood content was varied using 15, 20, 25 parts per weight of the polymer matrix. Since both the PVC and the major components of the beech wood (cellulose and lignin) are polar, we did not use a coupling agent between the polymer and the natural filler. Thus, the properties of the composites as function of concentration can be examined. The usual mechanical properties, such as tensile strength, elongation at break and so-called compact tensile fracture mechanical analysis was also performed. It has been found that the composites could be safely processed with up to 25 parts by weight wood content. However, the properties of 25 parts by weight filler content sample already showed a decrease in strength properties. We also tried to improve the fracture mechanical properties with fibrous (1 cm long) natural filler. In this case 20 parts by weight of wood flour content with one, three and five parts by weight hemp fiber were used in the PVC polymer matrix. The tests results showed that these fibrous fillers neither decrease nor improve the mechanical properties. The raw measurement results were evaluated according to the critical stress intensity factor (K_c) and the critical crack propagation energy (G_c). Ultrasound propagation measurements are in progress trying to determine the effect of structural changes and interactions at increased hemp fiber content.

Keywords: PVC composite, mechanical properties, K_c, G_c, LEFM.

1 INTRODUCTION

Poly (vinyl chloride) polymers are used in a variety of commercial products, such as pipes, windows profiles, cables, rigid sheets, insulators etc., due to their unique properties [1], [2]. Fiber additives are often applied to increase the mechanical strength of the materials. The natural fibers and flours made by grinding have been frequently applied to the polymers [3]. The natural fiber/flour-reinforced composites have attracted the attention of many scientists in the field of composite materials, because they have lot of advantages. These advantages are the lower density, high strength, high modulus and more biodegradable nature [4].

In fracture mechanical tests, the cracking behaviour of polymeric materials can be examined. Researchers reported that the composite's flexibility and strength are influenced by wood quantity and sizes. The high content of wood fillers resulted in poor adhesion between the polymer matrix and the wood filler. Studies have reported that a higher quantity of the filler reduces the impact strength by up to 60% [5]. To improve the fatigue behaviour of composites and to develop improved materials, it is essential to better understand the fatigue mechanism of unidirectional materials [6]. For investigating in the area of the linear elastic fracture mechanics (LEFM), the phenomenon of fatigue crack propagation analyzed by various analytical methods was evaluated, for example it can determine the K_c critical stress intensity factor [7].

WIT Transactions on Engineering Sciences, Vol 124, © 2019 WIT Press
www.witpress.com, ISSN 1743-3533 (on-line)
doi:10.2495/MC190111

The science of fracture mechanics has increasingly influenced the design and experimental work, which includes the prediction of lifetime of the products. However, the mathematical and physical models generally focus on a one axis stress crack area [8]. The time-dependent properties of the composite materials are very important, especially creep, stress relaxation and fatigue investigations. It was found that the rate of creep decreased with the wood content. The other parameter modifying the fatigue process is moisture. The moisture greatly influences the time-dependent properties of PVC composites, because the wood and also the hemp has hydrophilic nature [9].

According to recent literature, tests under cyclic loading are very rarely used. However, this type of investigations on building materials is very useful, in particular on products exposed to earthquakes or pedestrian bridges [10]. For testing fatigue, there exists a method, which examines the acceleration of fracture. Several studies deal with the nature of the spread of fatigue cracking in PVC. The change of the length of crack is expressed as a change in the cycle in the da/dN–ΔK period, which is usually described by the traditional Paris relation. The process is highly dependent on the composition; e.g. the molecular weight of the matrix polymer or the plasticizer content and type if used. It has been observed that the crack growth was reduced when the da/dN–ΔK period was decreased [11].

Fracture parameters are also tested using the three-point bending test based on the Jeng and Shah models [12]. The easiest way to analyze the fracture behaviour is to use Wöhler's rotating bending test for polymer materials. It is one of the most commonly used method for comparing the fatigue life expectancy of different materials. For thermoplastics the fatigue failure studies were originally based on the traditional technique of empirical analysis of metals [8].

2 THEORY

The fatigue is usually the most common structural failure of polymers; this problem is caused by cyclic or random loads. Fatigue failure begins with the formation of cracks in high stress regions [13]. Fig. 1 shows the flow chart of the fracture mechanism and the route of evaluation.

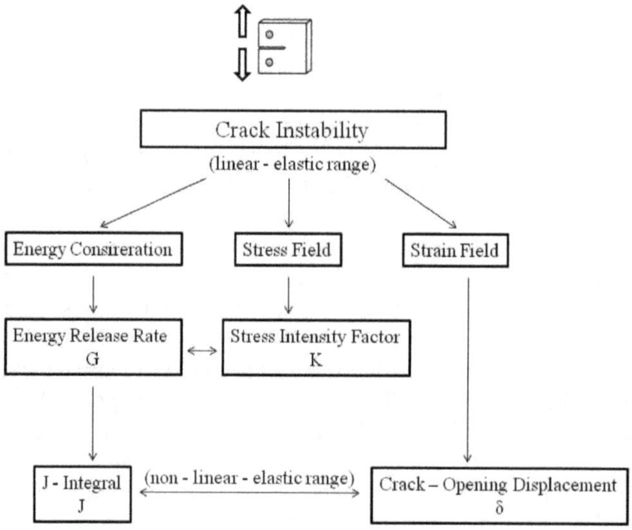

Figure 1: Outline of the fracture mechanism [14].

The tensile stress gradually increases during the linear fracture test while the size of the crack begins to increase in the specimens then the force is reduced. The intensity factors K_c give the stress-dependent term [14]

$$K_c = \sigma_i \sqrt{a}, \tag{1}$$

where σ_i = external stress, a = distance between the fixing and bottom of the notch (m). Since these values are evaluated by using finite samples, geometric factors (f) must be included, which leads to the final equations [14]

$$K_c = f \frac{F_{max}}{B\,W^{1/2}} \left[N/m\right], \tag{2}$$

where f = geometric factor, F_{max} = maximum load (N), B = thickness (m), W = width (m). The values of the "f" were determined from the geometry of the specimens.

We used critical crack propagation energy (G_c) to describe the instability of cracks material by stress and energy [14]

$$G_c = \frac{U}{B\,W\varphi} \left[J/m^2\right], \tag{3}$$

where U = deformation energy (Nm), W = width of the sample (m), φ = geometrical factor.

3 MATERIALS AND SAMPLES PREPARATION

3.1 Materials

The polymer matrix was poly (vinyl chloride) (PVC) of K value = 58. The wood flour was made of beech, with a particle size between 300 to 400 micrometers. The fiber size of the hemp is about 10 mm. All of the natural raw materials were purchased from a wholesaler in Hungary. The formulation contained the usual additives being necessary for PVC processing: Ca-Zn based stabilizer, paraffin based external, partial glycerol ester type internal lubricants, ground $CaCO_3$ filler and acrylic based processing aid.

3.2 Sample preparation

In the preparation of starting mixtures, first PVC powder and processing additive were added and mixed manually. This mixture was the reference of the measurement (its ID number is O-1). The other three mixtures contain different amounts of wood flour. The first composite mixtures contain 15 weight ratio of wood next to PVC. The second contain 20 and the last 25 weight ratio of wood. They will be referred to in the following line: C – 1, C – 2, C – 3. The last three composites also contain hemp fibers next to 20 phr wood and PVC. The amount of their application: 1, 3, 5 weight part per 100 PVC. Then they were referred to in the following: CH – 1, CH – 2, CH – 3.

The powder mixtures were roll milled at 170°C on at twin roll mill Schwabenthan 150UL electrical heated roll mill. The roll milled felts were compression moulded at 175°C about 100 bar pressure into 1 and 4 mm thick plates. Tests specimens were machined from these plates for the compact tensile tests: the specimens had nominal dimensions of 35 mm length, 5 mm thickness, 30 mm width (measuring parameter) and holes 5 mm diameter.

4 EXPERIMENTS AND METHODS

4.1 Determination of density, hardness, fracture and tensile mechanical tests

The density values were determined by Archimedes principle weight measurement in water. The hardness tests were carried out with a Zwich/Roell machine using Shore D duromer punching tool. The density and hardness values of the mixtures are shown in Table 1.

Table 1: Density and hardness measurement of PVC composites.

	Materials						
	O-1	C – 1	C – 2	C – 3	CH – 1	CH – 2	CH – 3
Density (g/cm^3)	1.41	1.42	1.40	1.42	1.42	1.43	1.43
Hardness (Shore D)	81±0.1	82±0.6	81±0.3	83±0.4	81±0.7	82±0.5	81±0.8

For fracture mechanics tests on polymers CT specimens are frequently used. The incision of the fracture mechanical specimen was further increased and sharpened with a blade, with which the front position of the initial crack can be fixed. After, these specimens were subjected to tensile stress. From the results of the tensile test the conventional fracture toughness (K_Q) can be determined. The conventional fracture toughness load parameters are the F_Q load at when the crack is starting. Fig. 2 illustrates the integral of force versus deformation curve and F_{max} value.

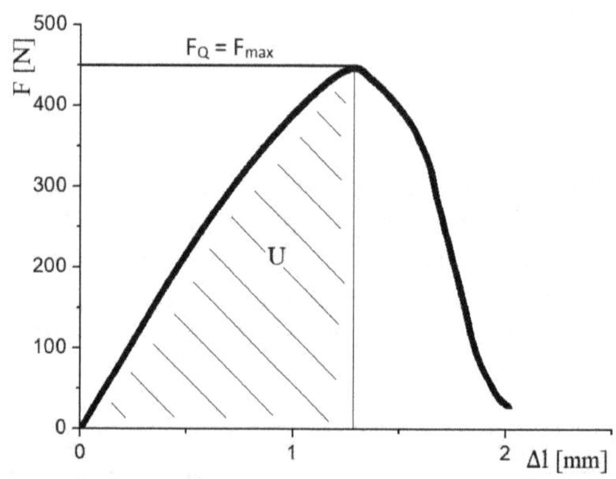

Figure 2: Evaluation of the tensile curves.

U is the integral of force until F_{max} i.e. the area under the curve from zero to F_{max}. The G_c values can be determined with the value of the energy associated with residual deformation.

According to the preliminary experiments, $F_Q = F_{max}$ was used as a critical break force to determine the fracture toughness. The final measurement results were determined from the average of five measurements. The tests were performed on INSTRON 5566 universal testing machine at room temperature (23±1°C). The cross-head speed was 10 mm/min.

The tensile tests were carried out with the same universal INSTRON 5566 machine at room temperature (23±1°C). The cross-head speed was 50 mm/min. When force–elongation data are converted to engineering stress and strain from the test results, a stress–strain curve that is identical in shape to the force–elongation curve can be plotted.

5 RESULTS AND DISCUSSION

The density values only slightly change regardless of the amount of wood flour and hemp fiber used. This is also experienced with the results of hardness tests. It is not surprising because the compact density both of the wood and the hemp fiber is about 1.5 g/cm^3 near to the density of matrix material.

It appears in Table 2 that the tensile strength is improved by the effect of hemp fiber and wood flour relative to the original PVC sample. However, the hemp fiber used as an additional reinforcement compared to the wood flour does not change the tensile strength values significantly compared to the 15 phr wood flour samples. The wood flour is micro fibrous filler so it has similar effects as hemps which are a long fibre compared to the flour. It is visible that using more than 15 phr wood flour start to weaken the composite as the interaction between the matrix and the filler is getting saturated. It is interesting that by using long fiber hemp we can improve the mechanical properties back to the 15 phr wood flour sample.

Table 2: Results of the tensile tests.

	Materials						
	O-1	C – 1	C – 2	C – 3	CH – 1	CH – 2	CH – 3
Tensile strength (MPa)	37.5±1.5	45±1.1	42±0.9	35.1±0.5	37.5±2.0	42.6±1.0	44.8±0.4

The higher the values of K_c and G_c in fracture tests, the more resistant is the material against crack propagation. Table 3 shows the critical stress intensity factor and the critical crack propagation energy values of the specimens. It is supposed that wettability of the wood flour and the hemp is not enough to obtain a good reinforcing effect; using a coupling agent would be advantageous.

Table 3: Results of the tensile test of the all materials.

	Materials						
	O-1	C – 1	C – 2	C – 3	CH – 1	CH – 2	CH – 3
K_c	2.4±0.1	2.5±1.1	2.9±0.1	2.4±0.0	2.8±0.2	2.6±0.2	2.6±0.2
G_c	11.8±2.0	7.5±0.9	8.7±1.3	7.2±0.8	6.9±0.7	5.9±0.7	6.4±1.0

No significant changes in the K_c values can be detected if the scatter values are taken into account. In contrast G_c values have visible changes. Forming a heterogeneous phase when using fiber and flour filler, can cause the structural deterioration. Fig. 3 shows K_c and G_c values with scatters. Since the values of G_c depends on the modulus of elasticity and maximum force, so the deformability and flexibility of the materials will also affect the crack propagation energy values.

Fig. 4 shows the process of propagation of crack. The second image of Fig. 4 shows the crack at the time of the maximum force values. Subsequently decreasing force was observed when the crack has grown.

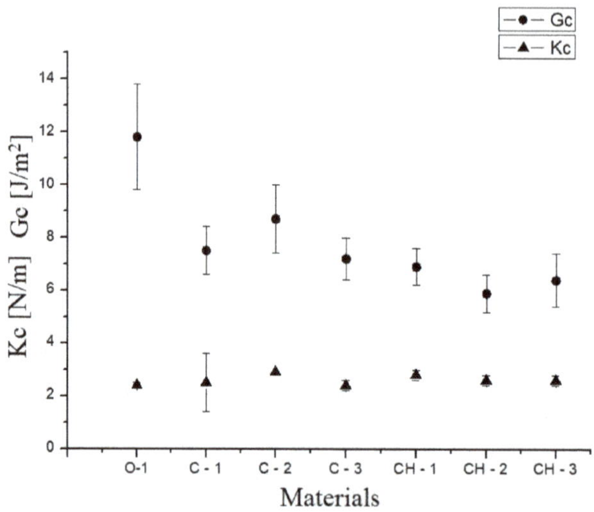

Figure 3: K_c and G_c values are the all samples.

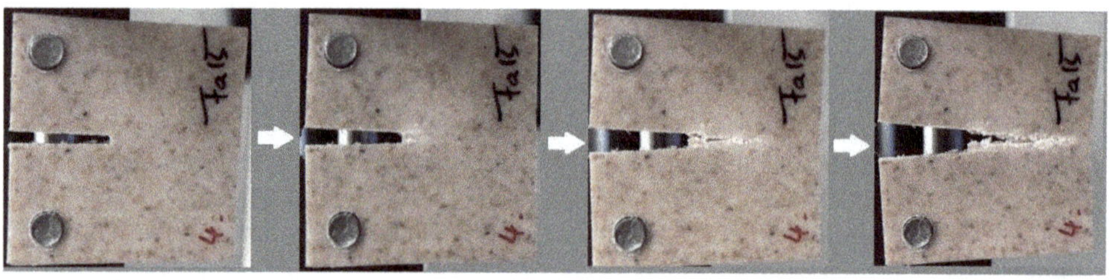

Figure 4: Measurement process for the fractural test.

6 CONCLUSION

The purpose of the fracture mechanics is to develop methods for characterizing the initial instability of microscopic cracks. Within the linear fracture mechanical test, the critical stress intensity factor and critical crack propagation energy have been shown to be suitable for characterization of the cracking process. The fracture toughness of PVC composites is hardly affected by the presence of additive. Only a small decrease in the fracture toughness was observed relative to the fracture toughness of the origin sample. The wood and hemp fillers leads to an increase of the craze dimension and decrease the loading capacity of the craze. The small variation in the G_c and K_c fracture parameters observed for PVC suggests that the amount of fillers gives almost balanced results. Therefore we continue the research with selected coupling agents.

ACKNOWLEDGEMENTS

The described article/presentation/study was carried out as part of the EFOP-3.6.1-16-00011 "Younger and Renewing University – Innovative Knowledge City – institutional development of the University of Miskolc aiming at intelligent specialisation" project implemented in the framework of the Szechenyi 2020 program. The realization of this project is supported by the European Union, co-financed by the European Social Fund. This research was supported by the European Union ant the Hungarian State, co-financed by the European Regional Development Fund in the framework of the GINOP – 2.3.4.-15-2016-00004 project, aimed to promote the cooperation between the higher education and the industry.

REFERENCES

[1] Titow, W.V., *PVC Technology*, 4th ed., Elsevier Applies Science Publishers: London and New York, 1984.

[2] Lee, J., Moon, J., Lee, H. & Choi, B., Investigation of fatigue and mechanical properties of the pipe grade poly (vinyl chloride) using recycled scraps. *eXPRESS Polymer Letters*, **9**(4), pp. 362–371, 2015.

[3] John K. & Naidu, S.V., Tensile properties of unsaturated polyester-based sisal fiber-glass fiber hybrid composites. *Journal of Reinforced Plastics and Composites*, **23**(17), pp. 1815–1819, 2004.

[4] Takagi, H., Nakagaito, A.N., Nishimura, K. & Matsui, T., Mechanical characterisation of nanocellulose composites after structural modification. *WIT Transactions on the Built Environment*, vol. 166, WIT Press: Southampton and Boston, pp. 335–341, 2016.

[5] Jiang, H. & Kamdem, D.P., Mechanical properties of poly (vinyl chloride)/wood flour/glass fiber hybrid composites. *Journal of Vinyl and Additive Technology,* **9**(3), pp. 138–145, 2003.

[6] Curtis, P.T., Tensile fatigue mechanisms in unidirectional polymer matrix composite materials. *International Journal of Fatique*, **13**(5), pp. 377–382, 1991.

[7] Radon, J.C., Physics fatigue of polymers: Crack growth in PVC. *Journal of Macromolecular Science, Part B Physics*, **14**(4), pp. 37–41, 2006.

[8] Leevers, P.S., Culver, L.E. & Radon, J.C., Fatigue crack growth in PMMA and rigid PVC under biaxial stress. *Engineering Fracture Mechanics*, **11**(3), pp. 487–498, 1979.

[9] Breen, J., Environmental stress cracking of PVC and PVC-CPE – Part I Crazing, *Journal of Materials Science*, **28**(14), pp. 3769–3776, 1993.

[10] Pulngern, T. et al., Effects of cross-section design and loading direction on the creep and fatigue properties of wood/PVC composite beams. *Journal of Vinyl and Additive Thechnology*, **16**, 2010.

[11] Hu, Y. & Summers, J., Correlation of fatigue and creep crack growth in poly (vinyl chloride). *Journal of Materials Science*, **38**, pp. 633–642, 2003.

[12] Xu, S. & Reinhardt, H.W., Determination of double-K criterion for crack propagation in quasi-brittle fracture. Part II: Analytical evaluating and practical measuring methods for three-point bending notched beams. *International Journal of Fracture*, **98**, pp. 151–177, 1999.

[13] Radon, J.C., Fatigue crack growth in polymers. *International Journal of Fracture,* **16**(6), pp. 533–552, 1980.

[14] Ramsteiner, F., Schuster, W. & Forster, S., Concepts of fracture mechanics for polymers. *Deformation and Fracture Behaviour of Polymers,* pp. 27–50, 2001.

SECTION 5
MECHANICAL AND THERMAL PROPERTIES

DIFFERENCE BETWEEN USING TABULATED AND EXACT VALUES OF THERMAL PROPERTIES OF MATERIALS IN NUMERICAL SIMULATIONS OF HEAT TRANSFER THROUGH A HIGH-PERFORMANCE WINDOW

MIHA JUKIĆ & SABINA JORDAN
Slovenian National Building and Civil Engineering Institute, Slovenia

ABSTRACT

The thermal properties of materials, primarily the thermal conductivity, are an essential input for numerical modelling of heat transfer in buildings and building components. When determining them according to relevant European standards, it is not uncommon to encounter materials for which the exact values are not appropriately specified and the tabulated values in standards are overly conservative. In such situations, the thermal conductivity of the material can be determined by measurement. However, this approach may prove inconvenient and too expensive, especially if the material in question turns out to have little influence on the overall thermal performance of the product. It is, therefore, of great interest to know how the thermal performance is affected by choosing either the accurate (measured) or the conservative (tabulated) value of the thermal conductivity. In this work, the two approaches are compared in a practical example – a high-performance window, Jelovica Jelofuture – using numerical simulations. Our study shows that modifying the thermal properties of individual materials generally leaves the thermal transmittances of the frame (U_f) and the window (U_W) almost unaffected. If all of the materials considered are modified simultaneously, U_f changes by 1–2% while the change in U_W remains below 1%. However, due to their small values, the calculated changes of U_f and U_W may be significantly affected (further increased or reduced) by the rounding of the results according to the relevant standards. In contrast, using the tabulated value of linear thermal transmittance (Ψ_g) of the junction with the glazing leads to an overestimation of U_W by up to 15%.

Keywords: thermal transmittance, numerical simulation, conductivity, tabulated and exact values, practical example.

1 INTRODUCTION

The EU Construction Products Regulation [1] requires the placement of construction products on the European market to be supported by assessment and declaration of their performance, with energy efficiency as one of the most important aspects. The thermal performance of many products, including doors and windows, is most conveniently assessed by performing numerical simulations of heat transfer. The latter is subject to national and international standards which, along with prescribing the methods, set restrictions on the input data for the analysis.

The present European standards which govern numerical modelling of heat transfer in doors and windows, e.g. [2], [3], allow several ways to determine the material properties used in the modelling. The preferred option is to use the tabulated values from the standards. Alternatively, the material properties may be taken from the declaration of performance of a specific product, or determined by measurement. The declared values are usually provided for the "important" products, e.g. thermal insulation, and the tabulated values usually suffice for the more generic materials, such as glass or steel. Still, it is not uncommon to encounter materials for which the thermal properties are not appropriately specified and the tabulated values are overly conservative. In such cases the properties can be determined by

WIT Transactions on Engineering Sciences, Vol 124, © 2019 WIT Press
www.witpress.com, ISSN 1743-3533 (on-line)
doi:10.2495/MC190121

measurement: however, this approach may prove inconvenient and too expensive, especially if the material in question only has a minor role in the analysed building or building component. It is therefore of great interest to know how the thermal performance is affected by using either the accurate (measured) or the conservative (tabulated) values of thermal properties. Studies that examined the influence of different parameters on the thermal performance of windows have been performed in the past, e.g. [4]–[6], but they mostly focused on the design aspect, i.e. optimising the window geometry, details and constituent materials.

In the present work, the focus was set on the assessment of the thermal performance of a finalised product in accordance with the relevant European standards [3], [7], [8]. The goal of the study was to investigate using a practical example – a high-performance window, Jelovica Jelofuture – how the choice between using the accurate or the tabulated values of thermal properties in numerical simulations of heat transfer influences the thermal performance of a window.

2 METHODOLOGY

The starting point of this study was the thermal performance assessment of a high-performance window, Jelovica Jelofuture. The assessment was based on numerical simulations of heat transfer, and comprised the calculations of the thermal transmittance of the window frame (U_f) the linear thermal transmittance at the junction of the frame and the glazing (Ψ_g), and the thermal transmittance of the window (U_W), all in accordance with the relevant European standards [3], [8]. In the original assessment, tabulated values of thermal conductivity had been used for several materials.

In the present study, the conservative, tabulated values of thermal properties used in the original assessment were modified in order to examine their influence on the thermal performance of the window and its elements. The modifications were applied in different combinations, individually and collectively. Numerical simulations of heat transfer were then repeated for each combination.

Two approaches were used for the subsequent calculations. The first one (referred to as "According to standard" in the tables below) strictly followed the procedures defined in standards [3], [8], which also specify the rounding of the results, i.e. thermal transmittances U_f, Ψ_g, and U_W. Since U_f is an input for the calculation of Ψ_g, and both are used to calculate U_W, the rounding may affect the results to a certain extent. In the second approach (referred to as "Without rounding" in the tables below) the calculations were carried out without intermediate rounding, in order to give more accurate and physically sensible results, and provide a better insight into the physical consequences of the applied modifications.

2.1 Description of the specimen

The window frame profiles of the high-performance window, Jelovica Jelofuture (the investment for the development of the window Jelovica Jelofuture in the scope of the project Jelofuture was co-financed by the Republic of Slovenia and the European Union – European Regional Development Fund) are designed as a combination of wood, PVC, and aluminium. The sill profile (D1) is depicted in Fig. 1(a). The central part of the frame consists of a PVC profile with highly insulative polyisocyanurate (PIR) foam inserts. The inner part of the frame and the casement are made of wood, and the outer cover is made of aluminium profile with a PIR foam insert. Silicone is used to seal the gap between the glazing unit and the casement. Expansion tape is used between the PVC and the wooden part of the frame. The gaskets are made of EPDM-based material. The head profile (D2) in Fig. 1(b) slightly differs from the sill profile – the casement has a different shape and two additional PIR foam inserts are used. The jamb profiles are identical to the head profile.

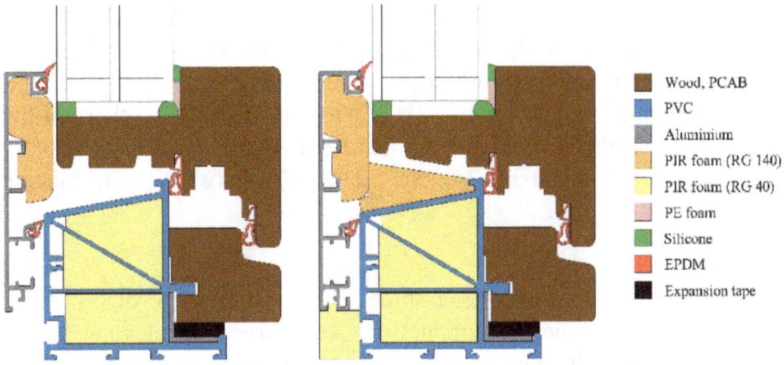

Figure 1: Window frame profiles. (a) Sill profile – D1; (b) Head and jamb profiles – D2.

Two different insulative glazing units (IGUs) were considered in this study. Both were triple-glazed, filled with argon, and had two low-emissivity coatings. The first one was 48 mm thick (4:/18/4/18/:4), with thermal transmittance $U_g = 0.5$ W/m^{2}K. The second was 44 mm thick (8:/10/8/10/:8), with thermal transmittance $U_g = 0.8$ W/m^{2}K. In the following text, they are referred to as IGU_{05} and IGU_{08}, respectively. In both cases a "Swisspacer Ultimate" glazing spacer was used.

Single casement windows of two standard sizes were analysed. Their overall frame dimensions (width × height) were 1.23 m × 1.48 m and 1.48 m × 2.18 m. According to [7], these may represent all windows with an overall area under/over 2.3 m^2, respectively.

2.2 Calculation of thermal transmittances U_f, Ψ_g, and U_W

Thermal transmittance of the window frame (U_f) was determined with numerical simulations of heat transfer according to the standard [3]. The simulations were performed using Physibel Bisco 11.0 software. Thermal properties (conductivity and emissivity) of materials were determined in accordance with the standard [3]. Their values and the details of the procedure are given in Section 2.3. The convective and the radiative heat transfer in cavities were treated using the single equivalent thermal conductivity method described in the standard [3]. The glazing was replaced by a 36 mm thick insulative panel extending 190 mm from the edge of the frame. The boundary conditions used in the simulations are listed in Table 1.

Table 1: Boundary conditions used in numerical simulations.

Surface	Air temperature (°C)	Surface resistance (m^{2}K/W)
External	0	0.04
Internal – normal	20	0.13
Internal – reduced (corners)	20	0.20

Linear thermal transmittance (Ψ_g) at the junction of the frame and the glazing was determined according to the standard [3] for all combinations of different frame profiles and different glazing units. For each case a numerical model was developed, similar to the one used for the calculation of U_f, but with the insulation panel replaced by the actual glazing. The argon filling was modelled with a replacement material with an equivalent thermal conductivity. The glazing spacers were modelled using the two-box method described in the ift guideline [9]. Linear thermal transmittance (Ψ_g) was then calculated from the total heat flow through the numerical model, the thermal transmittance of the glazing (U_g) and the previously determined thermal transmittance of the window frame (U_f).

Thermal transmittance of the window (U_W) was calculated from thermal transmittances U_g, U_f, and Ψ_g in accordance with the standard [8]. It was determined for two standard window sizes (123 cm × 148 cm and 148 cm × 218 cm) and two types of glazing.

2.3 Thermal properties

In the original thermal performance assessment, the thermal properties were determined as follows. Tabulated values of thermal conductivity, taken from the relevant standards [3], [10], were used for most materials, with two exceptions. The conductivity of the PIR foam (two types) was taken from the declaration of performance and the conductivity of the expansion tape was measured by an accredited institute. Due to the insufficient number of measurements, the latter was magnified by a factor of 1.25 in accordance with the standard [3]. Emissivities of all materials were taken from the standard [3]. The thermal conductivities and emissivities of materials are collected in Table 2.

Table 2: Thermal conductivities (original and modified values) and emissivities of materials used in numerical simulations.

Material	Conductivity (W/mK)	Modified conductivity (W/mK)	Emissivity
Aluminium (powder coated)	160	-	0.9
PVC	0.17	0.16 (A)	0.9
Wood, PCAB (spruce)	0.11	0.13 (B)	0.9
Puren PIR NE40-200 RG 140	0.036	-	0.9
Puren PIR NE40-200 RG 40	0.027	-	0.9
PE foam	0.05	0.04 (A)	0.9
Silicone	0.35	0.30 (A)	0.9
EPDM	0.25	0.20 (A)	0.9
Expansion tape Chemiefac 3 complete	0.0456 × 1.25 = 0.057	0.0456 (A)	0.9
Insulation panel	0.035	-	0.9

The linear thermal transmittance (Ψ_g) of the glazing spacer was calculated as described in the previous section (as opposed to using a tabulated value) and the thermal transmittance of the IGU was taken from the declaration of performance.

Modifications of the original thermal properties can be divided into two groups. The purpose of group (A) was to determine whether the thermal performance of a window could be substantially improved by utilising more accurate (non-tabulated) thermal conductivities of individual materials. A positive answer would justify the investment of time and resources into obtaining the appropriate documentation (measurement or declaration of performance).

In this group the thermal conductivities of PVC, PE foam, silicone, and EPDM were reduced by 5–20%. The modified values, denoted by (A) in Table 2, were chosen based on previous experience with such materials. The conductivity of the expansion tape was also reduced by omitting the factor 1.25, as would be the case if the measured conductivity was an average value of three or more measurements, instead of just one. The modifications in group (A) were applied individually and collectively to observe their impact on the thermal transmittances U_f, Ψ_g, and U_W.

Group (B) comprises two modifications of a different nature. Here, the original thermal properties were replaced with more conservative values. The conductivity of the wood was increased from 0.11 W/mK (tabulated value for the specific tree species) to 0.13 W/mK (general tabulated value for softwood). The modified value is denoted by (B) in Table 2. Also, the tabulated value, $\Psi_g = 0.06$ W/mK, was used for the linear thermal transmittance at the junction of the frame and the glazing. It corresponds to a thermally improved spacer in combination with triple glazing and low emissivity glass, according to the standard [3]. This value is roughly twice the calculated value of Ψ_g. The second modification only influences the calculation of U_W, while the first one also affects the calculation of U_f and Ψ_g.

3 RESULTS AND DISCUSSION

3.1 Thermal transmittance of the window frame, U_f

Thermal transmittances of the window frame profiles D1 (sill) and D2 (head and jambs), determined for the original and the modified material properties, are collected in Table 3. Each value is accompanied by the relative difference (Δ_{rel}) from the original U_f, i.e. thermal transmittance of the frame with the original thermal properties. The table shows the unrounded and rounded results.

Table 3: Thermal transmittance (U_f) of window frame profiles D1 and D2 – calculated according to standard, and without rounding, for different modifications of thermal properties.

Modified material	U_f (W/m²K)							
	D1 – sill profile				D2 – head and jamb profiles			
	Without rounding *	Δ_{rel}**	According to standard*	Δ_{rel}	Without rounding	Δ_{rel}	According to standard	Δ_{rel}
Original	0.8485		0.85		0.7223		0.72	
PVC	0.8418	-0.8%	0.84	-1.2%	0.7159	-0.9%	0.72	0.0%
Silicone	0.8464	-0.2%	0.85	0.0%	0.7206	-0.2%	0.72	0.0%
EPDM	0.8464	-0.2%	0.85	0.0%	0.7206	-0.2%	0.72	0.0%
PE foam	0.8468	-0.2%	0.85	0.0%	0.7210	-0.2%	0.72	0.0%
Expansion tape	0.8473	-0.1%	0.85	0.0%	0.7214	-0.1%	0.72	0.0%
All in group (A)	0.8350	-1.6%	0.84	-1.2%	0.7113	-1.5%	0.71	-1.4%
Wood	0.8837	4.1%	0.88	3.5%	0.7490	3.7%	0.75	4.2%

*See Section 2 for the description of the calculation procedures.
**Relative difference from the original value.

Modifying the thermal conductivity of an individual material from group (A) hardly affects the transmittance of the frame. In most cases U_f changes by a mere 0.2%, even with the reduction of the conductivity by 15–20%. A slightly bigger improvement of U_f is observed with PVC, despite the smaller reduction of its thermal conductivity (approx. 6%). This is because PVC constitutes part of the frame core which provides most of the thermal protection. Being surrounded by a superior material (PIR foam), PVC acts as a minor thermal bridge. Nevertheless, the improvement of U_f is still below 1%. Applying all modifications from group (A) simultaneously, improves the thermal transmittance of the frame by 1.5%.

Thermal conductivity of wood has a more noticeable influence on U_f. Increasing the thermal conductivity by 18%, which is comparable to the modifications in group (A), results in an approximately 4% larger U_f. Compared to group (A), this increase is an order of magnitude greater. This is not surprising, since wood covers virtually the whole inner surface of the frame in considerable thickness and thus creates a uniform thermally protective layer. On the contrary, the materials in group (A) mostly represent smaller elements scattered over the frame profile. The other modification in group (B) – application of a tabulated value for Ψ_g – does not affect the U_f calculation procedure.

The observations described above refer to the unrounded values of U_f. Subsequent rounding of those values has a significant impact on the differences between them. According to the standard [3], U_f has to be rounded to two significant figures, which implies a change in the order of 1%. Improvements of U_f, observed on the unrounded values, are generally much smaller and therefore irrelevant. This is confirmed by comparing the rounded results in Table 3. Individual modifications of thermal properties from group (A) have virtually no influence on the rounded value of U_f. Only when they are all combined is the improvement of U_f sufficiently big not to be annulled by rounding. The only relevant modification by this criterion is the increase of the thermal conductivity of wood.

Graphical results of the numerical simulation for the head/jamb profile (D2) with the original thermal properties are presented in Fig. 2 which displays the temperature field (a) and the heat flux field (b) in the frame profile.

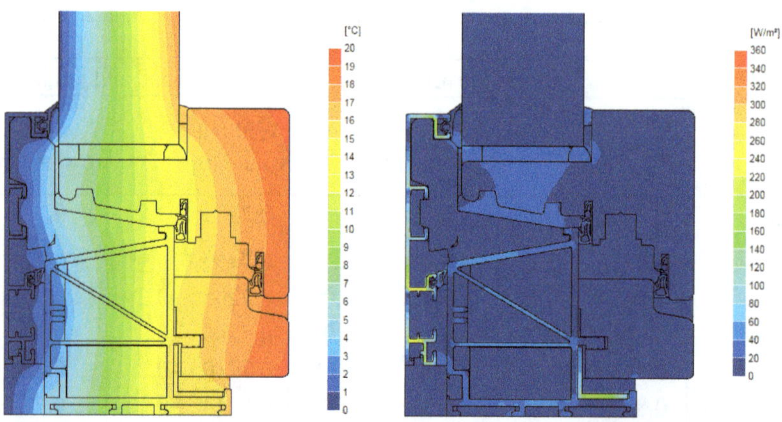

(a) (b)

Figure 2: Graphical output of the numerical simulation for the head/jamb profile (D2) with the original thermal properties. (a) Temperature field; (b) Heat flux field.

3.2 Linear thermal transmittance (Ψ_g) at the junction of the frame and the glazing

The linear thermal transmittances (Ψ_g) at the junction of the window frame and the glazing were determined for all combinations of the frame profiles (D1 and D2) and the glazing units (IGU_{05} and IGU_{08}), applying the different modifications of thermal parameters described in Section 2.3. The results are collected in Table 4 for IGU_{05} ($U_g = 0.5$ W/m^{2}K), and in Table 5 for IGU_{08} ($U_g = 0.8$ W/m^{2}K). Each Ψ_g is accompanied by its relative difference, Δ_{rel}, from the original value. Against intuitive expectation, Ψ_g takes smaller values for the glazing with lower U_g. The reason lies in the greater thickness of the glass panes (8 mm in IGU_{08} vs. 4 mm in IGU_{05}).

Table 4: Linear thermal transmittance (Ψ_g) at the junction of the window frame profile (D1 or D2) and the glazing IGU_{05} ($U_g = 0.5$ W/m^{2}K), calculated according to standard and without rounding, for different modifications of thermal properties.

Modified material	Ψ_g (W/mK)							
	D1 – sill profile				D2 – head and jamb profiles			
	Without rounding*	Δ_{rel}**	According to standard*	Δ_{rel}	Without rounding	Δ_{rel}	According to standard	Δ_{rel}
Original	0.02242		0.022		0.02292		0.023	
PVC	0.02242	0.0%	0.023	4.5%	0.02297	0.2%	0.022	-4.3%
Silicone	0.02237	-0.2%	0.022	0.0%	0.02287	-0.2%	0.023	0.0%
EPDM	0.02232	-0.4%	0.022	0.0%	0.02292	0.0%	0.023	0.0%
PE foam	0.02252	0.4%	0.022	0.0%	0.02302	0.4%	0.023	0.0%
Expansion tape	0.02242	0.0%	0.022	0.0%	0.02292	0.0%	0.023	0.0%
All in group A	0.02242	0.0%	0.022	0.0%	0.02287	-0.2%	0.023	0.0%
Wood	0.02287	2.0%	0.023	4.5%	0.02317	1.1%	0.023	0.0%

*See Section 2 for the description of the calculation procedures.
**Relative difference from the original value.

A distinction is made between the values of Ψ_g calculated in accordance with the standard [3], and the values calculated without rounding. The former were calculated using the rounded values of U_f. The latter were calculated using the unrounded values of U_f to provide more accurate and physically sensible results.

The unrounded results for IGU_{05} in Table 4 reveal that the modifications of material properties have no significant influence on the linear thermal transmittance of the glazing spacer. The relative differences (Δ_{rel}) do not exceed 0.4% except in the case of wood. The increased thermal conductivity of wood produces a 1–2% larger Ψ_g. In Table 5, the relative differences (Δ_{rel}) for IGU_{08} are even smaller.

Some modifications from group (A) increase the value of Ψ_g, while others decrease it. If they are all applied simultaneously, their effects balance out.

Rounding of Ψ_g to two significant figures plays an even more important role than it did in the case of U_f. It can change the typical Ψ_g value of an advanced thermally improved spacer (around 0.03 W/mK) by approximately 3%. This is confirmed by inspecting the results in Tables 4 and 5, calculated according to standard. Most of the small changes of U_f, caused by the modified thermal properties are voided. Some, on the other hand, are notably increased.

Table 5: Linear thermal transmittance (Ψ_g) at the junction of the window frame profile (D1 or D2) and the glazing IGU_{08} ($U_g = 0.8$ W/m^{2}K), calculated according to standard and without rounding, for different modifications of thermal properties.

Modified material	Ψ_g (W/mK)							
	D1 – sill profile				D2 – head and jamb profiles			
	Without rounding*	Δ_{rel}**	According to standard*	Δ_{rel}	Without rounding	Δ_{rel}	According to standard	Δ_{rel}
Original	0.03547		0.035		0.03547		0.036	
PVC	0.03547	0.0%	0.036	2.9%	0.03547	0.0%	0.035	-2.8%
Silicone	0.03552	0.1%	0.035	0.0%	0.03547	0.0%	0.036	0.0%
EPDM	0.03537	-0.3%	0.035	0.0%	0.03537	-0.3%	0.035	-2.8%
PE foam	0.03557	0.3%	0.035	0.0%	0.03552	0.1%	0.036	0.0%
Expansion tape	0.03552	0.1%	0.035	0.0%	0.03542	-0.1%	0.036	0.0%
All in group A	0.03547	0.0%	0.035	0.0%	0.03542	-0.1%	0.036	0.0%
Wood	0.03567	0.6%	0.036	2.9%	0.03592	1.3%	0.036	0.0%

*See Section 2 for the description of the calculation procedures.
**Relative difference from the original value.

The slightly "random" values of Δ_{rel} (e.g. in the case of PVC: positive for the D1 profile and negative for the D2 profile, while virtually non-existent in the calculation without rounding) are a consequence of calculating Ψ_g with the rounded values of U_f.

3.3 Thermal transmittance of the window, U_W

Thermal transmittance of the window (U_w) was calculated for two standard sizes of single casement windows and two types of glazing (IGU_{05} and IGU_{08}), applying the different modifications of thermal parameters described in Section 2.3. The results are presented in Table 6 for IGU_{05} and in Table 7 for IGU_{08}. The relative difference (Δ_{rel}) from the original U_W, i.e. determined with the original material properties, is given as well.

Again, calculation according to the standard [8] and calculation without rounding are distinguished in Tables 6 and 7. In the first case, the values of U_w were calculated using the rounded values of U_f and Ψ_g, which were determined in accordance with the standard [3]. The unrounded values of U_w were calculated using the unrounded U_f and Ψ_g to give more precise results.

The considered modifications of thermal properties only apply to materials in the window frame, while the thermal transmittance of the glazing (U_g) remains unchanged. The glazing represents approximately two thirds of the smaller window examined herein (123 cm × 148 cm), and three quarters of the bigger one (148 cm × 218 cm). It is therefore logical that the thermal transmittance of the window is less sensitive to the modifications than U_f and Ψ_g.

Individual modifications from group (A) have no significant influence on U_W. Even combining all of them only reduces U_W by circa 0.5%. The only noteworthy change of U_W (1–1.5%) is caused by increasing the thermal conductivity of wood. These effects are most notable on the smaller window (123 cm × 148 cm) with IGU_{05}. A larger window size means that a greater portion of the window is insensitive to modifications (glazing). And a larger value of U_W, caused by a larger U_g, means that the same absolute changes have a smaller relative value.

Table 6: Thermal transmittance (U_W) of two single casement windows (123 cm × 148 cm and 148 cm × 218 cm) using the glazing IGU_{05} (U_g = 0.5 W/m^{2}K) – calculated according to standard, and without rounding, for different modifications of thermal properties.

Modified material	U_W (W/m^{2}K)							
	123 cm × 148 cm				148 cm × 218 cm			
	Without rounding*	Δ_{rel}**	According to standard*	Δ_{rel}	Without rounding	Δ_{rel}	According to standard	Δ_{rel}
Original	0.6367		0.64		0.6073		0.61	
PVC	0.6349	-0.3%	0.63	-1.6%	0.6058	-0.2%	0.61	0.0%
Silicone	0.6361	-0.1%	0.64	0.0%	0.6067	-0.1%	0.61	0.0%
EPDM	0.6362	-0.1%	0.64	0.0%	0.6068	-0.1%	0.61	0.0%
PE foam	0.6366	0.0%	0.64	0.0%	0.6071	0.0%	0.61	0.0%
Expansion tape	0.6365	0.0%	0.64	0.0%	0.6070	0.0%	0.61	0.0%
All in group (A)	0.6330	-0.6%	0.63	-1.6%	0.6043	-0.5%	0.60	-1.6%
Wood	0.6467	1.6%	0.65	1.6%	0.6150	1.3%	0.61	0.0%

*See Section 2 for the description of the calculation procedures.
**Relative difference from the original value.

Table 7: Thermal transmittance (U_W) of two single casement windows (123 cm × 148 cm and 148 cm × 218 cm) using the glazing IGU_{08} (U_g = 0.8 W/m^{2}K), calculated according to standard and without rounding, for different modifications of thermal properties.

Modified material	U_W (W/m^{2}K)							
	123 cm × 148 cm				148 cm × 218 cm			
	Without rounding*	Δ_{rel}**	According to standard*	Δ_{rel}	Without rounding	Δ_{rel}	According to standard	Δ_{rel}
Original	0.8718		0.87		0.8572		0.86	
PVC	0.8698	-0.2%	0.87	0.0%	0.8556	-0.2%	0.86	0.0%
Silicone	0.8713	-0.1%	0.87	0.0%	0.8568	0.0%	0.86	0.0%
EPDM	0.8710	-0.1%	0.87	0.0%	0.8566	-0.1%	0.86	0.0%
PE foam	0.8715	0.0%	0.87	0.0%	0.8570	0.0%	0.86	0.0%
Expansion tape	0.8714	0.0%	0.87	0.0%	0.8569	0.0%	0.86	0.0%
All in group (A)	0.8680	-0.4%	0.87	0.0%	0.8542	-0.3%	0.85	-1.2%
Wood	0.8819	1.2%	0.88	1.1%	0.8651	0.9%	0.87	1.2%

*See Section 2 for the description of the calculation procedures.
**Relative difference from the original value.

The values of U_W, calculated according to the standard [8], are mostly unaffected by the modifications from group (A), unless all of them are combined. Even then, a non-zero Δ_{rel} is largely a consequence of the original U_W being near the border of being rounded up or down. Modifying the thermal conductivity of wood increases U_W by a little more than 1%.

Thermal transmittances of the window (U_W), calculated using the tabulated value of Ψ_g, are collected in Table 8. They were determined in accordance with the standard [8]. The table shows the results for two standard window sizes and two types of glazing. Here, Δ_{rel} represents the relative difference from U_W, calculated using the original thermal properties and the calculated value of Ψ_g (the results with the column tag "According to standard" and row tag "Original" in Tables 6 and 7).

Table 8: Thermal transmittance of the window (U_W) calculated using the tabulated value of $\Psi_g = 0.06$ W/mK – results for two window sizes (123 cm × 148 cm and 148 cm × 218 cm) and two IGUs ($U_g = 0.5$ W/m²K and 0.8 W/m²K).

IGU	Modified material	U_W (W/m²K)			
		123 cm × 148 cm		148 cm × 218 cm	
		According to standard	Δ_{rel}*	According to standard	Δ_{rel}
$U_g = 0.5$ W/m²K	Original	0.73	14.1%	0.68	11.5%
	Wood	0.74	15.6%	0.69	13.1%
$U_g = 0.8$ W/m²K	Original	0.93	6.9%	0.91	5.8%
	Wood	0.94	8.0%	0.91	5.8%

*Relative difference from U_w calculated using the original thermal properties and the calculated Ψ_g.

Using the tabulated $\Psi_g = 0.06$ W/mK proves most costly for the smaller window with the IGU_{05} glazing unit, for which the actual value of Ψ_g is almost three times smaller (0.022 W/mK for the D1 profile and 0.023 W/mK for D2). As a consequence, U_W grows by 14%. A lesser increase of U_W (by 7%) is observed in the case of IGU_{08}, because the actual Ψ_g is slightly larger here (0.035 W/mK for the D1 profile and 0.036 W/mK for D2). The values of Δ_{rel} are a little lower for the larger window, regardless of the type of glazing.

If thermal conductivity of wood is increased as well, the thermal transmittance of the window grows by additional 1–1.5%. Small deviations of this increase from the results in Tables 6 and 7 are caused by rounding.

4 CONCLUSIONS

The goal of this study was to examine how using the conservative, tabulated values of thermal properties in the numerical simulations of heat transfer, instead of obtaining the appropriate documentation which would allow the use of exact ones, affects the calculated values of the thermal transmittance of the window frame (U_f), the linear thermal transmittance at the joint of the frame and the glazing (Ψ_g) and the thermal transmittance of the window (U_W).

The study was based on the thermal performance assessment of a high-performance window, Jelovica Jelofuture, whose frame is designed as a combination of wood, PVC, and aluminium with PIR foam inserts, and which uses highly insulative glazing units. The tabulated values of thermal properties, used in the original assessment, were modified individually and collectively before repeating the numerical simulations and comparing the results to the original ones. The modifications comprised reducing the thermal conductivity of PVC and minor elements of the window frame (silicone, EPDM, PE foam, expansion tape); using the general thermal conductivity for softwood, instead of the value for the specific tree species; and using the tabulated value of Ψ_g in the calculation of U_W, instead of the exact one. U_W was calculated for two standard sizes (123 cm × 148 cm and 148 cm × 218 cm) of a single casement window.

The study showed that reducing the thermal conductivity of the minor frame elements by 15–20% had virtually no effect on the thermal transmittances U_f, Ψ_g, and U_W. Reducing the thermal conductivity of the PVC profile by 6% lowered U_f by 1%, but did not really influence Ψ_g and U_W. Combining these modifications had no surprising outcome – an improvement of

U_f by a bit more than 1%, and hardly any improvement of Ψ_g and U_W. Thermal conductivity of wood, which forms the inner part of the frame and the casement, proved to have a more noticeable influence. If the tree species had not been specified and the general tabulated value for softwood (18% higher value) had to be used, U_f would grow by approximately 4%, and Ψ_g and U_W by roughly 1–2%. Using the tabulated value of Ψ_g, instead of calculating the actual one, had by far the greatest consequences, increasing U_W by 6–14%. The increase depended partly on the window size, but mostly on the type of the insulating glazing unit which determined the actual value of Ψ_g. In our case, the latter was 2–3 times smaller than the tabulated Ψ_g.

The observations noted above exclude the effects of rounding the thermal transmittances U_f, Ψ_g, and U_W according to the relevant standards. Rounding them to two significant figures can usually change their values by an order of 1–2%, or a bit more for Ψ_g. This further confirms that most of the considered modifications are irrelevant, as shown by this study. Nevertheless, if the examined thermal transmittance is on the border of being rounded up or down, they may tip the scale.

The results of the study indicate that, in the numerical simulations of heat transfer, tabulated values of thermal conductivity can be used for the minor components of the window frame, such as gaskets, sealants, etc., without significant risk of compromising the (calculated) thermal performance of the analysed product. Attention should be given to materials that occupy larger, continuous areas in the frame cross section (wood in our case) and elements that can represent a thermal bridge. Even small adjustments (such as defining the tree species) may affect the thermal performance considerably.

Using a tabulated value of the linear thermal transmittance (Ψ_g) at the junction of the frame and the glazing should be avoided, especially if high-end glazing spacers are used. With the advance in technology, many high-performance spacers have been developed which allow values of Ψ_g several times lower than the conservative, tabulated values.

ACKNOWLEDGEMENT

The authors would like to thank the company Jelovica d.o.o. for their contribution to the research.

REFERENCES

[1] The European Parliament and the Council of the European Union, Laying down harmonised conditions for the marketing of construction products and repealing Council Directive 89/106/EEC. Regulation (EU) No. 305/2011 of the European Parliament and of the Council (Construction Products Regulation), 2011.

[2] International Organization for Standardization, Thermal bridges in building construction – Heat flows and surface temperatures – Detailed calculations. International standard SIST EN ISO 10211:2017, 2017.

[3] International Organization for Standardization, Thermal performance of windows, doors and shutters – Calculation of thermal transmittance – Part 2: Numerical method for frames. International standard SIST EN ISO 10077-2:2017, 2017.

[4] Gustavsen, A., Grynning, S., Arasteh, D., Jelle, B.P. & Goudey, H., Key elements of and material performance targets for highly insulating window frames. *Energy and Buildings*, **43**(10), pp. 2583–2594, 2011.

[5] Van Den Bossche, N., Buffel, L. & Janssens, A., Thermal optimization of window frames. *Energy Procedia*, **78**, pp. 2500–2505, 2015.

[6] Kalinović, S.M., Djoković, J.M. & Nikolić, R.R., Influence of windows geometrical parameters on calculations of the heat conduction coefficient. *Procedia Engineering*, **192**, pp. 404–409, 2017.
[7] International Organization for Standardization, Windows and doors – Product standard, performance characteristics – Part 1: Windows and external pedestrian doorsets. International standard SIST EN 14351-1:2006+A2:2016, 2016.
[8] International Organization for Standardization, Thermal performance of windows, doors and shutters – Calculation of thermal transmittance – Part 1: General. International standard SIST EN ISO 10077-1:2017, 2017.
[9] ift-Guideline WA-08engl/3, Thermally improved spacers – Part 1: Determination of representative Ψ-values for profile sections of windows. Report, 2015.
[10] International Organization for Standardization, Building materials and products – Hygrothermal properties – Tabulated design values and procedures for determining declared and design thermal values. International standard SIST EN ISO 10456:2008, 2008.

SECTION 6
COMPOSITE MATERIALS

INTERFACIAL SHEAR STRENGTH OF CARBON FIBER-REINFORCED POLYPROPYLENE USING UNEVEN CARBON FIBER UNDER A HOT–WET ENVIRONMENT FOR LONG-TERM SAFETY OF VEHICLES

KENICHI TAKEMURA[1], MAO MOCHIZUKI[1] & HIDEAKI KATOGI[2]
[1]Department of Mechanical Engineering, Kanagawa University, Japan
[2]Department of Human Environmental Sciences, Jissen Women's University, Japan

ABSTRACT

We have previously reported on the improvement of the interfacial shear strength of carbon fiber-reinforced polypropylene (CFRPP) using carbon fiber with added resin particles (uneven carbon fiber). This study was conducted to assess the interfacial shear strength of CFRPP with uneven carbon fiber under a hot–wet environment for long-term safety of vehicles. The matrix was maleic anhydride grafted polypropylene. The matrix concentration was 10%. Reinforcement was done with uneven carbon fiber. Water absorption tests of CFRPP were conducted using uneven carbon fiber. The range of water temperature was from room temperature (RT) to 80°C. The maximum immersion time was 112 days. After water absorption testing, a micro-debonding test of CFRPP using uneven carbon fiber was conducted at RT. Fracture surfaces of carbon fibers were observed using scanning electron microscopy (SEM). Results show that when the maximum immersion time was 112 days, interfacial shear strength of water-immersed CFRPPs using uneven carbon fiber at RT, 50°C and 80°C decreased by 54%, 94% and 96% respectively, compared with that of non-immersed CFRPP using uneven carbon fiber. However, interfacial shear strengths of water-immersed CFRPPs using even and uneven carbon fiber under 50°C and 80°C showed almost no change at 112 days. SEM observations revealed large resin particles on the fracture surface of carbon fiber at temperatures higher than 50°C. Large resin particles and matrix on fracture surface of carbon fiber were found at 80°C. The viscoelasticity and interfacial adhesion property of water absorbing polymer matrix probably decreased when CFRPP using uneven carbon fiber was immersed at temperatures higher than 50°C. Therefore, the interfacial shear strength of CFRPP using uneven carbon fiber was strongly affected because of water absorption by polymer matrix and water penetration into interface between fiber and polymer matrix under a hot–wet environment.

Keywords: CFRPP, hot–wet environment, interfacial shear strength, long term, maleic anhydride grafted polypropylene, uneven carbon fiber.

1 INTRODUCTION

FOMM Co. Ltd. announced a new concept for a microelectric vehicle in 2016 with a flotation system as a countermeasure against flood damage [1]. The microelectric vehicle is used under hot, rainy, and moist environments. The maximum temperature of the dashboard in vehicle reaches about 85°C [2]. Recently, control of motor temperatures in electric vehicle has been studied as a countermeasure against high temperatures [3], [4]. Mechanical properties of structural materials of microelectric vehicles under high temperature and water environments should be clarified for long-term safety. Carbon fiber-reinforced polypropylene (CFRPP) is used as a structural material for microelectric vehicles because carbon fiber has high specific strength and modulus. Therefore, many researchers have reported on CFRPP mechanical properties and molding methods [5]–[13]. Nevertheless, no report of the literature describes a study of CFRPP mechanical properties under temperature and water environments.

WIT Transactions on Engineering Sciences, Vol 124, © 2019 WIT Press
www.witpress.com, ISSN 1743-3533 (on-line)
doi:10.2495/MC190131

Katogi and Takemura [14] reported the effects of water temperature on the interfacial shear strength of CFRPP when using uneven carbon fiber for improvement. When the immersion time was a single day, the interfacial shear strength of CFRPP using uneven carbon fiber was affected by water temperature.

Nevertheless, few research papers describe interfacial shear strength of CFRPP using uneven carbon fiber under a long-term hot–wet environment. This study examined the interfacial shear strength of CFRPP using uneven carbon fiber under a hot–wet environment to ensure long-term safety.

2 MATERIALS AND SPECIMENS

2.1 Materials

Matrix was maleic anhydride grafted polypropylene (MAPP, Umex 1010; Sanyo Chemical Industries Ltd.). The MAPP concentration was 10%. Reinforcements were even and uneven carbon fibers (Sekisui Chemical Co. Ltd.).

2.2 Specimens

Specimens for micro-debonding test were made using a soldering iron. The molding temperature was 180°C. The CFRPP specimen shape using uneven carbon fiber was U-shaped [14]. Matrix was adhered to surface of carbon fiber like a spherical shape. The embedding length of specimen for micro-debonding tests was from 100 μm to 300 μm.

3 TESTING METHODS

3.1 Water absorption testing

Water absorption tests of CFRPP were conducted in distilled water using a constant drying oven. The water temperature range was from room temperature (RT) to 80°C. The maximum immersion time was 112 days.

3.2 Micro-debonding testing

Before and after water immersion, micro-debonding tests of CFRPPs using non-immersed and uneven carbon fiber were conducted using a tensile testing machine (RTC-1250A; A&D Co. Ltd.). A jig made in our laboratory was used for micro-debonding tests. The environmental temperature was RT. The crosshead speed was 0.1 mm/min. After micro-debonding test, the fracture surface of carbon fiber was observed using scanning electron microscopy (SEM, S-400; Hitachi High-Technologies Corp. and VE7200; Keyence Co.).

4 RESULTS AND DISCUSSION

4.1 Interfacial shear strength of CFRPP using uneven carbon fiber under a hot–wet environment

Figs 1–3 show the interfacial shear strength of CFRPP using uneven carbon fiber under a hot–wet environment. Interfacial shear strengths of CFRPPs using even and uneven carbon fibers decreased rapidly under all conditions after water immersion for one day. Interfacial shear strengths of CFRPPs using even and uneven carbon fibers decreased slowly under all

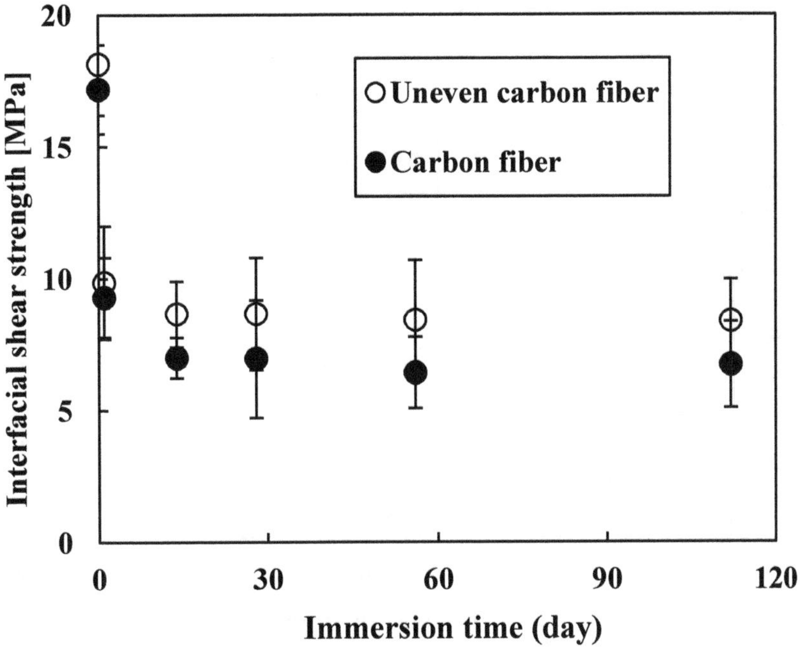

Figure 1: Interfacial shear strengths of CFRPPs using even and uneven carbon fibers (RT).

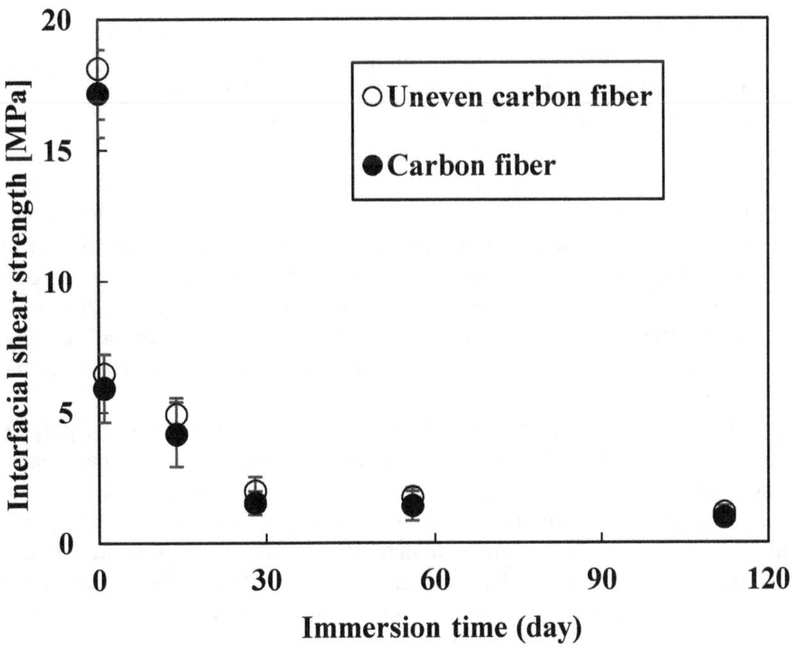

Figure 2: Interfacial shear strengths of CFRPPs using even and uneven carbon fibers (50°C).

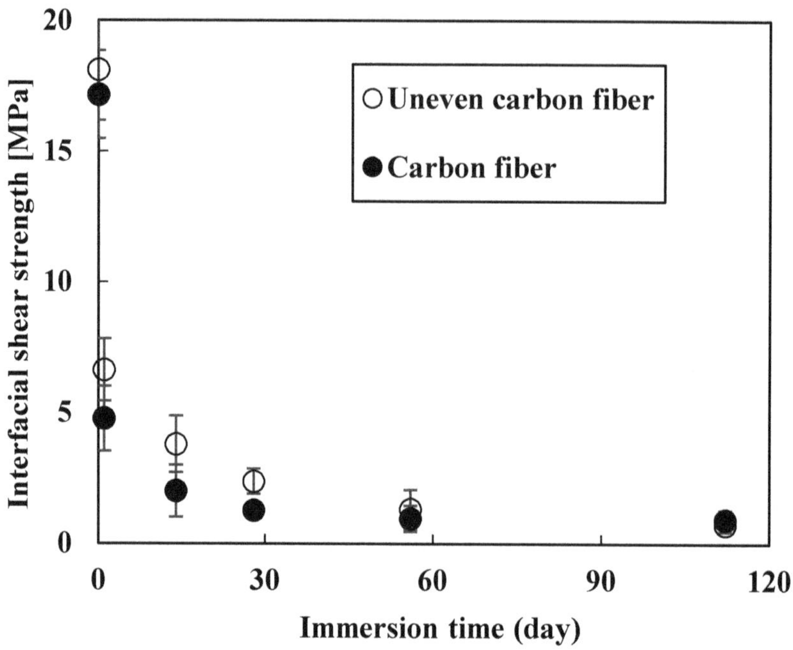

Figure 3: Interfacial shear strengths of CFRPPs using even and uneven carbon fibers (80°C).

conditions with an increase of immersion time until 28 days. Subsequently, the interfacial shear strengths of CFRPPs using even and uneven carbon fibers under all conditions remained almost unchanged with an increase of the immersion time.

When the water temperature was RT, the interfacial shear strength of CFRPP using uneven carbon fiber was larger than that of CFRPP using even carbon fiber under all immersion times. When the water temperature was higher than 50°C, the interfacial shear strength of CFRPP using uneven carbon fiber was greater than that of CFRPP using uneven carbon fiber after water immersion until 28 days. The interfacial shear strength of CFRPP using uneven carbon fiber was almost no different from that of CFRPP using even carbon fiber after water immersion over 56 days. For maximum immersion time of 112 days, interfacial shear strength of water-immersed CFRPPs using uneven carbon fiber at RT, 50°C and 80°C decreased by 54%, 94% and 96%, respectively, compared with that of non-immersed CFRPP using uneven carbon fiber.

It is possible that water penetration into the interface between fiber and polymer matrix occurred easily because of the water temperature [13]. The polymer matrix has a hydrophilic segment. Therefore, the interfacial adhesion property of the polymer matrix probably decreased with increased immersion time because of the chemical reaction of the hydrophilic segment of the polymer matrix under water temperature. However, the interfacial shear strength of CFRPPs using uneven carbon fibers after water immersion for the long term was not changed because debonding at the fiber and polymer matrix mainly occurred at temperatures higher than 50°C.

4.2 Hot–wet environment effects on the CFRPP fracture surface using uneven carbon fibers

Fig. 4 shows fracture surfaces of CFRPPs using even and uneven carbon fibers before water immersion. Before water immersion, the CFRPP fracture surface using even carbon fiber became smooth. However, small resin particles and the matrix on the fracture surface of CFRPP using uneven carbon fiber were found before water immersion.

Figs 5–7 depict the CFRPP fracture surface using even carbon fiber under a hot–wet environment. Figs 8–10 show fracture surfaces of CFRPP using uneven carbon fiber under a hot–wet environment. The fracture surface of water-immersed CFRPP using even carbon fiber almost became smooth under all conditions except for water immersion for one day at RT. When the water temperature was higher than 50°C, large resin particles were found on the fracture surface of water-immersed CFRPP using uneven carbon fiber at one day and 112 days. Much matrix material on the fracture surface of CFRPP using uneven carbon fiber at 80°C was often found compared with that of CFRPP using uneven carbon fiber at 50°C. Furthermore, matrix residue on the CFRPP fracture surface using uneven carbon fiber at 112 days was more than that of CFRPP using uneven carbon fiber at one day when the water temperature was 80°C.

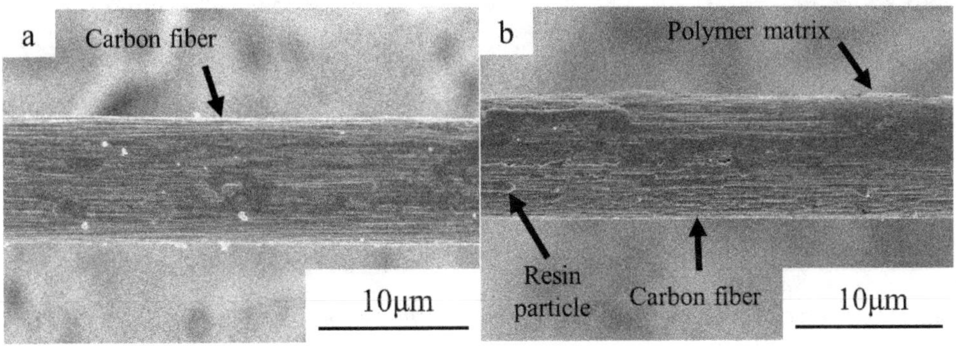

Figure 4: Fracture surface of CFRPP before water immersion. (a) Carbon fiber; and (b) Uneven carbon fiber.

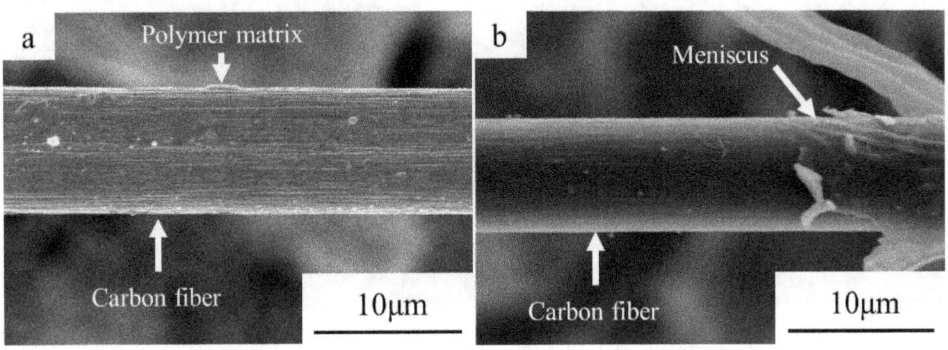

Figure 5: Fracture surfaces of CFRPPs using even carbon fibers (RT). (a) 1 day; and (b) 112 days.

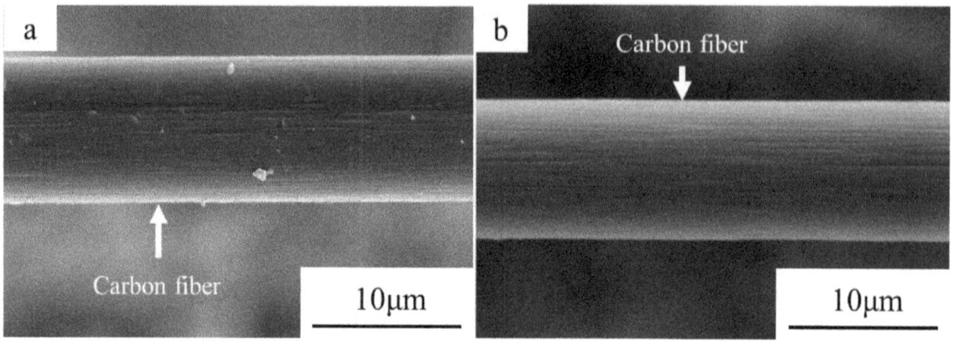

Figure 6: Fracture surfaces of CFRPPs using even carbon fibers (50°C). (a) 1 day; and (b) 112 days.

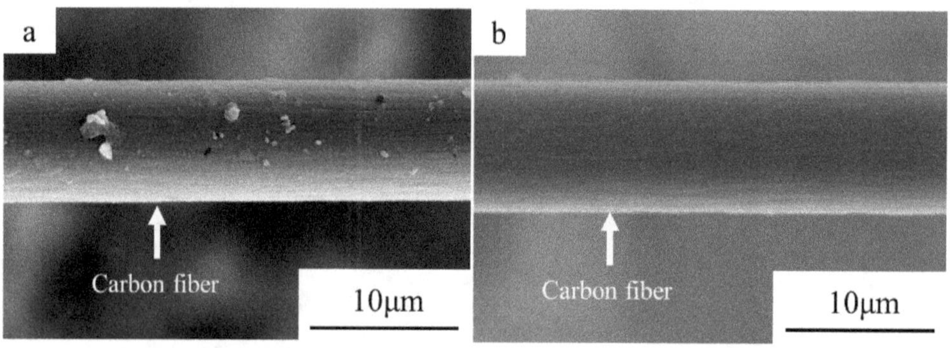

Figure 7: Fracture surfaces of CFRPPs using even carbon fibers (80°C). (a) 1 day; and (b) 112 days.

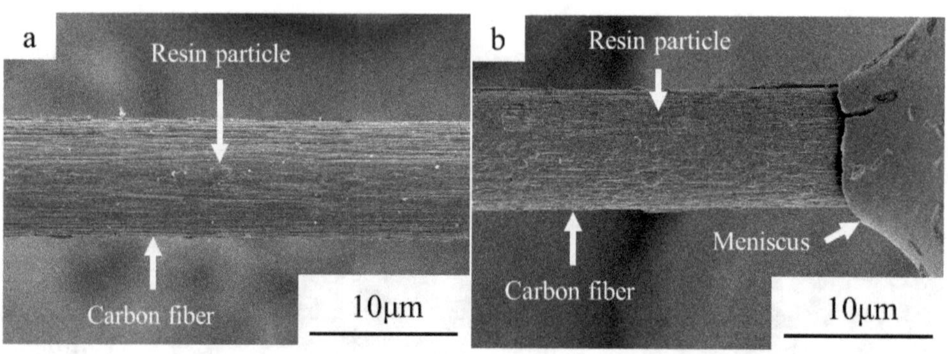

Figure 8: Fracture surfaces of CFRPPs using uneven carbon fibers (RT). (a) 1 day; and (b) 112 days.

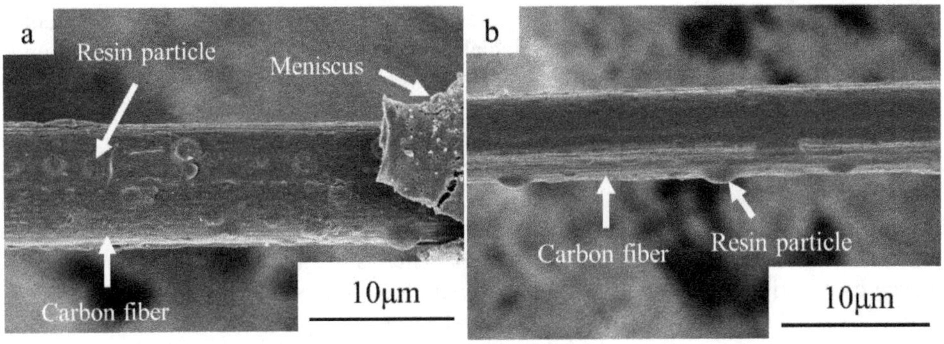

Figure 9: Fracture surfaces of CFRPPs using uneven carbon fibers (50°C). (a) 1 day; and (b) 112 days.

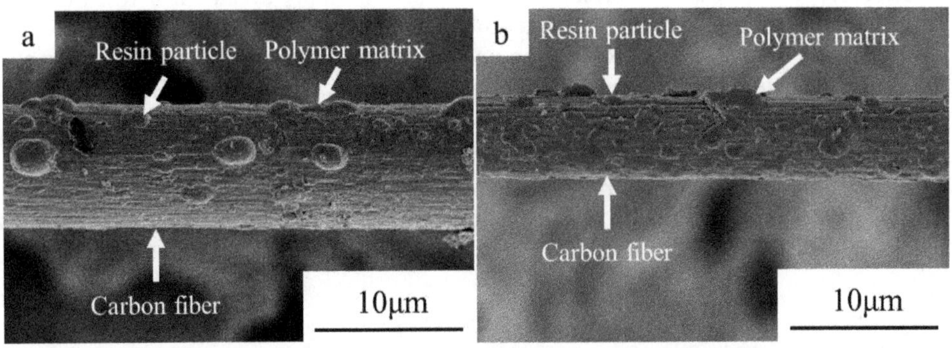

Figure 10: Fracture surfaces of CFRPPs using uneven carbon fibers (80°C). (a) 1 day; and (b) 112 days.

Water at the interface between the fiber and polymer matrix penetrated deeply to the center of CFRPP when the water temperature was 50°C. The surface roughness of carbon fiber was increased by resin particle addition. The interfacial shear strength of CFRPP using uneven carbon fiber after immersion until 28 days might be increased because of anchor effect [13] and increase of surface area of carbon fiber [14] by resin particle addition.

When the water temperature was 80°C, water absorption of polymer matrix near the surface of carbon fiber in CFRPP occurred rapidly after water penetration. Viscoelasticity and interfacial adhesion property of the polymer matrix near the surface of carbon fiber in CFRPP were decreased extremely by water absorption at 80°C. Therefore, interfacial shear strength of CFRPP was very improved by the anchor effect of resin particle after water immersion until 28 days at 80°C. However, interfacial shear strengths of CFRPPs using non-immersed and uneven carbon fibers almost did not change because decrease of viscoelasticity and interfacial adhesion property of polymer matrix under a hot–wet environment for a long term strongly dominated the CFRPP interfacial shear strength.

The results described above showed that the interfacial shear strength of CFRPP using uneven carbon fiber was affected strongly by water absorption of polymer matrix and water penetration into the interface between the fiber and matrix under a hot–wet environment up to 28 days. Viscoelasticity and the interfacial adhesion property of polymer matrix under hot–wet environment must be investigated further in future studies.

5 CONCLUSIONS

This study examined the interfacial shear strength of CFRPP using uneven carbon fiber under hot–wet environment for the long term. As a result, the interfacial shear strength of CFRPP using even and uneven carbon fiber under all conditions decrease rapidly after immersion for one day. Subsequently, interfacial shear strengths of CFRPP using even and uneven carbon fibers under all conditions decreased slowly with an increase of immersion time until 28 days. However, their interfacial shear strengths changed very little with increased immersion time beyond 56 days. From SEM observations, large resin particles on the carbon fiber fracture surface were found over 50°C. A lot of matrix material on the fracture surface of CFRPP using uneven carbon fiber at 80°C was often found compared with that of CFRPP using uneven carbon fiber at 50°C. The viscoelasticity and interfacial adhesion property of water absorbing polymer matrix probably became low when CFRPP using uneven carbon fiber was immersed at temperatures higher than 50°C. Therefore, the interfacial shear strength of CFRPP using uneven carbon fiber strongly affected the water absorption of matrix and water penetration into the interface between the fiber and matrix under a hot–wet environment.

ACKNOWLEDGEMENT

This study was supported by Mr. M. Nakamura (Sekisui Chemical Co. Ltd.), who supplied uneven carbon fiber.

REFERENCES

[1] FOMM Co. Ltd., www.fomm.co.jp/research-en.
[2] Eddy, D.S. & Sparks, D.R., Application of MEMS technology in automotive sensors and actuators. *Proceedings of the IEEE*, **86**, pp. 1747–1755, 1998.
[3] Katsuta, T., Yoshida, H., Kitaori, K., Ono, H., Iwabe, H. & Kasa, N., Structure and heat management of the motors developed in Okayama vehicle engineering project for the next EV. *Journal of the Japan Society of Applied Electromagnetics and Mechanics*, **26**, pp. 224–230, 2018. (In Japanese.)
[4] Takahashi, T., Ishikawa, R., Tominaga, R., Aizawa, S. & Minato, J., Development of epowertrain for small electric vehicles. *Panasonic Technical Journal*, **64**, pp. 63–68, 2018. (In Japanese.)
[5] Arao, Y., Yumitori, S., Suzuki, H., Tanaka, T., Tanaka, K. & Katayama, T., Mechanical properties of injection-molded carbon fiber/polypropylene hybridized with nanofillers. *Composites Part A: Applied Science and Manufacturing*, **55**, pp. 19–26, 2013.
[6] Enoki, S., Kojima, K., Mizuno, S., Katayama, K. & Tanaka, K., High-speed compression molding of continuous carbon fiber reinforced polypropylene. *WIT Transactions on the Built Environment*, vol. 137, WIT Press: Southampton and Boston, pp. 311–315, 2014.
[7] Kim, J.W. & Lee, J.S., The effect of the melt viscosity and impregnation of a film on the mechanical properties of thermoplastic composites. *Materials*, **9**, pp. 448–463, 2016.

[8] Tanaka, K., Ohno, K. & Katayama, T., Effects of modification and processing time on fiber/matrix interfacial strength for carbon fiber reinforced polypropylene. *WIT Transactions on the Built Environment*, vol. 166, WIT Press: Southampton and Boston, pp. 329–334, 2016.

[9] Tang, J. et al., Discontinuities as a way to influence the failure mechanisms and tensile performance of hybrid carbon fiber/self-reinforced polypropylene composites. *Composites Part A: Applied Science and Manufacturing*, **107**, pp. 354–365, 2018.

[10] Kada, D., Koubaa, A., Tabak, G., Migneault, S., Garnier, B. & Boudenne, A., Tensile properties, thermal conductivity, and thermal stability of short carbon fiber reinforced polypropylene composites. *Polymer Composites*, **39**, pp. E664–E670, 2018.

[11] Zao, J., Lu, C., Guo, S., Wang, K. & Fu, Q., Polymorphic structures phase diagram of shear-induced isotactic polypropylene/carbon fiber cylindrites. *Materials and Design*, **150**, pp. 40–48, 2018.

[12] Nguyen-tran, H.D., Hoang, V.T., Do, V.T., Chun, D.M. & Yum, Y.J., Effect of multiwalled carbon nanotubes on the mechanical properties of carbon fiber-reinforced polyamide-6/polypropylene composites for lightweight automotive parts. *Materials*, **11**, pp. 429–441, 2018.

[13] Katogi, H., Takemura, K. & Mochizuki, M., Effect of water temperature on interfacial shear strength of resin particles added CFRTP. *Key Engineering Materials*, **774**, pp. 7–12, 2018.

[14] Katogi, H. & Takemura, K., Interfacial shear strength of resin particles added to carbon fiber/maleic anhydride grafted polypropylene in a hot–wet environment. *WIT Transactions on Engineering Sciences*, vol. 116, WIT Press: Southampton and Boston, pp. 229–307, 2017.

DYNAMIC MICRO-TESTING OVER A LARGE RANGE OF STRAIN RATES FOR HOMOGENOUS AND HETEROGENEOUS LOCAL MATERIAL CHARACTERIZATION

FRANK HUBERTH, JOERG LIENHARD, BALAJI RAGUPATHI & MAX HAUBER
Fraunhofer-Institute for Mechanics of Materials IWM, Germany

ABSTRACT

The dynamic material characterization shows different macroscopic strain rate effects. The causal mechanisms cannot be identified at this level in most cases. Micro-tests allow a local transient analysis, which is illustrated in this article using the example of a long-fibre reinforced thermoplastic (LFRT). After a general introduction, the development and validation of a micro-test for a large strain rate range is presented. The validation explained for steel shows the advantage of the small sample for the dynamic characterization, if a homogenous material behaviour of this magnitude still exists, especially in the case of low-vibration force measurement. For a heterogeneous LFRT material, the micro-test shows strongly scattering test results that are no longer representative of the homogenized mechanical material behaviour, but reflect the local characteristics. These local properties are directly caused by the injection moulding process. Further SEM analyses of the samples indicate different macromolecular deformation mechanisms of the matrix at the different strain rates.

Keywords: material characterization, strain rate effect, micro-testing, crash, impact.

1 INTRODUCTION

The loading speed can influence the mechanical behaviour of a material. The influence of the load speed is usually described by the strain rate $d\varepsilon/dt$ as the time derivative of the strain resulting from the load. In the standard tensile test for metallic materials according to DIN EN ISO 6892-1 [1] quasi static loads are applied to the sample at very low load speeds (from strain rates $d\varepsilon/dt = 0.00007\ s^{-1}$). In many load cases, in particular at crash and impact, significantly higher strain rates are achieved in the components. The experimental procedure for the dynamic tensile test was presented in a FAT guideline "Dynamic Material Characteristics for Crash Simulation" (AiF No. 14205) [2] and based thereon was implemented in a standard "BS ISO 26203-2 Metallic materials Tensile testing method at high strain rates Part 2: Servo-hydraulic and other test systems" [3].

For these applications, first and foremost fast servo hydraulic test systems are used. For higher loading rates and resulting strain rates, other experimental equipment has to be used then, like e.g. gas pressure accelerators or so-called bar systems, such as the Split Hopkinson Bar. The various test facilities allow partially overlapping areas of loading speeds with different challenges and restrictions. During the dynamic test for crash and impact, in particular the force measurement is an essential issue. For this purpose, a patented force measurement (DE10201861) was developed at the Fraunhofer IWM for the middle strain rate range.

Another way is not to increase the strain rate, but to adjust the sample geometry to achieve higher strain rates. The strain rate range for the discussed experiments ranges from $10^{-3}\ s^{-1}$ to more than $1000\ s^{-1}$.

WIT Transactions on Engineering Sciences, Vol 124, © 2019 WIT Press
www.witpress.com, ISSN 1743-3533 (on-line)
doi:10.2495/MC190141

2 MICRO-TEST CONCEPT, METROLOGY AND VALIDATION

In order to realize material characterization over a large strain rate range on a single testing facility, two approaches, variable loading speed up to several meter per second and reduced sample design were realized for a new experimental setup. The extreme reduction of the sample size is motivated not only by the thus achievable strain rates, but also by locally focused observation and measurement in small scale ranges for local material characterization. The requirements were implemented in the testing facility IWM DynMicro [4], [5].

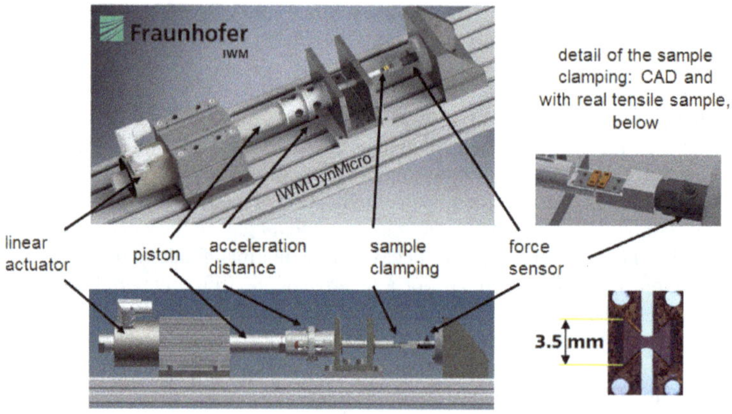

Figure 1: Test facility IWM DynMicro (CAD representation) for tests on small samples over a large strain rate range [5].

The IWM DynMicro has been tested on dual phase steel DP 500 (HCT 490X) and verified with macroscopic investigations on approved IWM mini tensile specimens with 14 mm gauge length at comparable nominal strain rates. The micro samples had a test area of 1.5x0.8 mm with an original sheet thickness of 0.5 mm. The strain rate range considered in the micro-tests included nominal strain rates of 0.001 s^{-1} to 1000 s^{-1}. The strains were determined optically on the sample in both test series for a defined measuring length with the aid of digital image correlation, DIC (gom ARAMIS). Local strain rates were even higher than 1000 s^{-1}.

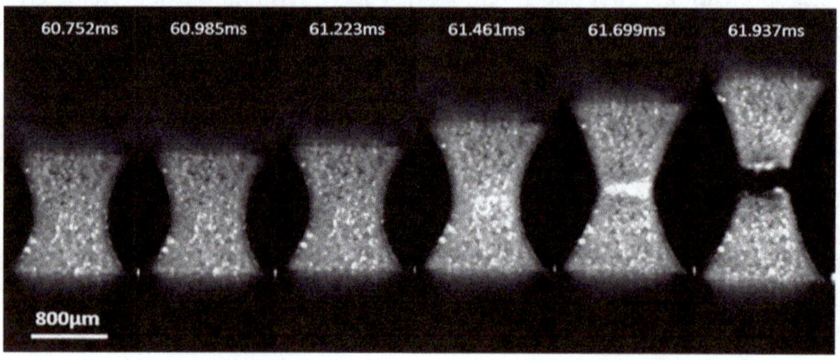

Figure 2: High speed video series micro tensile test steel at 1 ms^{-1} load speed, 300,000 frames per second (fps) [4].

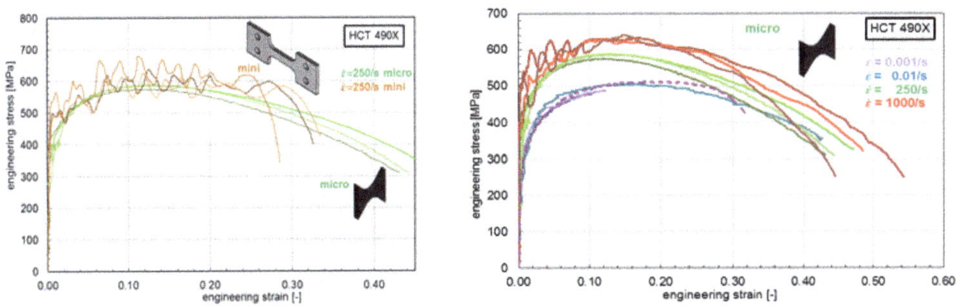

Figure 3: Technical stress–strain diagrams of the mini and micro samples for strain rates 250 s^{-1} in comparison (left) and the overview of the micro experiments for the different strain rates with a supplemented dashed curve for a quasi-static micro-test (right) [6].

Since the chosen material shows a very fine microstructure, it was possible to reproduce the material behaviour in the micro-tests with a measuring length of 1 mm as well as the tensile tests with a larger sample and a measuring length of 14 mm.

The measurement shows in the micro tensile test with a nominal elongation rate of 250 s^{-1} significantly lower vibrations in the dynamic force measurement, even in comparison with the patented IWM load cell measurement.

The experiments with 1000 s^{-1} provide usable results without additional optimization of the force measurement. Tests at higher rates showed significant oscillations in the force measurement. Adapting the quasi local force measurement is actually under development.

The verification test series showed the reliability and the benefit of the micro tensile tests and the main influencing factors like the sample preparation, the clamping system and the sensitivity to misalignment which are to be considered. An essential point for the successful introduction of the rate dependent tests was the application of the high speed camera system for the DIC at the micro scale for the deformation analysis on the sample. For such small samples the results are influence significantly by every setting or additional elasticities in the measuring length.

3 LOCAL STRAIN RATE EFFECTS ON DEFORMATION AND FAILURE FOR LFRT AND RESULTING GLOBAL MATERIAL RESPONSE

The influence of the strain rate for thermoplastics, reinforced thermoplastic materials and other composites was investigated by conventional large samples. The observation of the strain rate effect changed significantly from the pure thermoplastic matrix material and for the composites. The following diagrams directly show the difference for the pure matrix material (PP), the same thermoplastic matrix with long glass fibre reinforcement (LFRT) and a carbon fibre composite (CFRP).

For the pure PP matrix stress increases with increasing strain rate and failure strain decreases. For the other two materials a revers strain rate effect was observed for the failure strain. This effect was measured as shown in the diagrams and transferred into phenomenological material models for FEM simulations [7], [8] but the fundamental material behaviour of the responsible fibre matrix interaction is not assessable by these tests and neglected in homogenous phenomenological material models. For this purpose local material investigations were performed.

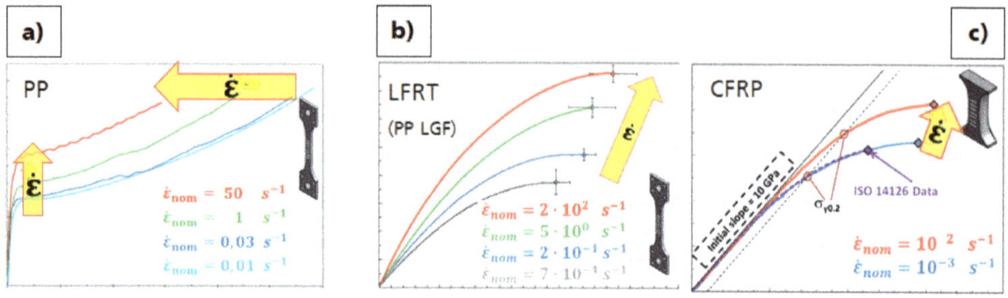

Figure 4: Tensile strain rate effect on the stress level and the failure strain for (a) thermoplastic (PP matrix) material; (b) long glass fibre reinforced PP (PP-LGF) [7], [8]; and (c) a unidirectional carbon fibre laminate [9].

Micro tensile tests allow not only high strain rates but also the transient investigation of the local heterogeneous material behaviour, depending on the microstructure. This is demonstrated using the example of a long glass fibre reinforced material (LFRT) with PP matrix (PP-LGF) [10]. This LFRT is used in many areas, especially in the automotive, which are exposed to high loads and load. These investigations of micromechanics of fibres and matrix at different strain rates were part of a research project. Here, the local fibre-matrix behaviour, in particular the orientation of the fibres and the damage of fibres and matrix were considered.

The samples were taken at different positions and variably over the wall thickness and prepared to observe the then partially visible fibres can. The local structure for each sample was influenced by the injection moulding process. For this reason, the tests show a large scatter, but represent the locale material behaviour as given in the component.

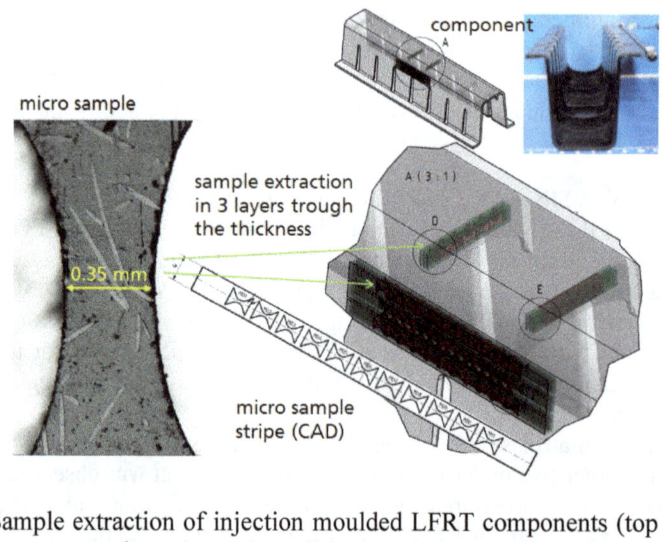

Figure 5: Sample extraction of injection moulded LFRT components (top right) in three layers as stripes (centre), polished and then cut as micro tensile samples (left) [10].

The sample extraction was done by a special micro water jet cutting technology and a finishing process. In the first step segments were prepared from components. These were used for preparing thin strips of 0.5 mm thickness in different position of the wall thickness of the segments. The sample geometry was cut in the stripes after a polishing process. By this special adapted process a very good bonding of surface fibres was achieved.

Micro-tests depend on a customized sample preparation, adapted for each material. This is essential for the tests.

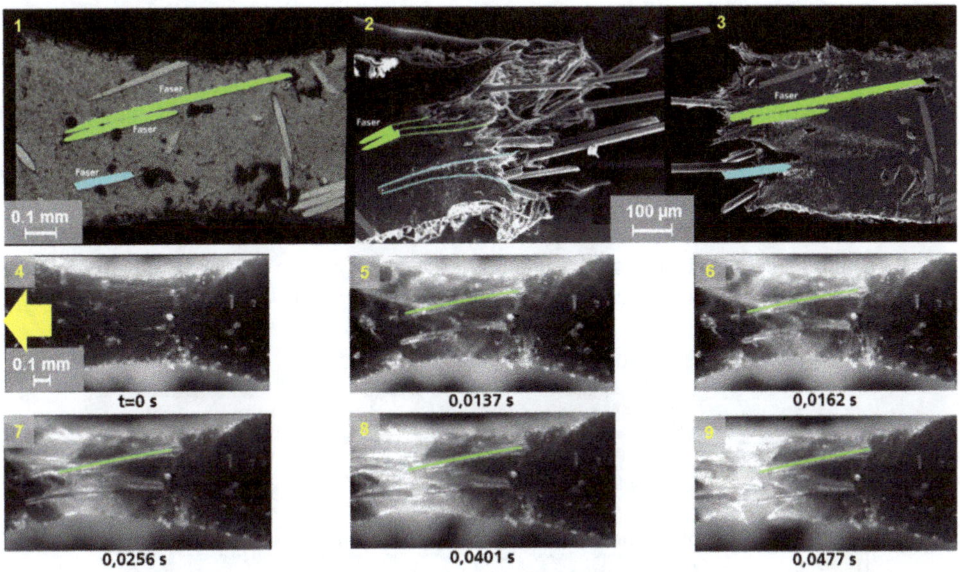

Figure 6: Micro tensile test PP-LGF; Micrograph polished sample (1), SEM images of the fractured ends (2) and (3); and high-speed image series (4 to 9 at 10,000 fps) of the experiment. The deformation zone at quasi-static loading is more localized around the fibres compared to the dynamic load. The local deformations indicate highly stretched polymer matrix at low strain rate and increased craze formation at higher strain rates.

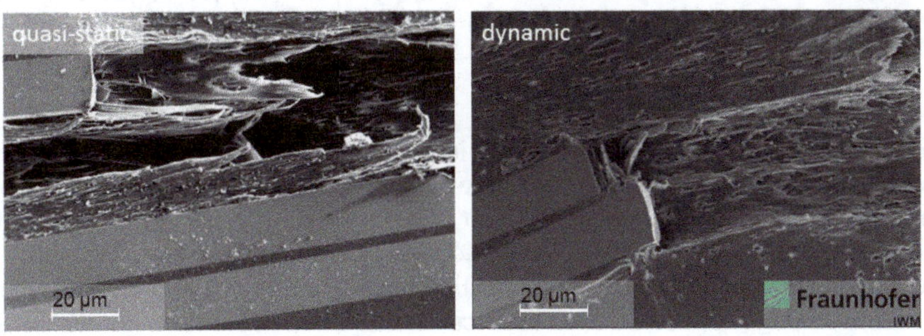

Figure 7: SEM micrograph PP LGF after quasi-static loading, left and dynamic loading, right, with fiber break and fiber pullout rates [10].

The results show different fibre-matrix interaction depending on the strain rate. This could be proven by the SEM examinations of the tested micro samples. The advantage here is to map the entire micro specimen and then zoom locally into the damage area. The SEM images clearly show local differences in the fibre environment with respect to the local deformations and their propagation around the fibres for different strain rates. This observation is an essential step of understanding the local strain rate effect, influencing the macroscopic LFRT reaction on different strain rates.

4 APPLICATION OF MICRO-TESTING FOR LOCAL BONDING STRENGTH CHARACTERIZATION

These applications of the micro tensile test are further developed and optimized at Fraunhofer IWM. The possibilities are expanded for other material groups and material composites, in particular joining zones. MPW is a joining process that allows the combination of heterogeneous material combination like steel and aluminium what is essential for mixed lightweight constructions [11].

As shown for the LFRT the samples for the joint characterization were extracted in stripes and then the shape of the micro tensile samples was cut in the stripes. Here a combined process of spark erosion and micro waterjet cutting was applied.

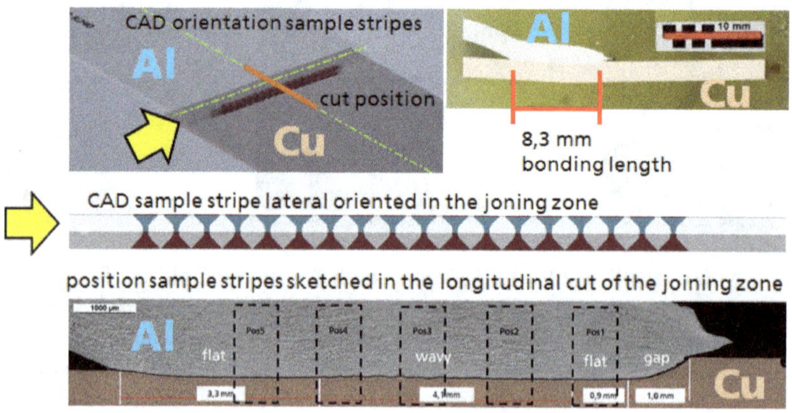

Figure 8: Micro tensile samples prepared out of a MPW joint of copper and aluminum, CAD sketch of the location in the coupon samples prepared in stripes, metallographic cut of the coupon sample overview and zoomed in with marked positions of the stripe extraction.

The analysis of this local joint is actually in progress by micro tensile tests. First results for a copper-aluminium combination are shown in the following image series. As shown in the series of images in Fig. 9, the MPW joint is stronger than the aluminium material. In a next step the sample shape will be adapted to strengthen the aluminium part and to force failure in the copper and the very small welding line pictured for two different samples in Fig. 10. This testing procedure allows the local determination of all strength values for the simulation.

This characterization concept can be transferred to other joints, like adhesives. The shape of the sample will be adapted to the geometries of the joined components and the bonding layer.

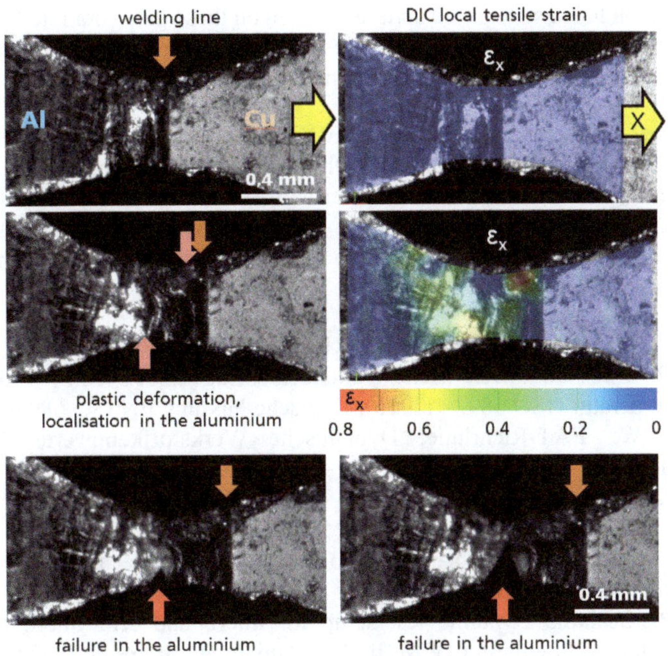

Figure 9: Micro tensile test for a MPW joint of aluminum (Al) and copper (Cu), initial frame at t0, localization without overlay (left) and with strain field overlay of the tensile strain (right) and the failure preparation in the aluminium.

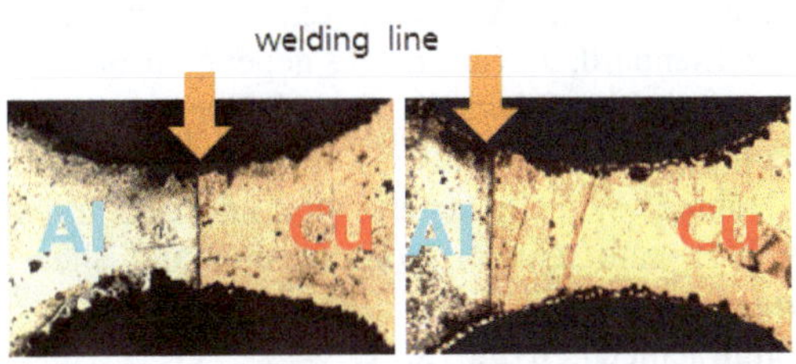

Figure 10: Micro tensile samples with centred welding line (left) and minimum cross section in the copper part (right) for shifting the failure.

5 CONCLUSIONS AND OUTLOOK

The micro-testing is a promising tool for local investigations of transient material behaviour in a wide range of strain rates. The application depends on the size of the representative structures in the material. If the sample size includes a quasi-representative structure, the tests can substitute the material characterisation with conventional large samples, especially for the strain rate dependent characterization. Here the main benefit is the wide range of feasible strain rates with only one testing device.

The main field of application will be the local investigation of mechanical performance and the mechanisms influenced by the strain rate. This investigations performed in a sufficient number will allow new correlations of local deformation processes influence by the local process induced state of the material. Based on these data combined with innovative data analysis methods and approaches for consideration of uncertainty and scatter the digital material twin could be realized for the mechanical performance analysis.

ACKNOWLEDGEMENTS

The work presented was partially funded by the Federal Ministry for Economic Affairs and Energy on the basis of a resolution of the German Bundestag within the framework of the IGF projects No. 19532 N/1 of the research associations FAT (Research Association for Automotive Technology, Germany) and No. 18943 N/1 in cooperation with the Research Associations for Steel Application (FOSTA).

REFERENCES

[1] ISO 6892-1:2017-02: Metallische Werkstoffe – Zugversuch, Teil 1: Prüfverfahren bei Raumtemperatur (ISO 6892-1:2016); Deutsche Fassung EN ISO 6892-1:2016, 2017.

[2] Böhme, W., FAT-Richtlinie: Dynamische Werkstoffkennwerte für die Crash simulation. *Mat. Testing*, **50**(4), pp 199–205, 2008. DOI: 10.3139/120.100865.

[3] ISO 26203-2:2011: Metallic materials – Tensile testing at high strain rates, Part 2: Servo-hydraulic and other test systems, 2011.

[4] Huberth, F. & Bimi, P., Dynamic micro testing over five decades of strain rates. *43rd MPA-Seminar*, Stuttgart, Germany, 11–12 October, 2017.

[5] Bimi, P., Scalable Dynamic test setup for micro and semi-micro tensile testing. Master's thesis, Fraunhofer IWM, IMTEK University of Freiburg, Freiburg, 2017.

[6] Huberth, F., Bimi, P., Klitschke, S. & Lienhard, J., Dynamic micro testing over more than five decades of strain rates. *ICILSM 2018*, Xi'an, China, 7–11 May, 2018.

[7] Sun, D. et al., IGF-Vorhaben 17334N: Entwicklung einer methode zur crashsimulation von langfaserverstärkten thermoplast (LFRT) bauteilen auf basis der faserorientierung aus der formfüllsimulation (IGF-Vorhaben 17334 N). Forschungsvereinigung Automobiltechnik e.V. (FAT), FAT-Schriftreihe 284, 2016.

[8] Schulenberg, L. & Lienhard, J., Strain rate dependent multiaxial characterization of long fiber reinforced plastic. *Composite Part B Engineering*, **141**, S164–S173, 2018. DOI: 10.1016/j.compositesb.2017.12.031F.

[9] Lienhard, J. & Böhme, W., Characterization of resin transfer molded composite laminates under high rate tension, compression and shear loading. *Engineering Fracture Mechanics*, **149**, S338–S350, 2015. DOI: 10.1016/j.engfracmech.2015.07.012.

[10] Huberth, F., Lienhard, J. & Klitschke, S., *Rateneffekte und Skaleneffekte bei der Werkstoffprüfung auf der Mikroskala, Tagungsband DVM Tagung Werkstoffprüfung 2018*, Günther Moninger (Hrsg.) Deutscher Verband für Materialforschung und prüfung e.V.: Berlin, S261–S266, 2018.

[11] Huberth, F., Klitschke, S., Gall, M., Sommer, S., Schnabel, K. & Baumgartner, J., Crash and durability of aluminum and mixed steel aluminum joints made by electro-magnetic pulse welding. *Proceedings ICHSF 2018*, Columbus, OH, 2018. OAI ID: eldorado.tu-dortmund.de:2003/369 62.

NEW ANTIFRICTION COMPOSITE MATERIALS BASED ON TOOL STEEL GRINDING WASTE

KRZYSZTOF JAMROZIAK[1] & TETIANA ROIK[2]
[1]Department of Mechanics, Materials Science and Engineering, Wrocław University of Science and Technology, Poland
[2]National Technical University of Ukraine, Igor Sikorsky Kyiv Polytechnic Institute, Ukraine

ABSTRACT

This paper studies the effect of technological parameters to the manufacture of new effective antifriction composite materials based on 7HG2VMF tool steel grinding waste with CaF_2 additions as solid lubricants in structures, having mechanical and tribological properties for high temperature friction (bearings perform at a temperature of up to 550°C, a sliding speed of 1 m/s, and a pressure up to 5.0 MPa, in air). This paper illustrates the mechanism of structure formation of the new materials and its effect on properties after using developed technological modes. Such technology can ensure a micro-heterogeneous fine-grained structure. The composite structure consists of a matrix based on tool steel waste and the uniformly distributed CaF_2 solid lubricant. The metal matrix is pearlite and consists of an α-solid solution based on iron with hard grains of alloying elements of carbides. Such heterogeneous structure promotes a high level of antifriction properties under severe operating conditions. The behavior of CaF_2 in the friction area under high temperature operating conditions has been shown. The CaF_2 solid lubricant is evenly distributed over the entire friction surface. Calcium fluoride and chemical elements of the contact pair form an antifriction film, which provides self-lubricating conditions. The possibility of predicting and controlling the behavior of antifriction materials at a high temperature, by the selection of initial metal grinding waste, for ensuring the high level of functional properties, has been demonstrated. The use of grinding waste to produce effective antifriction materials makes it possible to partially solve the global problem of environmental protection.
Keywords: antifriction composite material, grinding waste, tool steel, solid lubricant, technology, structure, properties, friction films.

1 INTRODUCTION

Materials that operate under severe operating conditions, such as high speeds, pressures, and aggressive environments, occupy a special place among the existing antifriction ones and are studied by experimental or numerical methods [1]–[3]. Bearing materials based on iron are often used under working conditions in the air at temperatures of 500–600°C and high stresses (of up to 5 MPa), since other materials, primarily based on non-ferrous metals, show either low tribological properties or are not able to be suitably operated under such conditions. In addition, many cast materials significantly increase the cost of the friction unit and the mechanism as a whole [1], [4].

The iron-based materials do not satisfy the growing needs of modern machines for withstanding severe friction conditions, for example high temperatures. In some cases, cast materials have insufficient performance (high friction and wear coefficients) or are generally inoperable. Powder materials have a high cost due to the high price of raw materials (starting powders) [4], [5].

At the same time, large potential sources of cheap valuable raw materials exist at plants. These are grinding wastes of ferrous and non-ferrous metals and alloys of machine-building, metallurgical, tool-making industries, which are taken to the dump because of their abrasive contamination and are not used in the repeated cycle of production. Such specified

WIT Transactions on Engineering Sciences, Vol 124, © 2019 WIT Press
www.witpress.com, ISSN 1743-3533 (on-line)
doi:10.2495/MC190151

waste contains a significant number of valuable alloying elements, such as W, Mo, V, Co, Al, Ti, Cr, Nb, etc. After proper processing, these wastes can be used in the further production cycle in particular for the production of effective slide bearings, as initiated by the authors [1], [4].

Therefore, the creation of new composite antifriction materials which are made of cheap and valuable raw materials with high operational properties under severe conditions is very relevant and requires a scientific and practical approach. This is especially important when it comes to searching for cheap raw materials and the implementation of tasks related to developing new resource-saving technologies.

The friction processes analysis shows that the wear rate of the friction surfaces is determined by the properties of the formed separating antifriction films (so called secondary structures) and depends on their nature [5]–[7]. The films on the contact surfaces can have a positive effect, minimizing the frictional forces and preventing the adhesion of the bearing and shaft, and a negative impact, when the friction film acts as abrasive depending on the working conditions. This is particularly important for friction under self-lubricating conditions. Moreover, during the friction films forming process, not only the separate property of the composite is significant, but the whole complex of properties affects the long and reliable operation of the antifriction material. These properties depend on the presence of alloying elements in the material, which are capable of forming such a structure that will ensure the effective formation of antifriction films and provide the required level of the functional properties.

The issue of using industrial metal waste in the recycling cycle is extremely important. Many scientists around the world devote their research to the problem of recycling and reuse of ferrous and nonferrous metal waste. Typically, such waste is a metal chip, scrub, etc. Such waste is the raw material for the production of new structural parts and tools which are manufactured using different technologies [8]–[10]. However, information on the reuse of grinding metal waste is somewhat limited.

Research in recent years has shown the suitability of using some steel and alloy grinding waste for the qualitative structural parts production [1], [8], [11]. The authors [8], [11] carried out experiments and received positive results on the use of some non-ferrous alloys and steel grinding waste for the manufacture of new bearing composite materials. However, studies on the use of grinding waste from a wide range of steels and alloys for composite antifriction materials were carried out in a limited number. Only individual publications have shown the prospect of this scientific direction.

The indicated arguments were the reason for comprehensive studies on the structural features, properties, and the composition of friction films of new antifriction composite materials based on the tool steel 7HG2VMF grinding waste depending on the synthesis modes. This is a significant problem both from a scientific and a practical point of view.

Solving this problem will allow the wide use of valuable grinding waste for the production of effective composite materials for various functional applications. The application of industrial grinding waste in the further production cycle will contribute to the partial solution of the environmental protection problem.

The objective of this research is to study structural features, properties, and the antifriction film composition of a self-lubricating antifriction composite material based on the tool steel 7HG2VMF grinding waste with CaF_2 solid lubricant additives for high-temperature friction units.

2 EXPERIMENTAL PROCEDURE

2.1 Preparatory procedures

The subject of study is a new antifriction composite material based on the tool steel 7HG2VMF grinding waste with the CaF_2 solid lubricant (Table 1) [12].

Table 1: Chemical composition of the materials based on the tool steel 7HG2VMF grinding waste.

Components, wt.%										
C	W	Cr	Mo	Si	Mn	V	S	P	Fe	CaF_2
0.68–0.76	0.55–0.9	1.5–1.8	0.5–0.8	0.2–0.4	1.8–2.3	0.1–0.25	0.02–0.03	0.02–0.03	basis	4.0–8.0

The 7HG2VMF steel is the closest analogue to the AM6F2 tool steel, AISI standard, USA. The steel 7HG2VMF grinding waste is formed during the grinding process of punches and dies. The waste is polluted with abrasive particles from grinding wheels. Therefore, such waste is not usually used in the subsequent production cycle. However, this steel contains valuable alloying elements (Table 1). Hence it is attractive for use as the basis for high-temperature antifriction material. The magnetic separation method was used to clean the grinding waste from the abrasive particles through using a magnetic separator [1]. The abrasive remains constituted 1–2% after the metal waste powders had been cleaned.

After cleaning, the metal powders were subjected to regenerative annealing at temperatures of 870–1000°C in a hydrogen atmosphere for removing excess oxygen. The regenerative annealing reduces the total concentration of oxygen in powders from 1.0% in the initial content to 0.4% after annealing. The above contributes to the formation of a homogeneous microstructure. The remains of the casting structure are observed in samples made from non-annealed powders in the form of heterogeneous carbide grain-boundary accumulations of a biographical character. However, the carbides are spheroidized after regenerative annealing, and the material's strength increases [13], [14]. A solid lubricant powder of calcium fluoride was added to the original mixture (Table 1). CaF_2 was chosen to be added to the antifriction material because it is a high-temperature and chemically stable solid lubricant. This solid lubricant is known to be effective at elevated and high temperatures and retains its properties at a temperature t up to 1300°C [15], [16]. The CaF_2 amount was chosen within the range of 4.0–8.0 wt.% because of the following reasons: at a quantity less than 4% of CaF_2, the function of the solid lubrication is not fully realized, and when the content exceeds 10% the strength and plasticity of the material are very low. Thus, in the experiments, the authors have examined the antifriction composite materials: powders of steel 7HG2VMF + (4.0–8.0)% CaF_2.

2.2 Preparation of powder charge and compaction

The initial components such as the regenerated powder waste of the tool steel 7HG2VMF and the solid lubricant (CaF_2) were mixed up for 4 hours. Then the powder mixture was subjected to pressing. The pressing process was carried out under a pressure of 700–900 MPa.

2.3 Sintering

The samples were sintered in an atmosphere of shielding gas (H_2) in a laboratory furnace at t = 1100–1150°C for 2 h.

2.4 Examination techniques

The composite structure was studied using optical and raster electron microscopes; calcium fluoride in the matrix was identified using scanning electron microscopy (SEM). The mechanical properties of the samples were determined as well. Tribological tests were performed on a VMT-1 friction testing machine (at temperature of up to 550°C, sliding speed V = 1.0 m/s, and load P of up to 5.0 MPa), the counterface was made of R18 cast tool steel (hardness 57–59 HRC). The counterface material R18 steel corresponded to the material of real shafts in the high-temperature friction units. The R18 steel has the following chemical composition, wt.%: 0.73–0.83 carbon, up to 0.5 silicon, up to 0.5 manganese, 3.8–4.4 chromium, 17.0–18.0 tungsten, up to 1.0 molybdenum, 1.0–1.4 vanadium, up to 0.03 sulfur, up to 0.03 phosphorus, and iron as the base.

3 EXPERIMENTAL RESULTS AND DISCUSSION

A heterogeneous composite material was formed after the sintering process. The new composite structure consists of a metal matrix and inclusions of CaF_2 solid lubricant (Fig. 1).

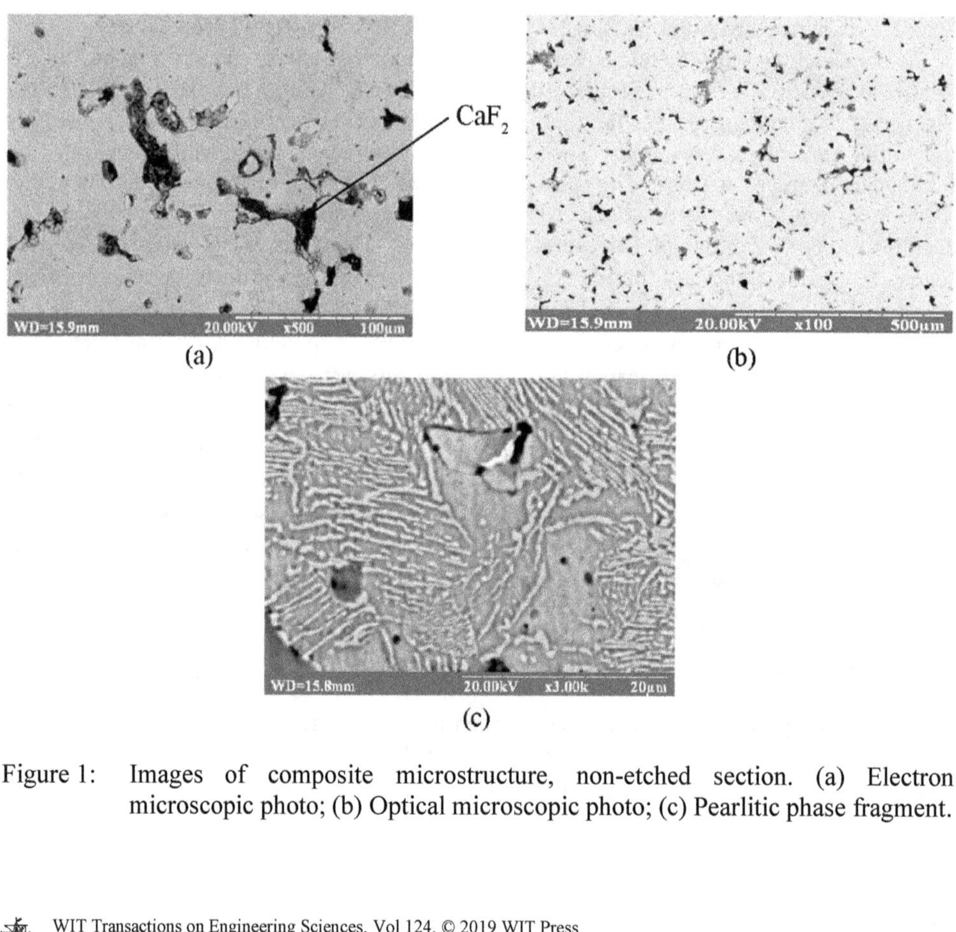

Figure 1: Images of composite microstructure, non-etched section. (a) Electron microscopic photo; (b) Optical microscopic photo; (c) Pearlitic phase fragment.

It can be seen on Fig. 1 that the CaF_2 solid lubricant is uniformly distributed over the entire composite material. The metal matrix is pearlite (Fig. 1(c)) where the carbide phases have been presented as alloying elements of carbides. These carbides are described by formulas $Me_{23}C_6$, Me_2C, Me_6C and MeC. Such phases are characteristic of a cast tool steel. In our case, a thin foil was prepared to identify the carbides in the new composite. Carbides presence was confirmed by electron diffraction using an electron microscope, followed by calculation of the lattice parameters (Figs 2 and 3).

The alloying elements in the waste powders contribute to the formation of the phases which are responsible for the formation of exploitation properties. It should be noted that the influence of alloying elements on the characteristics of the composite depends on the method of their introduction into the material.

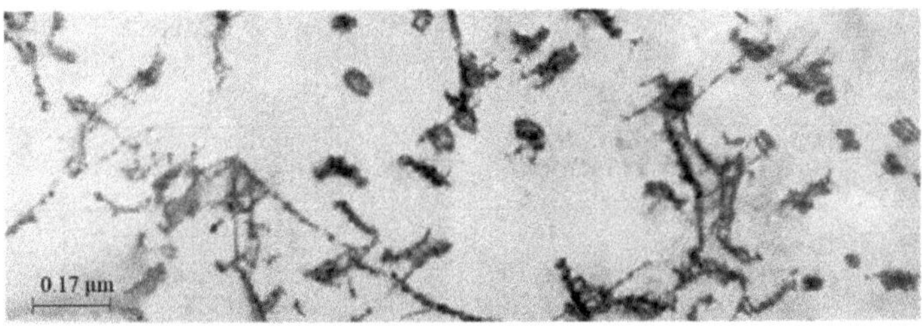

Figure 2: Mo_2C carbides in metal matrix.

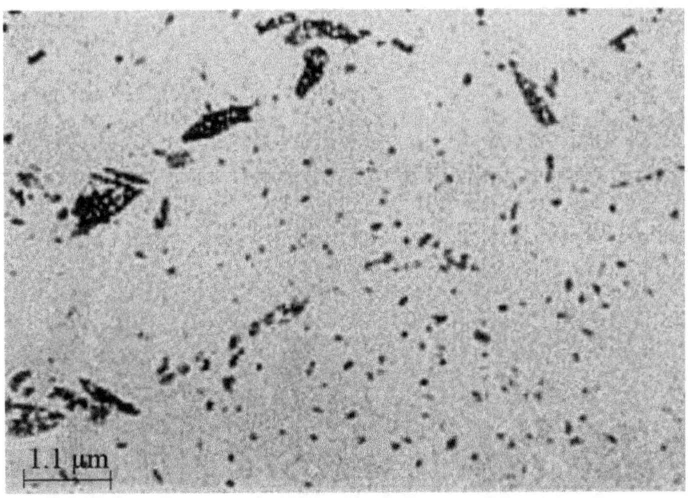

Figure 3: $(Fe, Cr)_{23}C_6$, VC and $(Fe, Cr,Mo)_6C$ carbides in metal matrix.

Addition of pure Cr powders leads to the formation of an extremely heterogeneous structure of the matrix. That is due to the deceleration of chromium dissolution processes on the iron base because of its high oxidation and ability of forming carbides [17]. However, the production of the composite made from the alloyed primary raw material provides a more homogeneous structure with higher mechanical properties. Such a structure has provided the development of the properties presented in Table 2.

Tribological and mechanical properties of the examined materials have been presented in Table 2 compared to the known powder materials based on Fe [1], [5], which are used under similar conditions. Table 2 shows that the new waste-based composite material demonstrates higher antifriction properties at friction in a self-lubricating mode at high temperatures and loads than the known Fe-based composite material, which showed unsatisfactory tribological characteristics under such operating conditions.

Table 2: Properties of the examined composite.

Composition, wt.%	Bending strength, σ_s, MPa	Impact resistance, KC, J/m²	Hardness HB, MPa	Friction coefficient/ wear rate, µm/km (at 5 MPa and 550°C)	Maximum allowable load, MPa/ temperature, °C
7HG2VMF + 4 CaF_2	590–630	650–680	2590–2640	0.18/41	5/550
7HG2VMF + 6 CaF_2	570–620	630–670	2570–2630	0.16/39	5/550
7HG2VMF + 8CaF_2	560–600	620–640	2550–2620	0.17/42	5/550
Fe+3C + 15 Mo (ZhGr3M15) composite material [1]	240–300	350–520	850–950	0.31/176	1.5/450

The dense and continuous antifriction films were formed on the contact surfaces, both on the surface of the examined material and counterface during the tribological tests. The scanning electron microscopy confirmed the presence of a thin dense antifriction layer on the friction surfaces after tribological tests (Fig. 4).

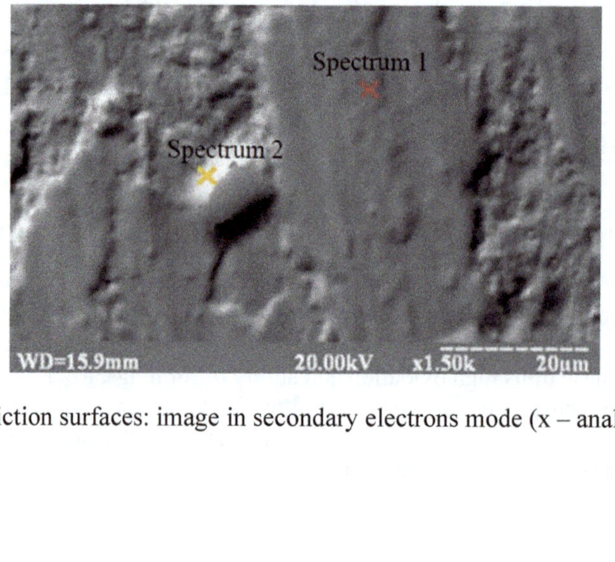

Figure 4: Friction surfaces: image in secondary electrons mode (x – analysis points).

Such films consist of contact materials alloying elements and the CaF_2 solid lubricant that was confirmed by the micro X-ray spectral analysis method (Fig. 5). The chemical elements of the composite material were determined after the analysis at points (the spectra 1 and 2). The data for spectra 1 and 2 (Fig. 5) show the presence of all components from the initial charge and counterface.

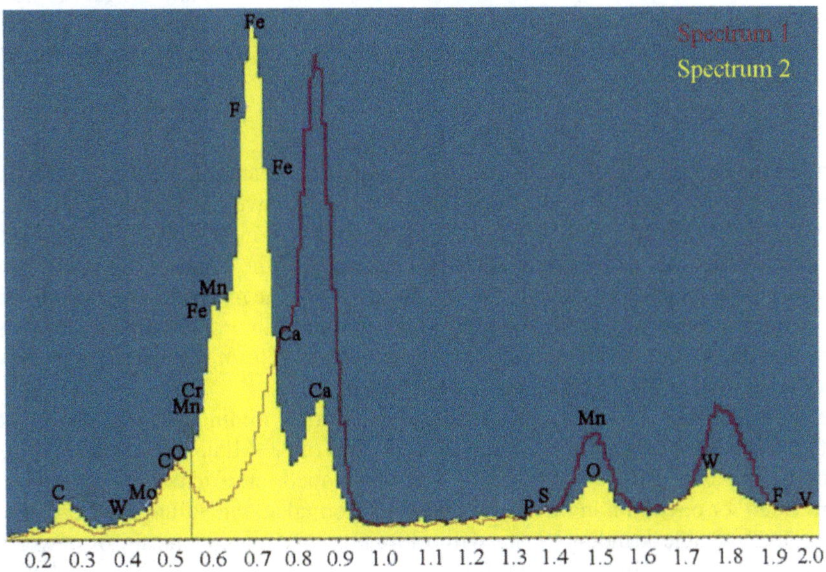

Figure 5: Spectra from selected areas of friction film.

The chemical elements ratio has been presented from the selected areas on the friction track (Fig. 5, spectra 1, 2) after tribological tests at 550°C, load 5.0 MPa and speed of 1.0 m/s (Table 3).

Table 3: EDS results of the wear track in selected areas.

Element, at.%	Spectrum 1	Spectrum 2
Fe	59.00	68.00
C	12.13	21.79
Ca	2.53	5.02
W	0.71	0.77
Mo	0.80	0.86
Cr	1.65	1.68
Mn	2.09	4.22
F	4.37	4.58
O	3.14	11.31
P	0.39	0.41
S	0.41	0.43
V	0.81	0.82

As can be seen from the spectral microanalysis data, oxygen is also present on the contact surfaces. This suggests that oxidation processes are activated at friction in the air at elevated temperatures.

Friction films probably consist of complex oxide and fluoride formations which cover the contact surfaces and prevent the composite from volume oxidation. Under these friction conditions, there is a balance between the wear rate of these films and the rate of new areas antifriction films formation. Thus, the self-lubrication mode is realized. Friction films (secondary structures) have smooth microtopography, which minimizes wear and stabilizes the friction pair operation.

4 CONCLUSIONS

The studies on the example of a new antifriction composite material based on the 7HG2VMF steel grinding waste with the CaF_2 solid lubricant have shown the possibility to determine the features of the formation of the material structure using grinding waste as the metal basis, taking into account the nature of the alloying elements and their phases in the metal matrix. The above allows predicting the nature of strengthening of the materials and their antifriction behavior. Such an approach opens the opportunity for controlling the composite structure and properties by the choice of initial alloyed grinding waste to form the necessary matrix basis and adding the required amount of the solid lubricant.

The analysis of tribological properties allowed determining the ranges of rational exploitation modes for the studied material: load of up to 5.0 MPa, slide speed of 1 m/s, and temperature of 500–550°C in the air. The developed antifriction composite can be recommended for operation in the friction units of thermal and metallurgical equipment.

Using the valuable grinding waste will allow not only manufacturing of new effective composites but also it will contribute to the partial solution of the environmental protection problem.

REFERENCES

[1] Kyrychok, P., Roik, T. & Gavrish, A. (eds), *New Composite Materials of the Printing Machines Friction Parts*, Publishing House of National Technical University of Ukraine, Igor Sikorsky Kyiv Polytechnic Institute: Kiev, 2015 (in Ukrainian).

[2] Baranowski, P., Damaziak, K., Malachowski, J., Sergienko, V.P. & Bukharov, S.N., Modeling of abrasive wear by the meshless smoothed particle hydrodynamics method. *Journal of Friction and Wear*, **37**(1), pp. 94–99, 2016.

[3] Baranowski, P. et al., Thermovision in the validation process of numerical simulation of braking. *Metrology and Measurement Systems*, **21**(2), pp. 329–340, 2014.

[4] Roik, T., Gavrish, A., Kirichok, P. & Vitsyuk, Y., Effect of secondary structures on the functional properties of high-speed sintered bearings for printing machines. *Powder Metallurgy and Metal Ceramics*, **54**(1–2), pp. 119–127, 2015.

[5] Migranov, M.S., Mukhamadeev, V.R., Migranov, A.M., Mukhamadeev, I.R. & Khazgalieva, A.A., The improvement of the tribotechnical properties of materials and coatings for metal cutting tool. *IOP Conference Series: Materials Science and Engineering*, **447**(012083), 2018.

[6] Roik, T.A., Kholyavko, V.V., Vitsyuk, Y.Y. & Melnyk, O.O., Influence of mechanism of secondary-structures tribosynthesis on properties of antifriction nickel-based composite materials. *Metallofizika i Noveishie Tekhnologii*, **38**(8), pp. 1111–1123, 2009.

[7] Roik, T.A. & Shevchuk, Y.F., Development of predictable secondary structures in materials for high-temperature bearings. *Powder Metallurgy and Metal Ceramics*, **45**(11), pp. 531–539, 2006.

[8] Shpak, P.A., Hrechaniuk, V.H. & Osokyn, V.A., Influence of electron-beam remelting on the structure and properties of high-speed steel R6M5. *Problems of special metallurgy*, **3**, pp. 14–17, 2002.

[9] Zhang, W., Zhu, Z. & Cheng, C.Y., A literature review of titanium metallurgical processes. *Hydrometallurgy*, **108**(3–4), pp. 177–188, 2011.

[10] Rao, S.R., *Resource Recovery and Recycling From Metallurgical Wastes*, Elsevier: Oxford and Amsterdam, 2011.

[11] Roik, T., Gavrish, O., Oliynik, V. & Vitsiuk, I., Analysis of the properties of antifriction composites based on aluminum alloy's grinding waste. *Eastern-European Journal of Enterprise Technologies*, **4**(12), pp. 16–22, 2018.

[12] Roik, T.A., Gavrish, A.P., Gavrish, O.A., Kyrychok, P.O., Vitsiuk, I.I. & Melnik, O.O., *Antifrictional Composite Material Based on Instrumental Steel*, Patent no. 102299, Bulletin no. 12, June 25, 2013 (in Ukrainian).

[13] Gheller, Y.A., *Tool Steels*, Metallurgy: Moscow, 1983 (in Russian).

[14] Christian, H., *Tool Steel*, Forskningscenter Risoe: Denmark, Risoe-R no. 1244(EN), 2001.

[15] Jamroziak, K. & Roik, T., Structure and properties of the new antifriction composite materials for high-temperature friction units. *Proceedings of the 7th International Conference on Fracture Fatigue and Wear*, pp. 628–637, 2019.

[16] Jianxin, D. & Tongkun, C., Self-lubricant mechanisms via the in situ formed tribofilm of sintered ceramics with CaF_2 additions when sliding against hardened steel. *International Journal of Refractory Metals and Hard Materials*, **25**(2), pp. 189–197, 2007.

[17] Antsiferov, V.N. & Akimenko, V.B., *Sintered alloy steels*, Metallurgy: Moscow, 1983 (in Russian).

DISPLACEMENT-BASED FAILURE ANALYSIS OF METAL MATRIX SYNTACTIC FOAM

ISTVÁN KOZMA, DÓRA HARANGOZÓ, HAJNALKA HARGITAI & IBOLYA ZSOLDOS
Department of Materials Science and Technology, Széchenyi István University, Hungary

ABSTRACT

Metal matrix syntactic foams are being used more and more widely thanks to their relative density as well as their perfect energy-absorbing characteristics. Multiple studies concern the economical production of these materials – particularly, how energy absorption as a physical characteristic can be increased. Many studies examined the effects of material choice as well as cell size and wall thickness of reinforcing materials on compressive strength. However, there are only estimated models about descriptions concerning decaying processes of samples from different material composition and geometrical variation. In this study, we introduced an "in situ" examination in order to model the decaying process. We upsetted aluminium-ceramic composite foams cyclically. We reconstructed the geometry of the sample with microcomputed tomography (μCT) technology and digital image processing at certain specified points of the compression test. During the complete decaying process, the process was evaluated with volume change and number of broken hollow particles, as well as elementary particle displacements, orientation and sphericity of the reinforced material.
Keywords: failure analysis, metal matrix syntactic foam, μCT, deformation analysis.

1 INTRODUCTION

A peculiar group of metal foams are metal foams reinforced with ceramic hollow particles. These are also known as metal matrix syntactic foams in academic literature. The two main components of metal matrix syntactic foams are the matrix material (usually technically pure aluminium, Al1050 or Al99.5) or alloys of aluminium and silicon, as well as the reinforcing material [1]. The main function of these reinforcing materials (almost all of them are hollow particles) is to implement porosity within the material, and thus, to create foam structure in the total volume. Hollow particle materials are usually ceramics, but academic literature also mentions iron alloys, glass, among others, as hollow particle materials.

The main advantage of cellular materials – including metal matrix syntactic foams – is their preferential mechanical properties, alongside the relatively small density. They are used widely in car bodies and in electromechanical industry. These applications are justified by the perfect energy-absorbing characteristics and the damping effects of cellular materials. Structural rigidity and compressive strength can be set by changing the volume percentage of hollow particles of the same size. The value of volume percentage theoretically can change between 0% and 74% [2]. This data can be precisely measured with industrial radiographical devices. During our examinations, we detected a volume percentage of 64% in the sample, which was manufactured with pressure infiltration. This value equals to that of the structure of non-probable densest fitting. Thus, the location of hollow particles of almost the same diameter in the material is not regular, but the densest. Alongside volume percentage, an important characteristic of metal matrix syntactic foam structures is the size of hollow particles, their outer radius (R) and their inner radius (r) [3]. These two values must be examined together, as their ratio ($\eta = r/R$) significantly effects the decaying behaviour of the hollow particles inside the material [4].

The most important characteristic of metal matrix syntactic foams is good compressive strength. Given the foam structure, these foams absorb energy perfectly, as the voids within the material do not collapse at the same time under the weight of an outside load, but

WIT Transactions on Engineering Sciences, Vol 124, © 2019 WIT Press
www.witpress.com, ISSN 1743-3533 (on-line)
doi:10.2495/MC190161

separately from each other in time and space; in this way, they take up the load gradually [5]. Many studies revolve around their decaying mechanism. Most of the research examined the decaying process under quasi-static loads [2]. These experiments usually tested the effect of the combination of matrix materials and reinforcing granule materials, in relation to compressive stress. During the quasi-static examination of particle geometry of the same material composition but varying geometry, it was revealed that the tested stress values increase as the wall thickness/diameter ratio also increases [6]. An obvious relation was detected between the first maximum stress peak and the matrix material strength: the bigger the material strength, the bigger the first maximum peak of the metal matrix syntactic foam made from this material. A general testing method to track structural changes in the deforming mechanism is by evaluating cross-sectional images of the metal foam. It has been determined that under the effect of upsetting performed in quasi-static state, the hollow particles remained intact on certain parts of the metal foam; however, a break line was detectable in the cross-section of the samples [7]. The presence of the break line is the consequence of the matrix material's malleable engineering strain. In case of the material under examination, this phenomenon precedes the decaying of ceramic hollow particles used as reinforcing material. Within the break line, hollow particles broke and their voids were compacted. However, hollow particles outside the break line were much less damaged or were not damaged at all. During the examination of syntactic foam decay, Wu et al. [8] stated that hollow particles located in conical parts close to the upper and lower pressure plates which shifted their shapes only to a minor degree, stayed practically intact, they did not break. The condition of minor engineering strain for these two parts is the friction on the contact surface of the pressure plates and the samples. This hypothesis is seconded by the fact that by a single layered MoS_2 lubrication, an acceptable measure of friction cannot be implemented; also, barrelling cannot be avoided. Furthermore, it was observed that in case of 90 μm diameter hollow particles which give a smaller porosity (58.1%), sample barrelling was smaller than in case of 150 μm diameter hollow particles which give bigger porosity (67.3%) [5]. Based on experiences, the authors have developed a schematic diagram about the assumed steps of decaying (Fig. 1).

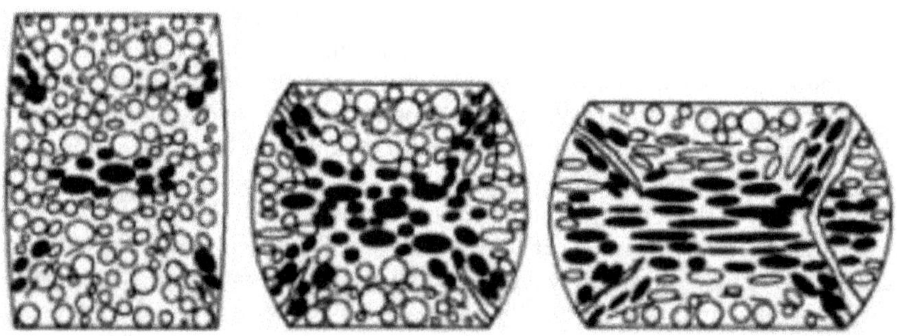

Figure 1: Schematic diagram of deformation metal matrix syntactic foam [8].

The first phase of decaying was the (shearing) decay of hollow particles found in corners in critical stress state. At the same time, hollow particle breakages occurred in the middle of the sample as well (under pressure) in the cross-sections of the weakest hollow particles. During the continuation of upsetting, damaged zones spread and a heavy barrelling occurred. This led to the detachment of material parts located in the outer volume of the sample.

2 MATERIALS AND TEST METHODS

During our tests, we used cylindrical samples with aluminium (Al 99.5) matrix, reinforced with ceramics (Globocer) hollow particles (R = 750 μm), height H/diameter D = 1. Pressure tests were performed according to the standard about the testing of cellular materials (DIN50134:2008). Quasi-static upsetting was performed sectionally, until a 30% of engineering strain has been reached with the samples. The test was interrupted after 200 μm of transformation in the flexible stage; and after 500 μm pressure plate displacement in the flatbed and compressing stage. Sample structures were tested with microcomputed tomography (μCT) technology in the starting state as well as between loads; in total, they were tested 18 times during the complete upsetting process.

X-ray Microcomputed Tomography tests were conducted with an Yxlon Modular device, with conical beam and flat panel detection. Geometric characteristics and technical parameters of the applied equipment ensured a magnification of M = 20.75. This resulted in a digital reconstruction of samples with 19 μm resolution.

During decay, change in matrix material and reinforcing material volume is negligible. However, the volumes of reinforced material voids decrease. These voids make up closed cells within the ceramic hollow particles. During shaping, cracks form on the hollow particles, then they break in certain damage zones (Figs 2 and 3). Internal hollow particle cells change in geometry as well as in topology. The decaying process was tested with these cell changes of volume, geometry and shape.

X-ray test data are the set of cubes consisting of volume and size per unit. Each of the voxels have a defined coordinate and grayvalue. We defined voxels of almost the same material with the threshold grayvalue, then we sorted them by adjacency (Fig. 4). The set of voxels defined in this way are the cell data of internal voids in the reinforced material. Data were processed with VolumeGraphics StudioMax software. In this program, we determined the geometrical and topological characteristics of each unique cell in the sampled steps of the decaying process.

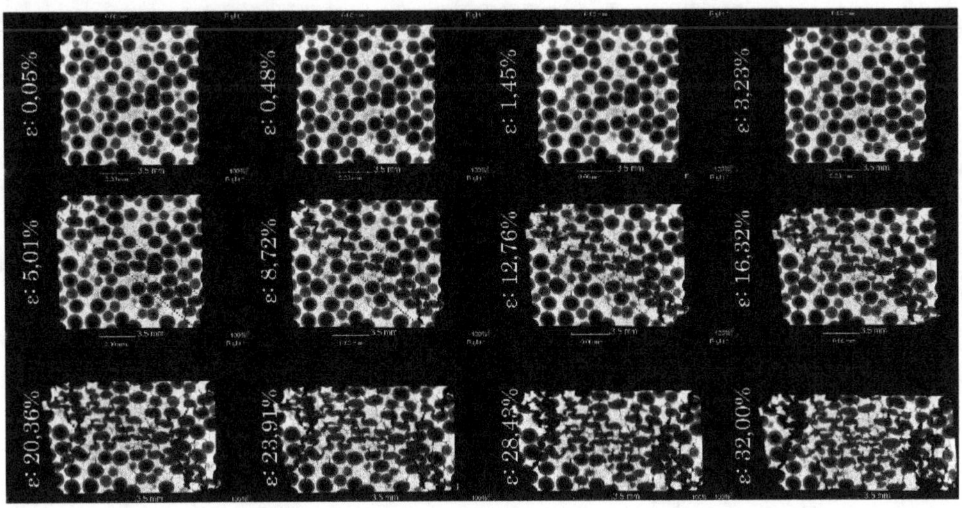

Figure 2: Structure change of metal foam on μCT 2D image slices, in relation to engineering strain.

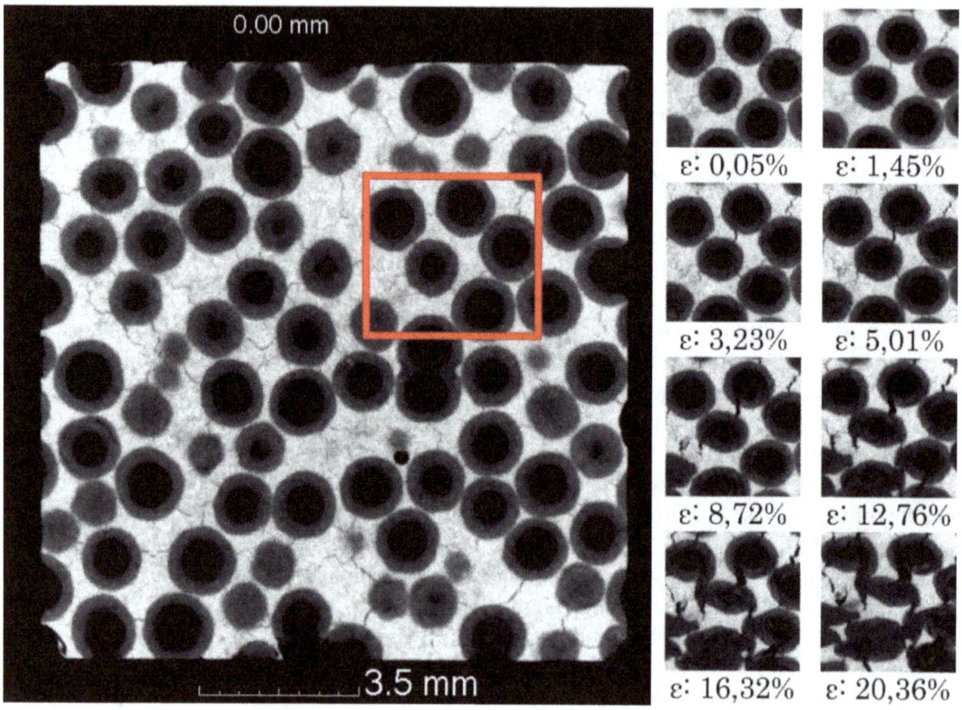

Figure 3: Decaying on μCT 2D image slices of metal foam, in relation to engineering strain.

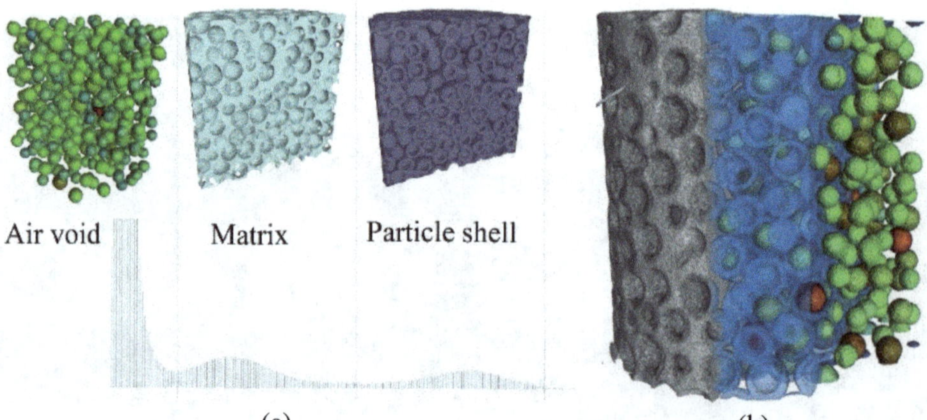

(a) (b)

Figure 4: Consisting elements of metal foam samples in relation to density.

Data were saved in a text file which includes x, y and z coordinates for the hollow particle voids, as well as their respective volume and diameter values, their spherical shape factors, and the number of hollow particle voids. Data assigned together were evaluated in GOM

Inspect software. Small, consecutive engineering strains induced slight displacements. Each coordinate of the consecutive states was fitted to the coordinates of the next phase state, based on the closest adjacent algorithm. Search distance was maximized in the value of the average hollow particle radius; thus we ensured the accuracy of the searching algorithm. We assigned an identifier to the hollow particle coordinates of the starting sample (Fig. 4(b)). This identifier marked that certain unique hollow particle during the whole transformation process, including its volume and diameter data (Fig. 5).

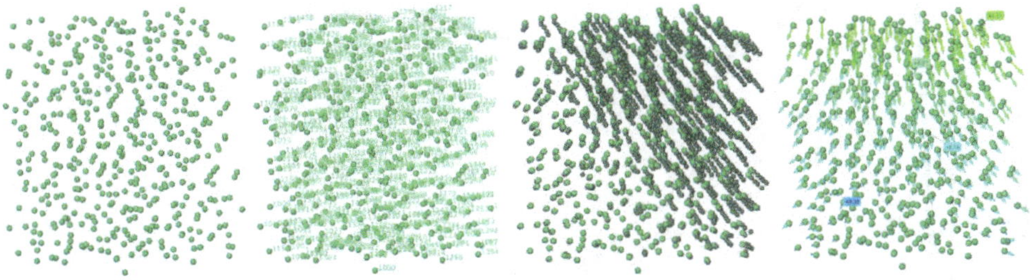

(a) (b) (c) (d)

Figure 5: (a) Hollow particle coordinates and (b) their identifiers, as well as Hollow particle displacements (c) as point cloud and (d) in vectorial display, in relation to engineering strain.

3 RESULTS

During the metal foam sample decay tests, we evaluated sectional images of X-ray tests as well as coordinate, volume and diameter data of unique hollow particle voids at the same time.

We inserted volume change data of reinforcing material into the push chart (Fig. 6). In the first phase of engineering strain, volume changes confirm literature references. Radius ratio of the hollow particles (η) affects volume change during the transformation process. Under load, the volume of hollow particles forming the reinforced material of samples – due to their radius direction – increase until the first stress peak of the transformation. Volume increase can be detected until an engineering strain of 2%; then, near the stress peak, it started to decrease steeply. The intensity of volume decrease changed at 4% engineering strain, and it showed near linear decrease in the flatbed stage.

The metal foam is significantly damaged during the transformation, within the 2%–5% engineering strain interval. After reaching the maximum stress, cracks form on ceramic hollow sphere walls under further load, and a 45° angle shear plane forms on the transformation direction. This range at the 2%–5% engineering strain load can be unequivocally determined based on both the sectional images and the volume change of samples. Along the 45° plane, the damage zone thickens under further pressure.

At 5% engineering strain it thickens to such a measure that a glide plane forms in this section of the sample. During further loading, the sample compresses in the complete volume of the test sample. Under the flatbed stage of the push chart, volume decrease is almost linear; it its intact after 30% engineering strain, defragmented hollow particles cannot be detected and hollow particle void volumes decrease to the detection threshold value.

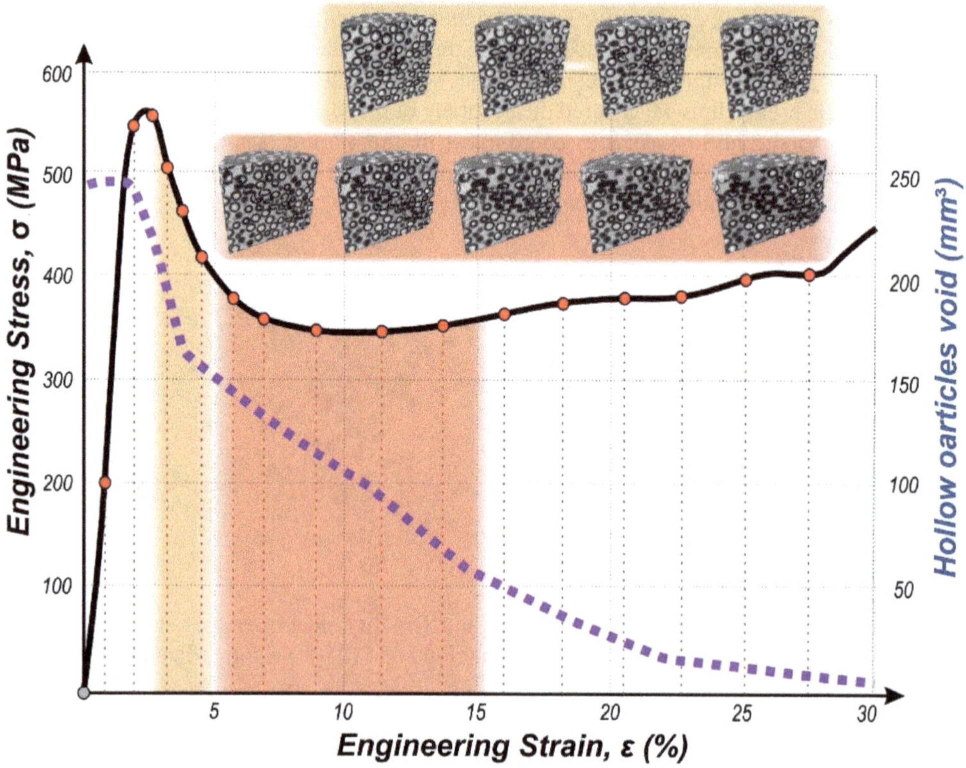

Figure 6: Push chart of tested samples and volume changes of hollow particle voids.

Hollow particle displacements induced by loads are depicted as vectors. The starting point of vectors is always the given, marked hollow particle gravity centre of the untransformed sample; the gravity centre of given state and position of the other hollow particle under the same identifier (Fig. 5). By this method, we determined not only the direction of displacement, but its size as well.

Our results show that, in the first stage of decaying which lasts up to the stress maximum of the push chart, engineering strain of the metal foam sample shows a similar process up to ~2% engineering strain as that of homogeneous materials (Fig. 7). Samples are barrelling, displacement vectors are small and they point to the lateral surface direction of the sample, almost equally from the axis.

The behaviour of metal foam can be explained with the malleable engineering strain of the aluminium matrix. At this stage, hollow particles did not take up critical load. However, at 2% engineering strain, there are detectable cracks in the matrix material (Fig. 3). Under further transformation, decaying starts on the hollow sphere wall, in thickness direction. At this stage, up to ~5% engineering strain, a ~45° angle break line forms in the loading direction. Hollow particles right next to the line start to break, then they completely detach. Hollow particle void volumes start to decrease significantly; this change correlates with the curve of the push chart.

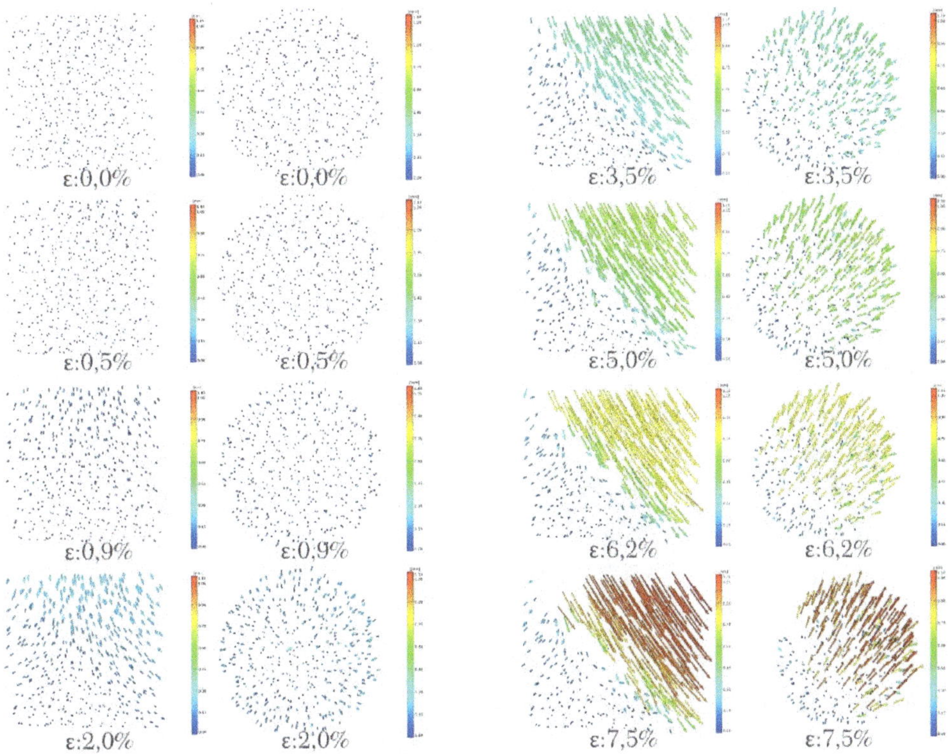

Figure 7: Hollow particle displacements, in relation to engineering strain.

Characteristics of displacement vectors also definitely change. The angle close to 45° with the load direction borders two different displacement structures (Fig. 8). In one of the segments, the barrelling characteristic remained with small homogeneous displacements. However, in the other area of the sample, big dislocations occur, usually in the same direction. Based on the vectorial map, the break line is also a shear plane, which is also a gliding plane during further transformations. Increasing the 5% engineering strain, the decaying process becomes homogeneous. By continuing upsetting, the shear plane thickens into a break line; the inner volume of hollow particles located on the line drastically decreases, then the line thickness increases further during decay; in the starting plane, the inner volume of the hollow particles reduces to a level where they no longer can be detected. Up to 20% engineering strain, hollow particles remote from the gliding plane stay intact. Above 20% engineering strain, decaying spreads to the whole sample. Fragmentation, decaying and volume change of hollow particles significantly increases in other segments of the sample as well.

We evaluated hollow particle displacements as only X, Y and Z directional components as well. Results have confirmed the previously described processes. The Z-directional displacement field shows the expected result. The largest displacement is on the load lateral surface; the length of vectors decreases towards the resting jaw. X- and Y-directional evaluation confirms the behaviour typical in case of homogeneous materials in the starting phase of transformation. Up to 2% engineering strain, displacements point symmetrically to the given tested (x and y) planes, or to the sample axis with the same vector size, only to the opposite direction.

Loading the samples further, the vector extent (that is, displacement) increases significantly. Next to increase, it shows an asymmetric image, as opposed to the starting symmetrical plane. It can be observed that this plane is not parallel to the load direction anymore, not even by increasing the transformation; it makes an angle with the load direction. The angle difference constantly grows above 2% engineering strain. Close to a 3.5% engineering strain, it makes a ~12° angle with the load direction (Fig. 8); above 4.5% engineering strain, it makes a ~30° angle (Fig. 9(b) and (c)); and at the measurement point above 5% engineering strain, it makes a 45° angle. Under the effect of shear power, the glide plane happens within the 2%–5% engineering strain range, based on hollow particle displacement.

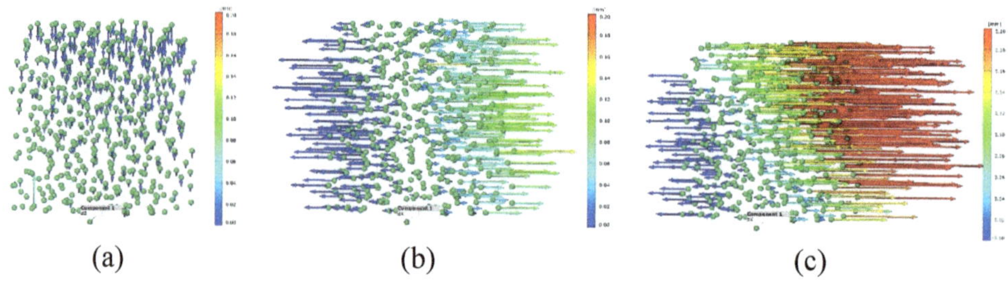

Figure 8: Displacement of hollow particles in the direction of transformation (Z): (a) Above 2% engineering strain. Displacement of hollow particles, perpendicularly to the transformation direction (X): (b) Above 2% engineering strain; (c) Above 5% engineering strain.

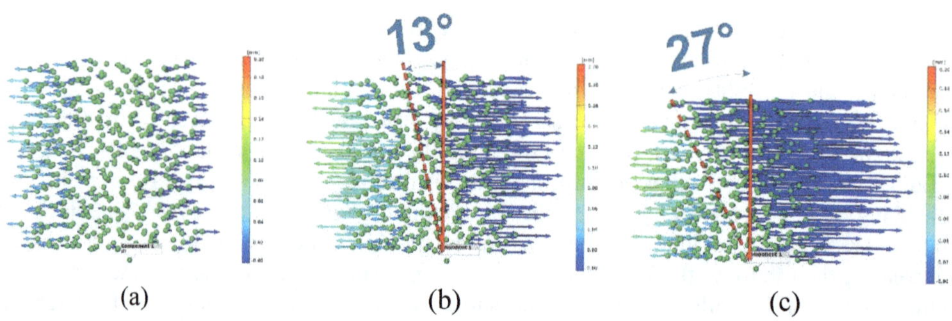

Figure 9: Displacement of hollow particles, perpendicularly to the transformation direction (Y): (a) Above 2% engineering strain; (b) Above 3.5% engineering strain; (c) Above 5% engineering strain.

The displacement histogram depicted in Fig. 10 also supports this. This figure quantitatively summarizes the numerical data of previously, graphically presented displacement vectors, in relation to engineering strain. On the X axis of the diagram, displacement values under given load can be read; on the Y axis, the number of displaced hollow particles can be read. It can be seen from the results that up to 2% engineering strain, the number of hollow particles changing position under load gradually increases; however, after further load, the quantitative factor does not change substantially, it becomes steady. At the same time, the distribution of displacement sizes during transformation supports earlier measurements and statements. The "barrelling", a homogeneous displacement characteristic, can be detected up to 2% engineering strain. The curve of the histogram does not show any local peaks; the size of displacements can be viewed as normally distributed. At 3.5% engineering strain, the homogeneity of the displacement field ceases to exist, 2 local peaks can be detected. The area under the two peaks – meaning, the number of hollow particles corresponding to given displacement measures – is approximately of the same magnitude. This similarity indicates two identical volumes in the same way as the linearly measurable distance between local peaks.

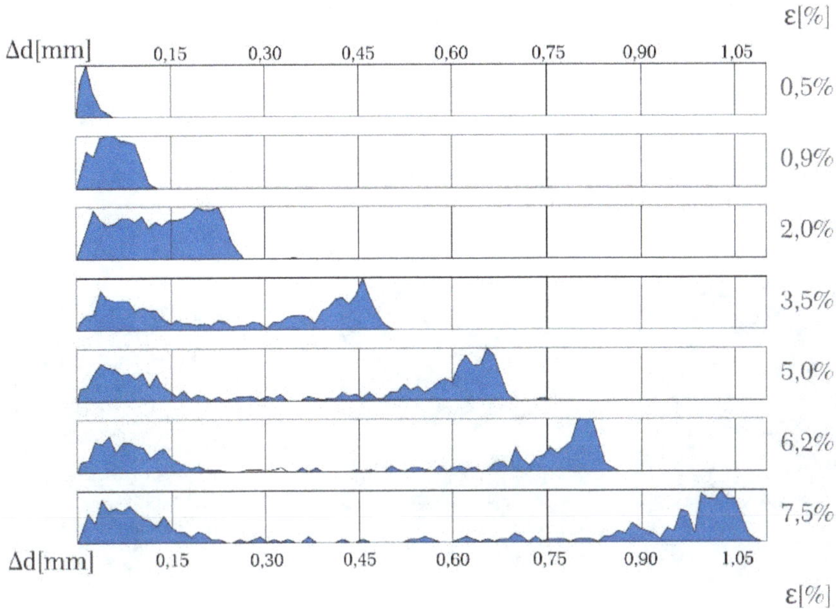

Figure 10: Displacement histograms in relation to engineering strain.

In Fig. 11, we can see the hollow particle displacement vectors of a sample above 2% engineering strain. We edited the excitation lines of vector displacement next to the figures – this is the characteristic referring to as "barrelling". In Fig. 11(b), the vector direction does not show only the engineering strain; arising forces can also be deduced based on the orderliness of directions. Based on this, it can be demonstrated that at this phase of transformation, the metal foam shows engineering strain typical to that of homogeneous materials; it deforms align a pressure cone. By increasing the load further, the characteristic shear plane forms.

Next to the displacement of gravity centres, we examined hollow particle surface deformation during transformation process. Volume data from microcomputed tomography scan were converted into surface models, then we evaluated the differences. In Fig. 12(a), the detail shows the starting geometry (blue) as well as geometry respective to 5% engineering strain (grey). In Fig. 12(b) and (c), details show differences between the two geometries, from different viewpoints. Red indicates additional material; blue indicates material loss. The evaluation confirms earlier deductions: parts can be localized in the sample structure where deformation cannot be detected. The big differences between geometries are detectable above the plane closing 45° degree with the axis. The detected surface differences usually do not show axial deformations either. Local geometry differences also occur in 45° degree direction.

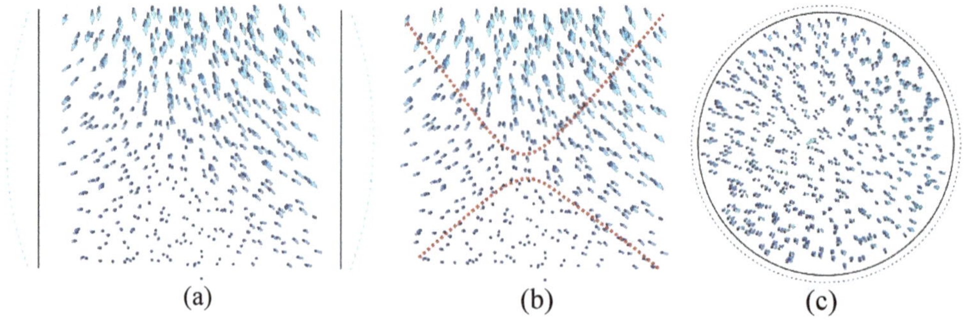

Figure 11: Hollow particle displacement vectors after 2% engineering strain histograms in relation to engineering strain.

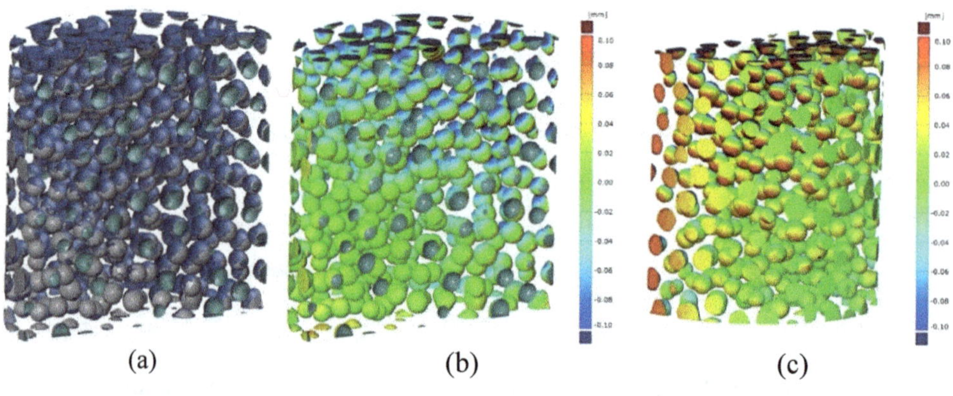

Figure 12: Hollow surface differences between the two geometries.

We have chosen 5 typically located hollow particle geometries from global surface data (Fig. 13), and we evaluated their local characteristics further. The SPH_A hollow particle is marked perpendicularly to the shear plane, right next to the upper pressure plate; the SPH_B hollow particle is marked on the decaying line border, closer to the upper pressure plate; the SPH_C hollow particle is marked in the centre of the decaying zone; the SPH_D hollow particle is marked on the decaying line border, closer to the lower pressure plate; and the SPH_E hollow particle is marked right next to the lower pressure plate, in the segment which was transformed the least.

Hollow particle displacements are presented in the diagram of Fig. 14. Displacement values can be stated linear in case of each hollow particle in relation to engineering strain; however, the scale depends on their position. There is no detectable, significant difference between the 2-2 hollow particles above and under the break line.

Figure 13: Individually evaluated hollow particles on the microcomputed tomography reconstruction.

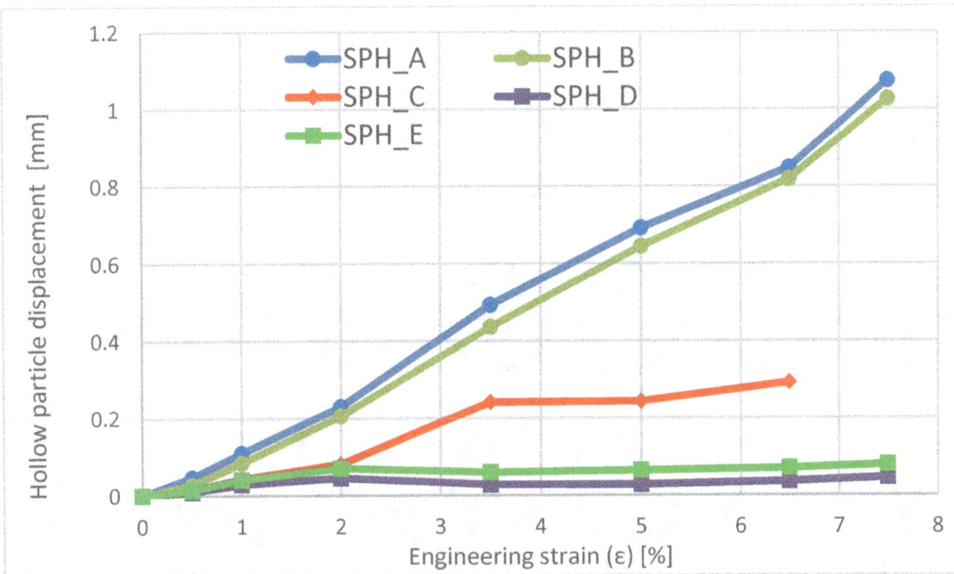

Figure 14: Local hollow particle displacements, in relation to the engineering strain.

During transformation, hollow particles deform, their shape characteristics change. The value of this phenomenon was characterised by the sphericity factor, based on the equation below

$$\Psi = \frac{\pi^{\frac{1}{3}}(6V_p)^{\frac{2}{3}}}{A_p}.$$

Measured values are presented in the diagram of Fig. 15. In the sphericity characteristic, the biggest difference was measured on the hollow particle marked in the break line. Change is depicted on the image series of Fig. 16. The deformation scale of further hollow particles is determined by their positions. Deformation was the biggest near the upper pressure plate; however, significant deformation cannot be detected near the lower pressure plate.

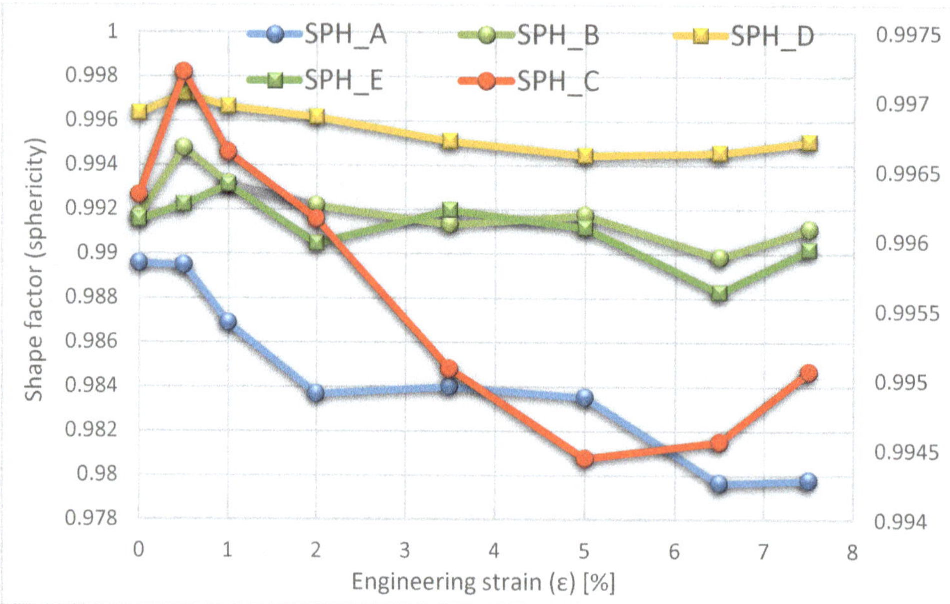

Figure 15: Change of sphericity factors in relation to transformation.

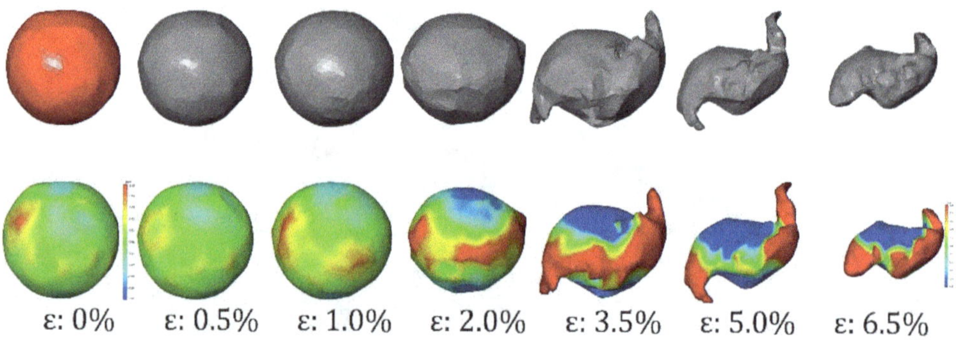

Figure 16: Deformation of hollow particle in the centre of the decaying line, in relation to engineering strain.

4 CONCLUSIONS

In this paper, we upset the metal matrix syntactic foam samples in stages. In order to model the decaying process, we reconstructed the geometry of the sample with X-ray Microcomputed Tomography technology and digital image processing at certain specified points of the compression test. During the complete decaying process, the process was evaluated with volume change, as well as elementary particle displacements, orientation and sphericity of the reinforced material.

During our tests, we concluded that at 2% engineering strain, there are detectable cracks in the matrix material. Under further transformation, decaying starts on the hollow sphere wall, in thickness direction. At this stage up to ~5% engineering strain, a ~45° angle break line forms on the load direction. Hollow particles right next to the line start to break, then they completely detach. Hollow particle void volumes start to decrease significantly; this change correlates with the curve of the push chart. We also evaluated the decaying process with hollow particle displacements. We concluded that the angle of 45° with the load direction borders two different displacement structures. In one of the segments, the barrelling characteristic remained with small homogeneous displacements. However, on the other area of the sample, big dislocations occur, usually in the same direction. Based on the vectorial map, the break line is also a shear plane, which is also a gliding plane during further transformations. Increasing the 5% engineering strain, the decaying process becomes homogeneous. By continuing upsetting, the shear plane thickens into a break line; the inner volume of hollow particles located on the line drastically decreases, then the line thickness increases further during decay; in the starting plane, the inner volume of the hollow particles reduces to a level where they no longer can be detected. Up to 20% engineering strain, hollow particles remote from the gliding plane stay intact.

ACKNOWLEDGEMENTS

We acknowledge EFOP-3.6.1-16-2016-00017 Internationalization, initiatives to establish a new source of researchers and graduates, and development of knowledge and technological transfer as instruments of intelligent specializations at Szechenyi University.

REFERENCES

[1] Orbulov, I.N. & Majlinger, K., Microstructure of metal-matrix composites reinforced by ceramic microballoons. *Materiali in Tehnologije*, **46**(4), pp. 375–382, 2012.

[2] Orbulov, I.N. & Májlinger, K., Compressive properties of metal matrix syntactic foams in free and constrained compression. *JOM*, **66**(6), pp. 882–891, 2014.

[3] Balch, D.K. et al., Plasticity and damage in aluminum syntactic foams deformed under dynamic and quasi-static conditions. *Materials Science and Engineering: A*, **391**(1), pp. 408–417, 2005.

[4] Vesenjak, M. et al., Impact behavior of composite hollow sphere structures. *Journal of Composite Materials*, **43**(22), pp. 2491–2505, 2009.

[5] Gupta, N. & Ricci, W., Comparison of compressive properties of layered syntactic foams having gradient in microballoon volume fraction and wall thickness. *Materials Science and Engineering: A*, **427**(1), pp. 331–342, 2006.

[6] Gupta, N. & Rohatgi, P.K., *Metal Matrix Syntactic Foams: Processing, Microstructure, Properties and Applications*, DEStech Publications, 2014.

[7] Zou, L.C. et al., Compression deformation behaviour of aluminium matrix syntactic foams under quasistatic loading: In situ SEM compressive tests. *Materials Research Innovations*, **18**(sup4), pp. S4-536–S4-539, 2014.

[8] Wu, G.H. et al., Compression behaviors of cenosphere–pure aluminum syntactic foams. *Scripta Materialia*, **56**(3), pp. 221–224, 2007.

SECTION 7
MATERIALS FOR BIOENGINEERING APPLICATIONS

ELECTROCHEMICAL METHOD OF FRETTING CORROSION TESTING IN MODULAR HIP PROSTHESES

CLAUDIO T. DOS SANTOS[1*], IEDA M. V. CAMINHA[1], WELLINGTON G. FERNANDES[1], MAURÍCIO J. MONTEIRO[1], PATRÍCIA O. CUBILLOS[2] & CARLOS R. M. ROESLER[2]
[1]National Institute of Technology (INT), Brazil
[2]Biomechanical Engineering Laboratory, Federal University of Santa Catarina (UFSC), Brazil

ABSTRACT

Modular hip prostheses allow better adjustment to the patient's anatomy. We used a combination of materials to make the prosthesis more resistant to wear and corrosion; however, modular connections are inevitably subjected to fretting corrosion, due to micromotion. Evaluation of this corrosion is generally based on two in vitro test methods: a long-term test method (fatigue-corrosion) for the simultaneous assessment of mechanical stability, particle release and analysis of damage on the surfaces of prosthesis components; and a short-term electrochemical test method, for current and potential generated by fretting and corrosion in the modular connection. In both methods, the prosthesis is mechanically loaded in a saline environment, to simulate human body conditions. Although these methods are complementary, there are still few studies using the electrochemical test method. This study aims to present the methodology used in an electrochemical test and the results obtained in our laboratory for a stainless steel commercial prosthesis. In addition, we discuss the mechanisms involved in the fretting corrosion process. We conclude that the results of the electrochemical test, together with those obtained in the fatigue-corrosion test, contribute to an effective evaluation of the quality of modular prostheses, in terms of their safety and performance.

Keywords: hip prosthesis, fretting corrosion, electrochemical test, modular interface, medical device safety, performance, head-stem combinations.

1 INTRODUCTION

The partial or total hip prosthesis is used to recover the function of the hip joint when functionality is lost due to the occurrence of trauma, disease or to relieve the pain resulting from a chronic inflammatory process that does not respond to conventional treatments [1].

The modular type of hip prosthesis has become widely used, because it provides a better fit to the patient's anatomy, since it is possible to combine components of different sizes, types and materials [2], [3]. Surgeons may also perform femoral head replacement on already fixed stems, for a less aggressive revision procedure [4], [5]. The head-stem modular connection of the hip prosthesis usually consists of a metal-on-metal conical junction where the femoral stem has a conical tip and the femoral head has the tapered bore for fitting, as can be seen in Fig 1.

Hip prostheses may vary from manufacturer to manufacturer, in terms of angle of the taper and bore of the head (one example can be seen in Fig. 1), tolerances, surface roughness, material, diameters of the taper and presence of "skirt" on the head [6]. Metallic materials used in these prostheses should have a set of properties, such as: good biocompatibility, mechanical strength, abrasion and corrosion resistance. The most widely used materials for manufacturing prostheses are in the AISI 316 stainless steel family, cobalt-based alloys, and titanium and its alloys.

* *ORCID: https://orcid.org/0000-0002-7343-4307*

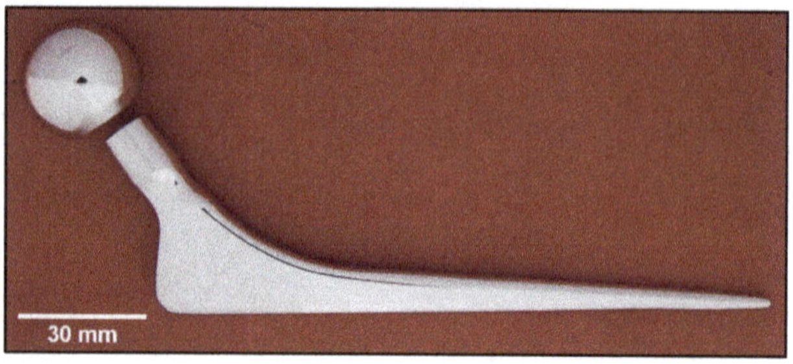

Figure 1: Hip prosthesis, head-stem modular connection. (Source: Authors, 2015 [19].)

These materials rely on the formation of a passive film that prevents the process of corrosion. The film consists of an oxide that forms spontaneously on the surface of the metal, avoiding the transport of ions and electrons [7]; however, when the components of this prosthesis are connected, they inevitably form a crevice in the connection, which can lead to so-called mechanically assisted crevice corrosion or fretting corrosion, a corrosion process that occurs due to the formation of crevices combined with the fretting in the interfaces that are submitted to small-scale relative motion [4], [6]. This corrosion is related to the combined action of the mechanical removal of the passive film due to fretting, and corrosion of the exposed metal surface with the consequent release of debris, as seen in the Fig. 2 example. It is not yet clear which design, material or surface modifications can minimize or eliminate the fretting corrosion without the need to eliminate the modular connection itself.

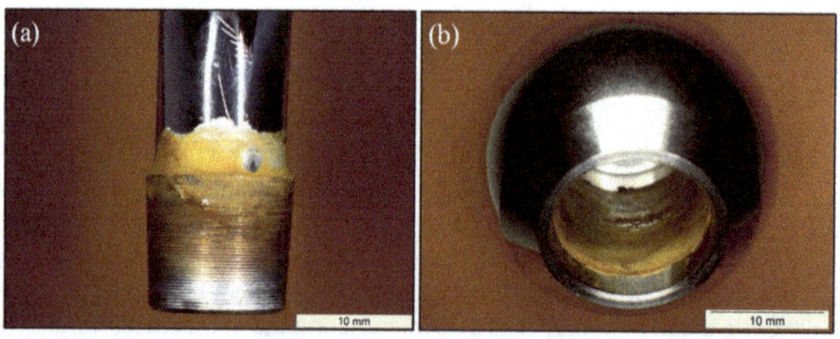

Figure 2: Corrosion in the trunnion (a) and head (b) of a stainless steel hip prosthesis, after long-term fatigue-corrosion test (10^7 cycles) in an aqueous 0.90% NaCl solution. (Source: Authors, 2015 [19].)

Corrosion of a metal prosthesis in the human body is undesirable for two main reasons: it can reduce the structural integrity of the prosthesis; and the release of ions and particles usually causes chronic allergies, inflammatory reactions or other toxic reactions in the body, which can lead to the need for revision surgery, thus increasing patient suffering and treatment costs [2]. The revision surgery may have double the cost of the primary surgery, due to the complications associated with the fixation of the material and the bone quality of the patient [8], [9]. Ideally, the release of ions and particles should be maintained at a satisfactory low level over the time in service, which is estimated at 15 or 20 years, depending on the material [10].

The analysis of the fretting corrosion of hip prostheses is extremely important, in order to evaluate their propensity to release ions and particles into the modular interfaces, allowing the comparison of prostheses with different characteristics, such as different manufacturers, designs, surface finishes and combinations of materials for the head and the stem. This analysis is usually performed by means of a short-term test using an electrochemically simulated corrosion chamber. According to a recent study [4], there are still few data in the literature that were obtained using electrochemical chambers, with most of the information pertinent to fretting corrosion damages based on retrieval studies [4], [11]–[13].

It is possible to detect the corrosion not only in the interface of the head-taper, but also in other regions of the prosthesis that are subject to the relative movement. The great advantage of this method is the possibility of real-time evaluation of the effect of compressive loads on the current generated at the head-taper interface in a controlled environment [4].

The purpose of this study is to present the electrochemical method used in the fretting corrosion test of modular hip prostheses and provide the results obtained in our laboratory for a commercial prosthesis. The results are discussed in terms of the mechanisms proposed in the literature for this corrosion process. At the end of the study, the importance of the electrochemical test, together with the fatigue-corrosion test, is highlighted for obtaining an evaluation of the quality of the prostheses.

2 MATERIALS AND THE METHOD OF ELECTROCHEMICAL TESTING

Our electrochemical chamber consists of a cylindrical acrylic vat filled with 1 L of 0.90% NaCl aqueous solution (Alphatec, 99.5%) containing the reference electrode (Ag/AgCl, Analion, R-684), the working electrode (WE), which in this case is the hip prosthesis, and a counter electrode (CE) of the same material as the prosthesis (Fig. 3). The apparatus was assembled in accordance with the recommendations of ASTM F1875:2014 [14]. The saline solution used in the test is called an artificial physiological environment, because it seeks to simulate the aggressive effect within the human body.

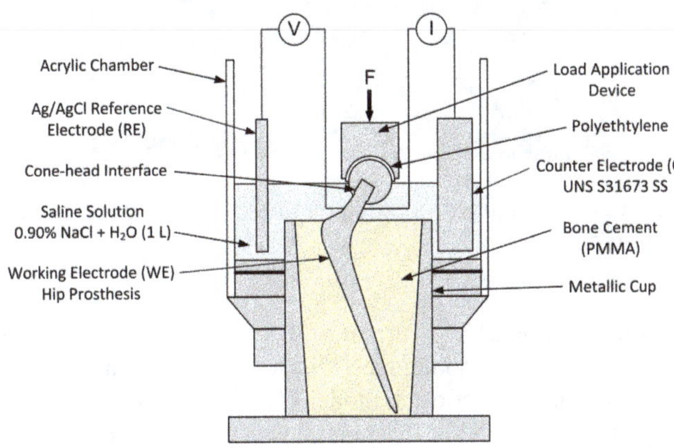

Figure 3: Schematic drawing of the electrochemical chamber for fretting corrosion testing of the head-taper connection of hip prostheses.

To perform the tests, the modular head was fitted to the stem with a load of -2000.0 N at a displacement rate of 0.05 mm/s, as recommended by ASTM F2009: 2011 [15]. For this, we used the Instron uniaxial mechanical testing machine, Model 3382. The head assembly is performed after prior wetting of the surfaces of the connection with the test solution, thus simulating the "worst-case" condition, whereby the aggressive environment penetrates into the cone gap during the service period, contributing to the process of corrosion.

To position and support the prosthesis for the application of cyclical mechanical loading of the prosthesis is obtained by embedding in polymethylmethacrylate bone cement (PMMA, Baumer). The prosthesis was embedded in an anatomical position, according to ISO 7206-6: 2013 [16], with angles $\alpha = 10°$ and $\beta = 9°$. Then, the chamber is mounted on the mechanical loading system of the dynamic test machine (Fig. 4). In this case, the Instron servo-hydraulics machine (model 8872), was used with a load cell of 25.0 kN.

The head of the prosthesis was mechanically loaded by means of a loading device with a semi-spherical surface of ultra-high molecular weight polyethylene (UHMWPE). The loading device has a ball bearing to enable lateral movement due to deformation of the prosthesis during the test. This measurement system permits the performance of tests with different levels of cyclic load and different oscillation frequencies, allowing the evaluation of the effect of these parameters on the fretting corrosion rate.

Figure 4: Electrochemical chamber mounted on the servo-hydraulic machine and the load application device.

For monitoring potential (OCP) and current of the prosthesis during the test, generally a potentiostat is used, connected to a computer dedicated to the data acquisition. In our study, we used the Metrohm Autolab potentiostat, model PGSTAT302N. For the OCP reading, the electrochemical noise module (ECN) of this potentiostat was used. The tests were performed at room temperature (~25°C) and the pH of the solution was measured at the beginning and at the end of the test: pH was maintained between 6 and 7 in all tests performed. The prosthesis was kept at rest for at least 1 hour prior to mechanical loading, to stabilize the potential and current reading.

We show examples of the results obtained using this apparatus on a UNS S31673 stainless steel hip prosthesis (stem-head pair similar to that of Fig. 1), composed of a 150 mm polished stem, 12/14 taper and a head diameter of 28 mm. Results were obtained with the application of a sinusoidal cyclic load with a minimum value of -2300.0 N, maximum of -230.0 N (R =

10) and a frequency of 1.0 Hz, according to ISO 7206-6: 2013 [16]. It is important to highlight that this is still the type of prosthesis that is most used in Brazil, due to the low cost and satisfactory properties of steel. Table 1 shows the nominal composition of UNS S31673 steel (AISI 316LVM) [17].

Table 1: Nominal composition (weight%) of UNS S31673 stainless steel [17].

Carbon (C)	Silica (Si)	Manganese (Mn)	Nickel (Ni)	Chromium (Cr)
0.030 max.	0.75 max.	2.00 max.	13.00 – 15.00	17.00 – 19.00
Mo	Sulphur (S)	P	Copper (Cu)	N
2.25 – 3.00	0.010 max.	0.025 max.	0.50 max.	0.10 max.

Note: The balance is Iron (Fe).

3 RESULTS AND DISCUSSION OF ASSOCIATED MECHANISMS

The results of the fretting corrosion test of the UNS S31673 (AISI 316LVM) stainless steel hip prosthesis are shown in Fig. 5.

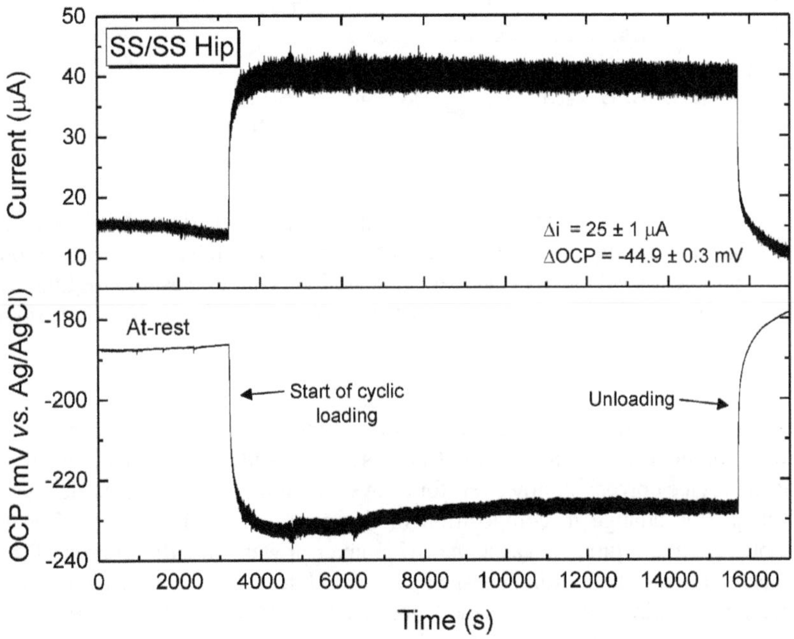

Figure 5: Current (top) and potential (bottom) profiles of the UNS S31673 stainless steel prosthesis, obtained before and after cyclic mechanical loading of -230.0 to -2300.0 N, f = 1.0 Hz.

With the start of cyclic loading, it is possible to observe an abrupt displacement in the curves of current and OCP. This displacement indicates the onset of fretting corrosion, which is the combined action of passive film abrasion and chemical changes within the crevice formed in the head-taper connection.

According to the mechanism described in [6], mechanical loading induces high shear and compression stresses on the taper and head surfaces, which are sufficient to break the passive film. With the load cycling, the film becomes constantly broken by the fretting process. The rupture of the film exposes the metal directly to the aggressive medium, causing its ionic dissolution: ($M \rightarrow Mn^{+} + n\acute{e}$). On the other hand, high driving force for the formation of passive oxide film induces the hydrolysis of the aqueous solution according to the following equation.: $M + H_2O \rightarrow MO + 2H^{+} + 2\acute{e}$ [6].

These two processes generate electrons and change the potential of the prosthesis, causing the observed displacements [6]. The increase in current with fretting is due to the potential difference that is created between the prosthesis that is suffering fretting and the counter electrode at rest. Fig. 6 illustrates the ionic dissolution and repassivation processes that occur after the oxide layer rupture.

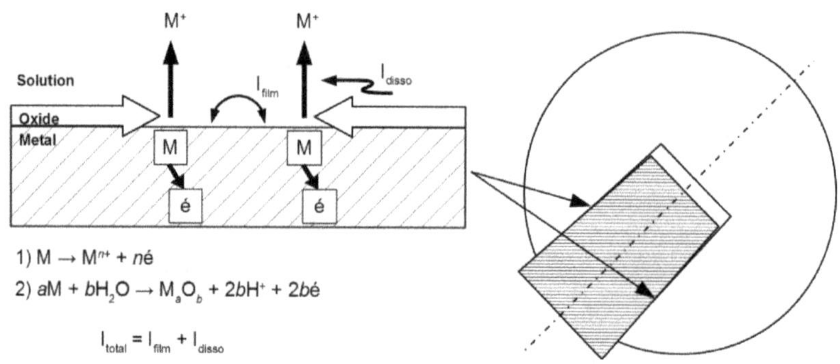

Figure 6: Schematic drawing that shows the repassivation reaction after the rupture of the oxide film and the ionic dissolution reaction (left); and the region of the connection where mechanically assisted crevice corrosion (fretting corrosion) occurs (right). *(Source: Adapted from Gilbert and Jacobs [6] and Hussenbocus et al. [18].)*

In fretting corrosion, the fluid enters the crevice and the crevice itself contributes to modify the chemical composition of the medium (increase of chloride ions [Cl^-] and local reduction of pH), as in crevice corrosion; however, for alloys which form a passive oxide layer such as stainless steel, this change in composition of the confined fluid is not sufficient for the corrosion process to continue, because the oxide layer prevents the dissolution of the metal. With the contribution of the fretting caused by the small-scale cyclic movement, the corrosion process becomes continuous, which is evidenced by the displacement observed in the current and OCP curves. According to Gilbert and Jacobs [6], as the fretting makes the potential more negative, it may have the effect of reducing the thickness and stability of the oxide layer that reforms on the surface of the entire prosthesis. Consequently, the oxide film becomes less able to withstand corrosive attack, resulting in large-scale corrosion and localized corrosion, which are generally evidenced by observations in retrieved implants.

Fig. 7 shows the magnified profile of current, load and OCP curves at the end of the test shown in Fig. 5. Note that there is a periodic oscillation with a frequency of 1 Hz in the curves, which corresponds to the frequency of mechanical loading, indicating that it has a direct influence on current and potential behavior. This behavior is indicative of the continuous process of breakdown and repassivation of the oxide film.

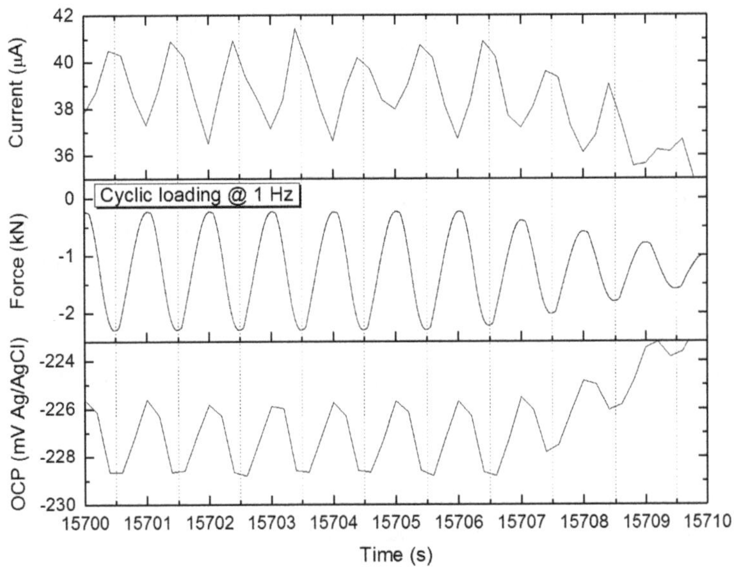

Figure 7: Magnification of the current (top) and potential (bottom) profiles at the end of the mechanical loading (middle). Note the change in current and potential behavior with the gradual onset of unloading.

The average current and potential in the initial resting regime were 14.8 ± 0.5 μA and -187.0 ± 0.2 mV, respectively. On the other hand, in the mechanical loading regime, the root mean square (RMS) values of current and potential were 40.1 ± 0.9 μA and -231.9 ± 0.2 mV, respectively. Current and potential have opposite profiles, which means that when the potential is at its minimum the current is at its maximum, while the profile of the potential is similar to that of mechanical load, which means that when the load reaches its minimum value (F_{min} = -2300 N), the potential is also minimal, while at this point the current is at its maximum.

The results of the electrochemical fretting corrosion test are important for understanding the effect of design changes, load application modes, and combination of components made from different materials. Recently, use of electrochemical tests by Rowan et al. [4] evaluated the effect of two different designs (V40 and 12/14) of titanium alloy taper and two different head materials (Co-Cr and ceramic) on the fretting corrosion. They showed that the V40 taper with ceramic head presented the lowest propensity to fretting corrosion than the same taper with a Co-Cr head. While in the 12/14 taper design, the ceramic and Co-Cr heads exhibited similar behavior, indicating that design also has an influence on fretting corrosion. In another recent study using a similar methodology, Panagiotidou et al. [12] examined fretting corrosion in several head-stem combinations as a function of the frictional torque and bending moment. They showed that the Co-Cr tapers presented the lowest current, when combined with a ceramic head.

Although the electrochemical test helps to identify the fretting corrosion, the use of complementary techniques is essential for a more effective assessment of this corrosion process. Analysis of chemical composition, microstructure, surface damage, roughness and disassembly force of head are fundamental actions for a more complete characterization and understanding of the differences in electrochemical behavior between the prostheses.

It is important to highlight that when the prostheses have good performance and safety, they improve the quality of life of the implanted patient and reduce the risk of rejection during the time in service, besides reducing the high hospital expenses associated with the purchase of a new prosthesis, revision surgeries and medical treatment.

4 CONCLUSION

The electrochemical method proved to be fast and efficient for the evaluation of fretting corrosion of modular hip prostheses. Results of this test, together with those obtained in the fatigue-corrosion test, allow for a consistent evaluation of the quality of modular prostheses, in terms of their mechanical and electrochemical safety and performance.

ACKNOWLEDGEMENTS

We are grateful to Metrohm do Brasil and its specialists Jonas Motte Nogueira, André Guimarães de Oliveira, Paula Gimenes Machado and Vinícius Santos for technical support. Thanks to Jeremy L. Gilbert of Clemson University (MUSC) and Ivan Bastos (UERJ) for information that contributed to the assembly and testing of our measurement system. Thanks to Javier A. Carreno Velasco and Rodrigo Vieira Landim (INT/LAH$_2$S) for their suggestions and help in the preparation of the method. Also, thanks to Jorge A. Hinostroza Medina (INT/LACPM) for contributing to this work.

REFERENCES

[1] Simioni, S., *Manual da Qualidade de Implante em Artroplastia de Quadril [Manual of Quality of Arthroplasty Implants]* (In Portuguese), 1st ed., Curitiba: Champagnat, p. 106, 2012.

[2] Urban, R. et al., Characterization of solid products of corrosion generated by modular-head femoral stems of different designs and materials. *Modularity of Orthopedic Implants*, ASTM International: West Conshohocken, PA, p. 33, 1997.

[3] Brown, S., Abera, A., D'Onofrio, M. & Flemming, C., Effects of neck extension, coverage, and frequency on the fretting corrosion of modular THR bore and cone interface. *Modularity of Orthopedic Implants*, ASTM International: West Conshohocken, PA, pp. 189–198, 1997.

[4] Rowan, F.E., Wach, A., Wright, T.M. & Padgett, D.E., The onset of fretting at the head-stem connection in hip arthroplasty is affected by head material and trunnion design under simulated corrosion conditions. *J Orthop Res*, **36**(6), pp. 1630–1636, 2018.

[5] De Thomasson, E., Conso, C. & Mazel, C., A well-fixed femoral stem facing a failed acetabular component: To exchange or not? A 5- to 15-year follow-up study. *Orthop Traumatol Surg Res*, **98**(1), pp. 24–29, 2012.

[6] Gilbert, J. & Jacobs, J., The mechanical and electrochemical processes associated with taper fretting crevice corrosion: A review. *Modularity of Orthopedic Implants*, ASTM International: West Conshohocken, PA, p. 15–45, 1997. Available from: https://www.astm.org/DIGITAL_LIBRARY/STP/PAGES/STP12020S.htm.

[7] Cohen, J., Current concepts review. Corrosion of metal orthopaedic implants. *J Bone Joint Surg Am,* **80**(10), p. 1554, 1998.

[8] Rodrigues, C.V.M., *Próteses Totais de Quadril: Diagnóstico do Panorama Nacional e Elaboração de Diretrizes para as Políticas Públicas*. Universidade Federal do Rio de Janeiro, 2006.

[9] Arndt, Â.B.M., Telles, J.L. & Kowalski, S.C., O custo direto da fratura de fêmur por quedas em pessoas idosas: análise no setor privado de saúde na cidade de Brasília,

2009 [The direct cost of femur fracture for…: analysis in the private sector in the city of Brasilia, for 2009]. *Rev Bras Geriatr e Gerontol*, **14**(2), pp. 221–231, 2011.

[10] Chen, Q. & Thouas, G.A., Metallic implant biomaterials. *Mater Sci Eng R Reports*, **87**, pp. 1–57, 2015.

[11] Goldberg, J.R. & Gilbert, J.L., In vitro corrosion testing of modular hip tapers. *J Biomed Mater Res*, **64B**(2), pp. 78–93, 2003.

[12] Panagiotidou, A. et al., The effect of frictional torque and bending moment on corrosion at the taper interface. *Bone Joint J*, **97B**(4), pp. 463–472, 2015.

[13] Dos Santos, C.T., Barbosa, C., Monteiro, de J.M., Abud, I.C., Caminha, I.M.V. & Roesler, C.R.M., Characterization of the fretting corrosion behavior, surface and debris from head-taper interface of two different modular hip prostheses. *J Mech Behav Biomed Mater*, **62**, pp. 71–82, 2016.

[14] ASTM F1875-98(2014), *Standard Practice for Fretting Corrosion Testing of Modular Implant Interfaces : Hip Femoral Head-Bore and Cone Taper Interface*, vol. i, ASTM International: West Conshohocken, PA, pp. 1–6, 2014.

[15] ASTM F2009-00, *Standard Test Method for Determining the Axial Disassembly Force of Taper Connections of Modular Prostheses*, ASTM International: West Conshohocken, PA, pp. 1–4, 2011.

[16] ISO 7206-6:2013. Implants for surgery. Partial and total hip joint prostheses. Part 6: Endurance properties testing and performance requirements of neck region of stemmed femoral components. ISO Web Site, 2013. www.iso.org/standard/51186.html. Accessed on 28 Mar. 2019.

[17] ASTM International, *ASTM F138-13a. Standard Specification for Wrought 18Chromium-14Nickel-2.5Molybdenum Stainless Steel Bar and Wire for Surgical Implants (UNS S31673)*, ASTM International, pp. 1–5, 2013.

[18] Hussenbocus, S., Kosuge, D., Solomon, L.B., Howie, D.W. & Oskouei, R.H., *Head-Neck Taper Corrosion in Hip Arthroplasty*, pp. 18–21, 2015.

[19] Dos Santos, C.T., Caminha, I.M.V., Fernandes, W.G., Monteiro, M.J., Cubillos, P.O. & Roesler, C.R.M., Evaluation of the technology of national and imported orthopedic prostheses used in total hip arthroplasty. Report for and funded by Brazilian Ministry of Health, 2015.

A NOVEL TECHNIQUE FOR TESTING OSTEOINTEGRATION IN LOAD-BEARING CONDITIONS

GIULIA PASCOLETTI[1], GIANCARLO DI CHIO[2], ANTONGIULIO MARMOTTI[3], ALESSANDRO BOERO BARONCELLI[4], PIERO COSTA[5], ANDREA TANCREDI LUGAS[2] & GIANPAOLO SERINO[2]
[1]Department of Industrial and Mechanical Engineering, University of Perugia, Italy
[2]Department of Mechanical and Aerospace Engineering, Politecnico di Torino, Italy
[3]University Hospital St. Luigi Gonzaga, University of Torino, Italy
[4]Department of Animal Pathology, University of Torino, Italy
[5]Intrauma S.r.l., Rivoli, Italy

ABSTRACT

When a new material for the realization of an implantable device in the bone is being studied, in addition to its chemical-physical-mechanical characterization, tests regarding osteointegration are performed. Usually, researchers evaluate the ability of biomaterials to bind to the bone under load-bearing conditions, through animal experiments in the phase of a preclinical study, provided the respective authorization by the ethics committee. In more detail, plugs made of the material under investigation are prepared and implanted into a weight-bearing portion of the skeleton of animals (typically into the knee joint of goats, pigs, rabbits or dogs); after a pre-set time, the animal is sacrificed, the bone element is extracted, it is tested mechanically – generally by means of a pull-out test – and finally it is examined histologically. Mechanical tests often require demanding specimen preparation, which could bias results. In the scope of a research regarding the interface behaviour of a ceramic plug (two different ceramic plugs) compared to a titanium one, the authors have suggested a novel testing technique which allows to perform 'push-in' tests, instead of the more common pull-out tests. This methodology has been followed here to compare titanium versus ceramic plugs at different times from implant (0, 3 months, 1 year) into goat knees. As a result, the study reports the shear resistance of bone–plug interfaces. The statistical analysis of the data allowed us to establish that titanium plugs systematically exhibit a higher resistance ($p<0.10$); this resistance undergoes a significant increment as time passes ($p<0.07$) due to progressive osteointegration.

Keywords: mechanical test, bone–biomaterial interface, osteointegration.

1 INTRODUCTION

When a new material for the realization of an implantable device is being investigated, first of all chemical-physical tests are performed, then biocompatibility is tested, and eventually, according to its destination (for example in the case of prostheses or bone synthesis devices), its bioactivity is tested. The osteointegration of an implantable device depends on its constitutive material and on surface properties. Bone remodelling, i.e. the structural reorganization of the bone not only at the interface but all around the device, depends above all on the state of strain and stress that the device induces into the bone when it is implanted.

The knowledge of the mechanical properties of the bone tissue and of the structural behaviour of the whole native bone both in physiological and pathological conditions is mandatory in order to identify main specifications for restorative materials, and to assess if and how the bone can be interfaced with other materials and devices.

The structural analysis of skeletal body elements and of biomechanical systems consisting of a bone element coupled to a prosthesis, an implant or a fracture synthesis device, can be performed both numerically and experimentally [1]. There are many examples of clinical problems which have moved from a qualitative assessment to a quantitative evaluation thanks to the respective modelling [2]–[9] or to the application of classical experimental methods of structural analysis to the evaluation of the efficacy of procedures or surgical techniques [10]–

WIT Transactions on Engineering Sciences, Vol 124, © 2019 WIT Press
www.witpress.com, ISSN 1743-3533 (on-line)
doi:10.2495/MC190181

[18] or to the evaluation of the mechanical characteristics of the materials used at different scales of investigation [19]–[24].

However, each approach has its own limitations: numerical models can be very complex and consequently need to be validated; experimental tests cannot faithfully reproduce the real conditions and they require simplifications, in most cases. On the basis of these assumptions, both approaches still remain necessary and in most cases complementary.

Preclinical studies can be performed as a validation of both *in vitro* tests and numerical models. Here researchers usually focus their attention on the ability of prosthetic materials to bind to the bone under load-bearing conditions, and resort to animal experiments. Synthetic plugs are implanted into the weight-bearing portion of the femoral condyles of animals (typically goats, pigs, rabbits or dogs); and the animals are sacrificed after a pre-set time, bone portions are tested mechanically and examined histologically.

Mechanical testing often requires demanding specimen preparation which could bias results: first of all bone cubes must be cut from the whole bone [25] (Fig. 1(a)); secondly, the plugs must bear an attachment area (like a transversal hole [26] or an axial threaded hole [27], Fig. 1(b)) that is not easily accessible; third, the mechanical attachment to the plugs for pull-out tests requires milling adjacent bone (in the transversal hole case), or the removal of ingrowth bone (in the case depicted in Fig. 1(b)).

A novel technique is here introduced, which allows to perform 'push-in' tests after performing two easy operations which do not affect the condylar area: planar cutting of the whole bone and end milling of an axial hole.

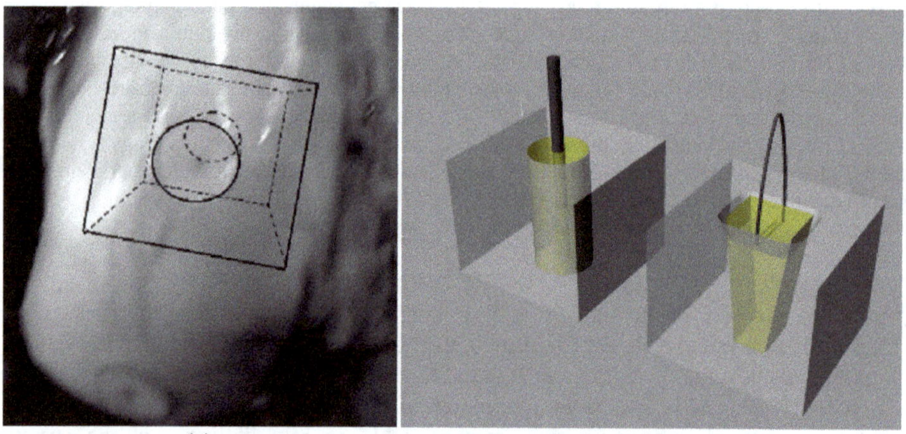

(a) (b)

Figure 1: Specimen preparation for pull-out tests. (a) Bone cube cutting [25]; (b) Types of attachment to the specimen.

2 MATERIALS AND METHODS

The right knees of eighteen adult Roccaverano goats were implanted with two types of ceramic plugs, Al_2O_3/ZrO_2 nanocomposites, produced starting from high purity powders, with different ratios among the components, respectively alumina toughened zirconia (16 wt% Al_2O_3, 84 wt% ZrO_2) surface treated with phosphoric acid in hydrothermal conditions and zirconia toughened alumina (20 wt% ZrO_2, 80% Al_2O_3) coated on the surface with bioglass and a titanium plug (Ti_6Al_4V, sandblasted with corundum on the surface) on the two condyles (Fig. 2(a)). The plugs were rectangular cuboids (8 mm x 4 mm x 8 mm, Fig. 2(b)). Experiments on animals were authorized by the Ethical Committee of Animal Care and Use Center (CISRA, Faculty of Veterinary Medicine, University of Torino, Grugliasco, Italy).

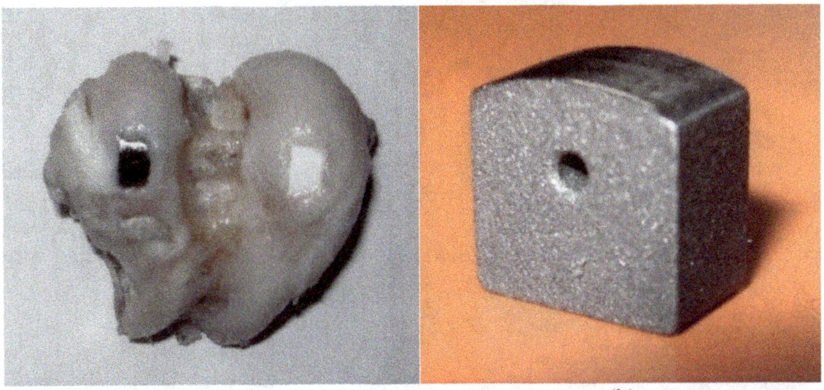

(a) (b)

Figure 2: "Push-in" test. (a): Initial specimen; (b) Example of titanium plug.

Goats were sacrificed respectively 3 months and 1 year following the implantation and the contralateral knee has been implanted with the same plugs, to be used as a reference (time zero). Following excision, bone specimens underwent histological and mechanical tests; the second ones are considered in this work. The authors set up a novel technique for the experimental evaluation of osteointegration, through the measurement of the maximum load withstood by the plug before starting to slide in the housing. This evaluation is revealing the shear stress which the interface can withstand.

The procedure suggested is the following:

- first of all, the femur bones are excised;
- a planar cut is performed, perpendicularly to the axis of the plug;
- a hole is drilled on the distal side: this hole and the plug should be coaxial, the hole should be larger than the plug, and its depth should be such as to contact the bottom of the plug (Fig. 3(a));
- the bone specimen is placed inside a grasp designed so as to receive and block the knees whose shape, might be slightly different from one another (Fig. 3(b));
- a "push-in" test is performed moving the machine crosshead downward (MTS QTest-10 Testing System), at a speed of 2 mm/min (Fig. 3(c));
- a punctual compression test is performed on intact bone to determine the quality of the surrounding trabecular and cortical bone;
- the whole curve of force versus displacement is recorded.

In "push-in" tests, force/displacement curves were analysed in order to establish the maximum load withstood and to evaluate the ultimate interface shear stress as the peak force divided by the lateral surface of the plug.

In bone compression tests (Fig. 4), the initial linear portion of the force/displacement curves [1] was used to assess bone quality since it is related to bone's Young modulus, according to Timoshenko and Goodier's equation [28].

(a) (b) (c)

Figure 3: Suggested procedure. (a) Drilled specimen; (b) Gripping and loading system; (c) 'Push-in' test.

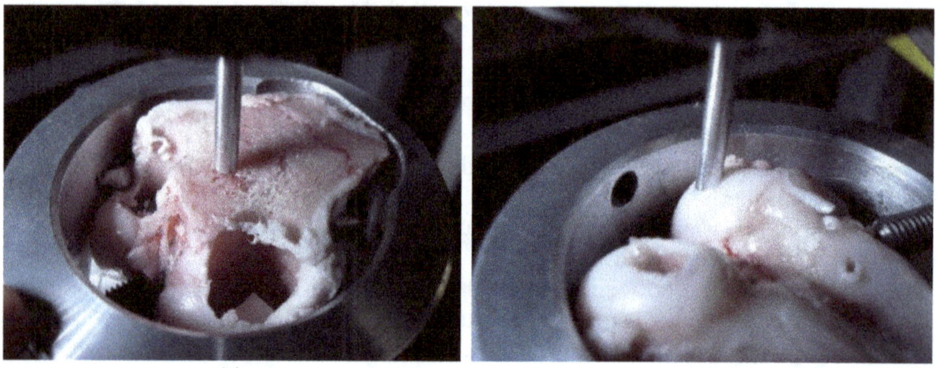

(a) (b)

Figure 4: Compression tests. (a) Trabecular bone; (b) Cortical bone.

3 RESULTS AND DISCUSSION

Bone compression tests results showed a high variance of stiffness both for cortical and trabecular bone (Fig. 5) and no significant differences between bone quality at 0 months, 3 months and 1 year have resulted.

Fig. 6 shows an example of force/displacement curves obtained during "push-in" tests on ceramic plugs at time zero. The test performed on the two types of ceramic plugs were put together because their behaviour was quite the same. In Fig. 7 the mean values of the peak loads recorded in the push-in test are shown, with the respective standard deviation, both for ceramic and titanium plugs, at different times from implant.

The statistical analysis of data (two-way ANOVA) allowed to establish that titanium plugs systematically exhibit a higher resistance ($p<0.10$); they osteointegrate and therefore the shear resistance undergoes a significant increment with time ($p<0.07$). The difference between titanium and ceramic plugs does not vary significantly at different times from implant.

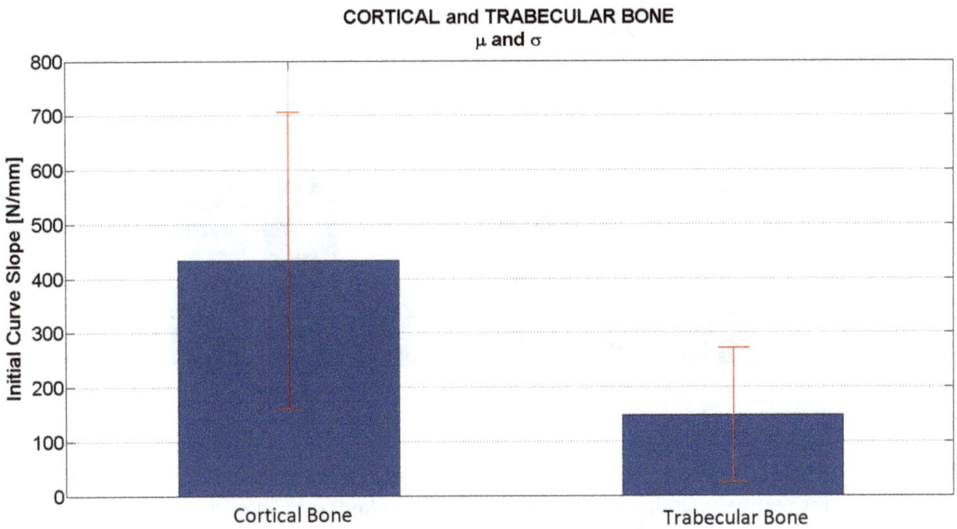

Figure 5: Mean values μ and standard deviation ±σ of cortical bone and trabecular bone stiffness.

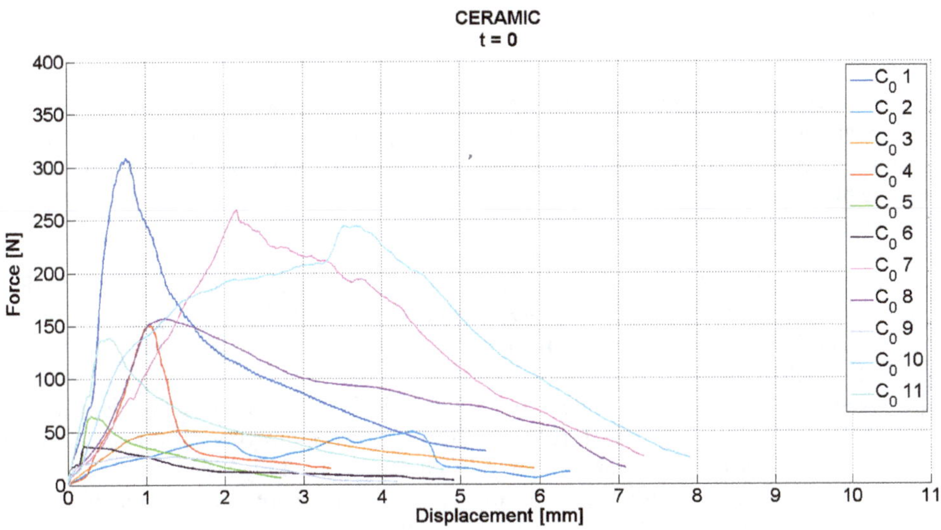

Figure 6: Force/displacement curves obtained during "push-in" tests on ceramic plugs at time zero.

The evaluation of ultimate interface shear stress resulted to range from 0.71±0.52 MPa (time zero, ceramic plug) to 3.05±0.54 MPa (1 year, titanium plug); Suzuki et al. [26] found higher values on goats: 4.43±0.45 MPa, however they used tapered implants; an ultimate shear stress equal to 0.62 MPa can be derived from Whiteside et al. [25], on cylindrically shaped implants, at time zero, on pigs.

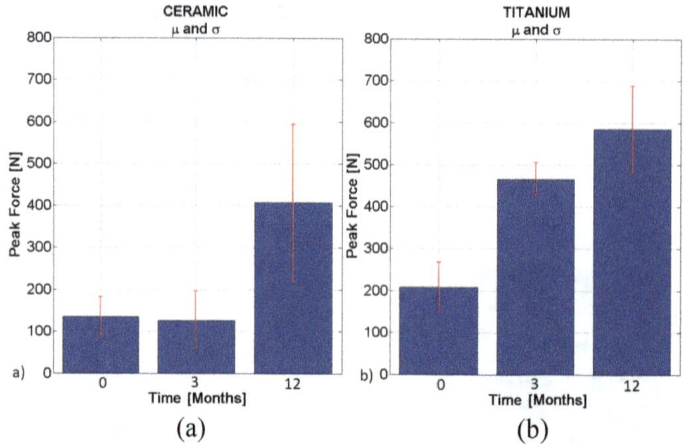

Figure 7: Mean values μ and standard deviation ±σ of peak loads recorded in the push-in, respectively, at the different times from implant. (a) Ceramic; (b) Titanium plug.

From the histological point of view, the titanium plugs proved to be more stable and at 3 months and 1 year than the ceramic ones, in fact they achieved greater peripheral osteointegration; ceramic plugs, on the other hand, were surrounded by a fibrous reaction and only few areas of new bone apposition.

4 CONCLUSIONS

The test has resulted to be sensitive to the plug constitutive material and the time passed since the implantation. Tests allowed to establish that the end milled hole should overpass the maximum plug dimension by at least 2 mm in order to avoid secondary impingement of the bone plug against bone walls.

REFERENCES

[1] Zanetti, E.M. & Bignardi, C., Structural analysis of skeletal body elements: numerical and experimental methods. *Biomechanical systems technology*, ed. by C.T. Leondes, World Scientific Publishing: University of California, Los Angeles, vol. 3, pp. 185–225, 2009.

[2] Vitale, M.C., Chiesa, M., Coltellaro, F., Bignardi, C., Celozzi, M. & Poggio, C., FEM analysis of different dental root canal-post systems in young permanent teeth. *European Journal of Paediatric Dentistry*, **9**(3), pp. 111–117, 2008.

[3] Zanetti, E.M. & Bignardi, C., Mock-up in hip arthroplasty pre-operative planning. *Acta of Bioengineering and Biomechanics*, **15**(3), pp 123–128, 2013.

[4] Zanetti, E.M. et al., A structural numerical model for the optimization of double pelvic osteotomy in the early. *Veterinary and Comparative Orthopaedics Traumatology*, **3**(4), pp. 256–264, 2017.

[5] Calì, M. et al., Influence of thread shape and inclination on the biomechanical behaviour of plateau implant systems. *Dental Materials*, **34**(3), pp. 460–469, 2018.

[6] Zanetti, E.M., Bignardi, C., Terzini, M., Putame, G. & Audenino, A.L., A multibody model for the optimization of hip arthroplasty in relation to range of movement. *Academy of Management Journal*, **11**(10), pp. 486–491, 2018.

[7] Putzer, D., Nogler, M., Terzini, M., Mannara, R. & Bignardi, C., A finite element analysis for a new short stem concept design with spherical bone interface for hip

resurfacing. *International Journal of Mechanical Engineering and Technology*, **9**(3), pp. 923–935, 2018.

[8] Aldieri, A. et al., Osteoporotic hip fracture prediction: is T-score based criterion enough? A Hip Structural Analysis based model. *Journal of Biomechanical Engineering*, **140**(11), 2018.

[9] Terzini, M. et al., Multibody modelling of ligamentous and bony stabilizers in the human elbow. *Muscles, Ligaments and Tendons Journal*, 7(4), pp. 493–502, 2018.

[10] Bresciano, M., Schierano, G., Manzella, C., Screti, A., Bignardi, C. & Preti, G., Retention of luting agents on implant abutments of different height and taper. *Clinical Oral Implant Research*, **16**, pp. 594–598, 2005.

[11] Menicucci, G., Ceruti, P., Barabino, E., Screti, A., Bignardi, C. & Preti G., A preliminary in vivo trial of load transfer in mandibular implant-retained overdentures anchored in 2 different ways: allowing and counteracting free rotation. *International Journal of Prosthodontics*, **19**(6), pp. 574–576, 2006.

[12] Zanetti, E.M. & Audenino, A.L., Differential thermography for experimental, full-field stress analysis of hip arthroplasty. *Journal of Mechanics in Medicine and Biology*, **10**(3), pp. 515–529, 2010.

[13] Zanetti, E.M., Bignardi, C. & Audenino, A.L., Human pelvis loading rig for static and dynamic stress analysis. *Acta of Bioengineering and Biomechanics*, **14**(2), pp. 61–66, 2012.

[14] Boero Baroncelli, A., Reif, U., Bignardi, C. & Peirone, B., Effect of screw insertion torque on push-out and cantilever bending properties of five different angle-stable systems. *Veterinary Surgery*, **42**(3), pp. 308–315, 2013.

[15] Manzella, C., Burello, V., Bignardi, C., Carossa, S. & Schierano, G., A method to improve passive fit of frameworks on implant-supported prostheses: An in vivo study. *International Journal of Prosthodontics*, **26**(6), pp 577–579, 2013.

[16] Manzella, C., Bignardi, C., Burello, V., Carossa, S. & Schierano, G., Method to improve passive fit of frameworks on implant-supported prostheses: An in vitro study. *Journal of Prosthetic Dentistry*, **116**(1), pp. 52 58, 2016.

[17] Zanetti, E.M., Pascoletti, G., Calì, M., Bignardi, C. & Franceschini, G., Clinical Assessment of dental implant stability during follow-up: What is actually measured, and perspectives. *Biosensors* (Basel), **8**(3), p. E68, 2018.

[18] Bignardi, C., Zanetti, E.M., Terzini, M., Ciccola, A.R., Schierano, G. & Audenino, A.L., Reliability, learnability and efficiency of two tools for cement crowns retrieval in dentistry. *The Open Biomedical Engineering Journal*, **12**(1), pp. 27–35, 2018.

[19] Peluccio, M.S., Bignardi, C., Lombardo, S., Montevecchi, F.M. & Carossa, S., Comparative study of nanomechanical properties of cements used in teeth restoration. *Journal of Physics: Condensed Matter*, **19**, p. 395003, 2007.

[20] Bignardi, C., Petraroli, M. & Pugno, N.M., Nanoindentation on conch shells of gastropoda and bivalvia molluscs reveal anisotropic evolution against external attacks. *Journal of Nanoscience and Nanotechnology*, **10**(10), pp. 6453–6460, 2010.

[21] Zanetti, E.M., Perrini, M., Bignardi, C. & Audenino, A.L., Bladder tissue passive response to monotonic and cyclic loading. *Biorheology*, **49**(1), pp. 49–63, 2012.

[22] Terzini, M., Bignardi, C., Castagnoli, C., Cambieri, I., Zanetti, E.M. & Audenino, A.L., Ex vivo dermis mechanical behavior in relation to decellularization treatment length. *The Open Biomedical Engineering Journal*, **10**, pp. 34–42, 2016.

[23] Terzini, M., Bignardi, C., Castagnoli, C., Cambieri, I., Zanetti, E.M. & Audenino, A.L., Dermis mechanical behaviour after different cell removal treatments. *Medical Engineering & Physics*, **38**(9), pp. 862–869, 2016.

[24] Aldieri, A., Terzini, M., Bignardi, C., Zanetti E.M. & Audenino A.L., Implementation and validation of constitutive relations for human dermis mechanical response. *Medical & Biological Engineering & Computing*, **56**(11), pp. 2083–2093, 2018.

[25] Whiteside, R.A., Bryant, J.T., Jakob, R.P., Mainil-Varlet, P. & Wyss, U.P., Short-term load bearing capacity of osteochondral autografts implanted by the mosaicplasty technique: An in vitro porcine model. *Journal of Biomechanics*, **36**(8), pp. 1203–1208, 2003.

[26] Suzuki, T., Fujibayashi, S., Nakagawa, Y., Noda, I. & Nakamura, T., Ability of zirconia double coated with titanium and hydroxyapatite to bond to bone under load-bearing conditions. *Biomaterials*, **27**(7), pp. 996–1002, 2006.

[27] Aebli, N. et al., In vivo comparison of the osseointegration of vacuum plasma sprayed titanium- and hydroxyapatite-coated implants. *Journal of Biomedical Materials Research A*, **66**(2), pp. 356–363, 2003.

[28] Timoshenko, S.P. & Goodier, J.N., *Theory of Elasticity*, New York: McGraw-Hill, 1970.

COLLAGEN CROSS-LINKER EFFECT ON THE MECHANICAL PROPERTIES OF THE RADICULAR HYBRID LAYER IN RESTORATIVE DENTISTRY: A NANOINDENTATION STUDY

GIANPAOLO SERINO[1], CRISTINA BIGNARDI[1], CLELIA BOCCAFOSCHI[1],
NICOLA SCOTTI[2], ELIO BERUTTI[2] & ALBERTO AUDENINO[1]
[1]Department of Mechanical and Aerospace Engineering, Politecnico di Torino, Italy
[2]Department of Surgical Sciences, University of Turin, Italy

ABSTRACT

Bond strength between the dentin and the restorative resins is deeply dependent on the nature of their interface. Resins impregnate dentin, which is rich in collagen fibers. This in vitro study aimed to evaluate the effect of collagen carbodiimide cross linker agent (EDC) on the mechanical properties of the adhesive interface in endodontically treated teeth. Twenty upper premolar teeth were selected and divided into two groups according to the dentin pretreatment procedures: no EDC application (group A) and EDC application (group B). Three typical zones, i.e. the dentin, the radicular hybrid layer and the resin, were analyzed using a Nanoindenter XP equipped with a diamond Berkovich indenter. The input curve was characterized by loading and unloading phases with a strain rate value of 0.1 s^{-1} and, an intermediate dumbbell phase of 30 s. The maximum indentation depth was set to be 200 nm. The load-displacement curves were analyzed by using the "Oliver and Pharr" method. The mean values of nanoindentation modulus were determined for the dentin, the radicular hybrid layer and the resin for both the samples with and without crosslinker. In general, the application of EDC was found to modify the mechanical properties of the radicular hybrid layer. The mechanical properties of the radicular hybrid layer could be related to the efficient infiltration of the adhesive systems and collagen crosslinker through dentin.

Keywords: quasi-static nanoindentation, hybrid layer, collagen crosslinker, mechanical properties.

1 INTRODUCTION

The discipline of dental science includes the diagnosis of diseases in the mouth and teeth, and the procedures involved in restoration of their integrity and function. Restoration of lost tooth structure with suitable materials play an integral part in the successful rehabilitation of oral tissue.

There are many examples of clinical problems that went through from a qualitative assessment to a quantitative evaluation thanks to their modelling [1], [2], furthermore, the application of classical experimental methods of structural analysis to the evaluation of the efficacy of procedures or surgical techniques [3]–[7] or to the evaluation of the mechanical characteristics of the materials used at different scales of investigation [8]–[10] pushed the dental science forward.

The study of Buonocore [11] was the first to report that etching the enamel, to expose the underling organic matrix of the tooth, improved the bond strength between the resin and the enamel. The ability of the resin to penetrate the organic collagen network and envelope apatite crystallites played a crucial role [12]. A transition zone defined hybrid layer, by Nakabayashi et al. [13], and better observed through scanning electron microscopy [14] confirms what was already found by Gwinnett and Matsui [12]. Exposed collagen matrix acts like a scaffold [15] where the resin can infiltrate and establish a strong and durable bonding with enamel or dentin.

WIT Transactions on Engineering Sciences, Vol 124, © 2019 WIT Press
www.witpress.com, ISSN 1743-3533 (on-line)
doi:10.2495/MC190191

As reported in literature [16], the adhesion process to dentin or enamel occurs in two different phases. In a first step the inorganic component is etched generating the structure that in a successive step, defined hybridization phase, will be perfused by the resin. Depending on the adhesion strategy, which Van Meerbeek classified in "etch and rinse" or "self-etch" [16] the adhesion process can be performed in one, two or three phases correlated to the number of steps required for the application of the resin. The adhesion strategy adopted in this study was the "etch and rinse", in this case the bonding lies mostly on the diffusion of the resin into the collagen scaffold. The weak affinity between the functional groups of the monomers and the exposed collagen make their chemical bonding very unlikely [16].

Although the findings of Buonocore pushed the use of dental resins in a new scenario, some issues still are not completely overcome. Dealing with these adhesion procedures imply a pre-treatment of dentin to eliminate the inorganic phase by using phosphoric acid solution. Acidic treatment of the dentin onset the activation of the metalloproteinases (MMPs) through an acid activation mechanism [17]. This procedure combined with the ionic and hydrophilic nature of dental adhesives currently adopted, leads to the hydrolysis of collagen fibers [18]–[20]. The MMPs activation and the water adsorption lead to a decline over time of the hybrid layer and to a consequent failure of the bonding between the dentin and the resin [21].

Integrity of the collagen network is essential to preserve the stability and ensure the durability of the hybrid layer over time. The use of crosslinkers is able to establish covalent bonding with the exposed site of the enzymes in order to inactivate the MMPs is one of the strategies currently applied [22]–[24]. The 1-ethyl-3(3-dimethylaminopropyl) carbodiimide (EDC) is able to reduce the hydrolysis of collagen fibers [21] and stabilize the collagen matrix [25], [26]. Moreover, treating the exposed collagen network with EDC for less than 1 minute increases its stiffness [25], [27], and preserves the dentin bond strength when aged over than one year in artificial saliva [28].

Bond strength between dentin and the resin composite is deeply dependent by the nature of their interface [29], [30]. Nanoscopic Dynamic Mechanical Analysis (nanoDMA) was applied to investigate the influence of the EDC [31] and, the effect of the biomimetic remineralization [32] on the hybrid layer, furthermore, combining this technique with fatigue test allowed to analyze the hybrid layer when undergone to cycling stress [33].

Effectiveness of the application of crosslinkers, such as EDC and chlorhexidine, on the durability of dentin-resin bond, was also evaluated through microtensile test [33], [34].
All the studies reported the good efficacy of the EDC crosslinker to preserve the bond strength over time.

Thus far, quasi-static nanoindentation test were performed to evaluate the nanomechanical properties of the enamel and the dentin [35]–[37] or to evaluate the performance of cements in dental restoration [10], and in the 1993 by Van Meerbeek's group to study the effect of 4 different adhesive systems and conditioning agent on the resin-dentin bonding area [29].

Even if the nanoDMA method permits to obtain a useful mapping of the mechanical properties of the hybrid layer with a very fine resolution, in order to obtain information from the single phases that compose it, in this in vitro study quasi-static nanoindentation method was adopted to evaluate the effect of EDC on the mechanical properties of the adhesive. When opportunely set, through quasi-static nanoindentation method is possible to identify the mechanical properties of the hybrid layer as a result of the contribution of all the phases that compose it.

2 MATERIALS AND METHODS

2.1 Restoration procedure

Twenty upper premolars teeth, extracted for periodontal reasons, were selected. The inclusion criteria were: sound teeth extracted within one month, with nearly similar root size (length > 12 mm) and no cracks under trans-illumination and magnification.

After 24 hours in water storage at 37°C, a standardized post space, 8 mm depth, was prepared in each sample with a dedicated bur, etched for 15 s (Ultraetch, Ultradent Products Inc.) with phosphoric acid, rinsed for 30 s and dried with air and paper point. Then, samples were divided in 2 groups (10 samples for each group) according to the dentin pretreatment procedure: no EDC application (group A) and EDC application (group B). In all samples, a universal adhesive system (Scotchbond Universal, 3M Science Applied to Life™) was employed in multiple application. A red stain was added to the adhesive system in order to better recognize the hybrid layer. Then, fiber posts (Rebilda Post ∅ 1.2 mm, Voco GmbH) were luted with a dual-curing luting cement (Rely-x Universal, 3M Science Applied to Life™) and cured for 180 s (Celalux LED lamp, Voco GmbH). The coronal cavity was restored with a nanohybrid composite (Filtek Supreme XTE, 3M Science Applied to Life™).

2.2 Sample preparation

The samples were embedded in resin and then sectioned with diamond saw in 1 mm thick slices perpendicularly to the longitudinal axis, from the occlusal surface until the post end. Each slice was flattened with ascending sandpapers till 2400 grit and, opportunely polished in order to reach a reasonable value of the mean surface roughness [38].

From the center of the slice three typical regions can be identified. The cement in the center and, in the outermost circular crown the dentin. The hybrid layer is the transition zone where the adhesive system infiltrates the etched dentin, with and without pretreatment.

2.3 Nanoindentation procedure

Tests were carried out with the Nanoindenter XP (MTS System Corporation), equipped with a diamond Berkovich indenter and characterized by a theoretical force resolution of 50 nN and a theoretical displacement resolution lower than 0.01 nm. The shape function of the indenter and the frame stiffness of the instrument were fine calibrated following the standard procedure [39]. Input curve is characterized by loading and unloading phases imposing a strain rate value of 0.1 s^{-1}. Before the onset of the unloading step the maximum load value was held for 30 s.

Depending on the etching agent used during the pretreatment procedure the width of the hybrid layer varies from 3 μm to 6 μm [30], [32]. Tests were performed in displacement control imposing a maximum value of indentation depth of 200 nm to assure that only the mechanical properties of the hybrid layer were analyzed. The cross-section value obtained from the calibration procedure at the set indentation depth is compatible with the hybrid layer width [40]. Imposing such limit of the indentation depth allows to analyze only the mechanical properties relative to the region of interest given the compatible dimensions of the area under the indenter and the indented portion of the tooth slice.

The loading-displacement curves were then analyzed by using the Oliver and Pharr method [39] in order to obtain the nanoindentation modulus.

2.4 Statistical analysis

For each slice 20 indentation points have been selected for the resin, the hybrid layer and, the dentin. A total of 800 indentations for group A and for group B for every region respectively were performed. The outliers were identified and discarded through the modified Thompson Tau Test before applying the one-way analysis of variance (ANOVA) test ($p<0.05$). In this study we recognize two levels: no application of EDC (group A) and application of EDC (group B). Nanoindentation modulus is the independent variable.

3 RESULTS AND DISCUSSION

Representative nanoindentation curves obtained for the dual-curing cement, the dentin and, the hybrid layer is shown in Fig. 1.

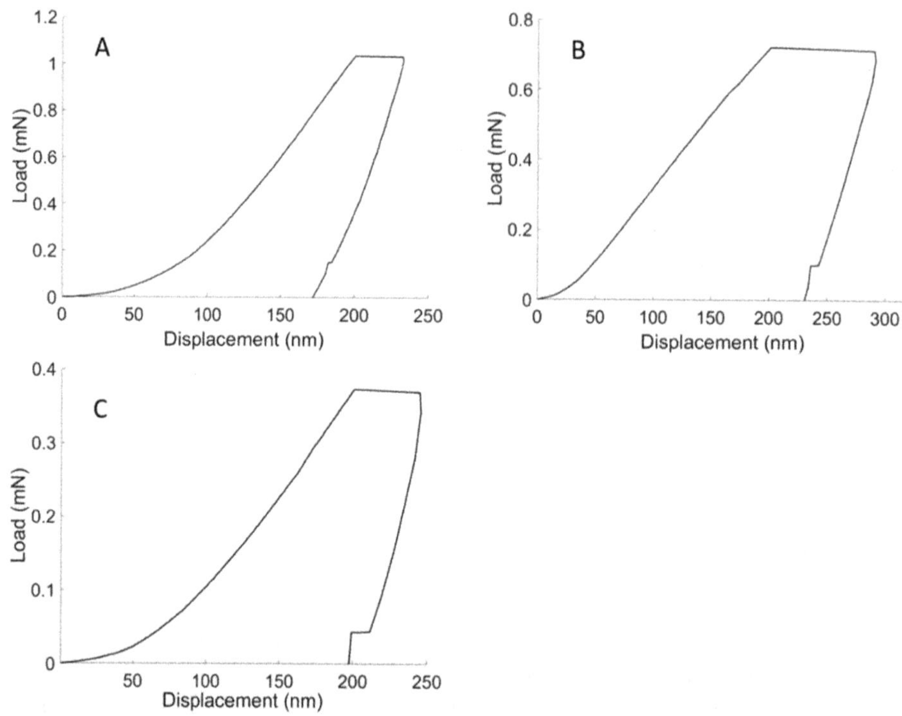

Figure 1: Representative nanoindentation curves obtained for: (a) Dentin; (b) Hybrid layer; (c) Resin.

The amount of residual deformation seems to indicate that plasticization phenomena were occurred. However, observing the indentation site at the end of the test no residual imprints were detected. In other words, the misleading plastic deformation is due to the viscous properties of the materials under investigation. When the load is totally removed the deformation recovery continues, and the imprint becomes undetectable, such phenomenon was already observed elsewhere [41]. No plasticization limit was reached to produce permanent deformation due to the magnitude of the applied load. The holding time defined before the onset of the unloading phase stabilize the time-dependent phenomena [42], [43], allowing to apply the Oliver and Pharr method and calculate the nanoindentation modulus and the hardness of the cement. The displacement registered during the dwell time reported in Fig. 1(A) indicates that the collagen fibers, which compose the dentin, heavily contribute to defining the behavior of the tissue. Such response, better known as creep, characterizes the class of polymeric materials. The displacement relative to the hybrid layer registered during the dwell period is the higher when compared with the other two indented materials, indicating a higher creep rate as summation of the effect due to the cement and to the collagen fibers. Quantitative analysis of the nanoindentation curves is reposted in Fig. 2.

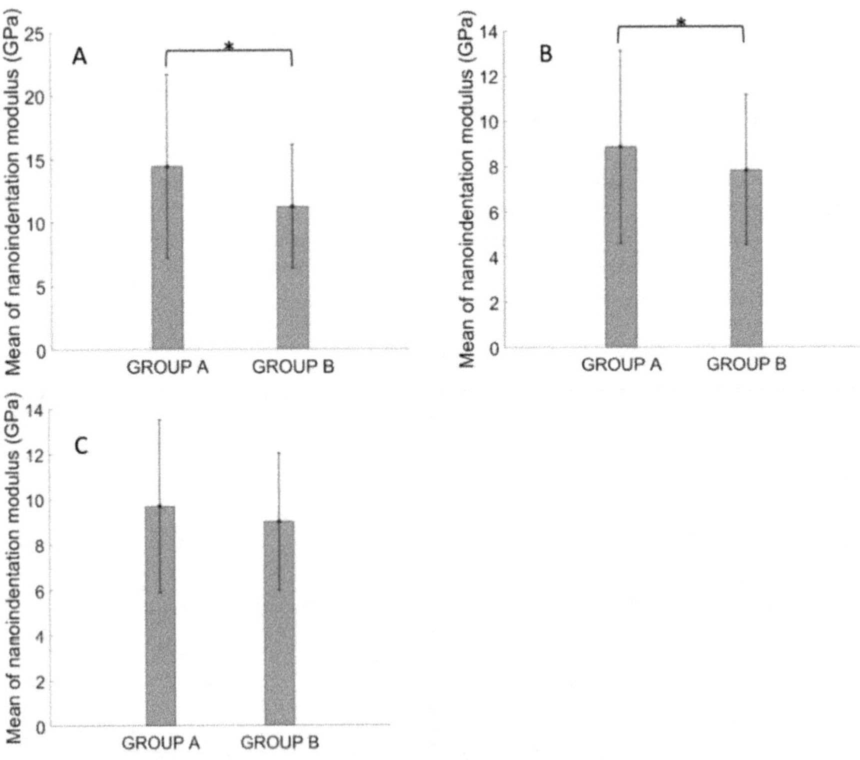

Figure 2: Nanoindentation Modulus obtained for: (a) Dentin; (b) Hybrid layer; (c) Resin. Values of nanoindentation modulus of the dentin and hybrid layer are significant different for group A and group B ($p<0.05$).

The higher values of nanoindentation modulus were registered for the dentin due to the presence of the inorganic material. Nanoindentation modulus mean values of dentin were respectively 11.255 (± 4.871) GPa and 14.430 (± 7.249) GPa for the samples with EDC and without.

As can be expected, the values obtained for the cement and the hybrid layer are very close to each other since the hybrid layer is generated by the impregnation of demineralized dentin by the cement.

Nanoindentation modulus mean values of cement, obtained for the samples with and without EDC, were respectively 8.998 (±3.804) GPa and 9.690 (±3.033) GPa. Nanoindentation modulus mean values of the radicular hybrid layer were respectively 7.828 (± 3.331) GPa and 8.855 (± 4.281) GPa.

The mechanical properties of the dentin evaluated for the sample obtained using the EDC are significantly lower than those obtained without EDC ($p<0.05$). The statistical difference, however, cannot be ascribed to the EDC application considering that, the indentation sites for the dentin were chosen far away from the region where the EDC was applied. As matter of fact the statistical difference is due to the microstructure features of the dentin composition, in other words the inter-variability between the two samples is due to the collagen fiber network that encloses the inorganic phase and not to the EDC application [44], [45].

No significant differences of the resin mechanical properties were found ($p>0.05$) between group A and group B, as can be expected. The statistical analysis revealed, furthermore, that the EDC application slightly modify the mechanical properties of the radicular hybrid layer ($p<0.05$). Counterintuitive reduction of the nanoindentation modulus values of the hybrid layer obtained for the spacemen of group B can be explained observing that the mechanical properties of the dentin sample of this group are, on average, minor than those registered for the control sample (group A). In other words, the slight, even if significant difference between the mechanical properties obtained for the hybrid layer indicates that the EDC crosslinker increases the links between the collagen fiber, [25]–[27], [46], [47] matching the mechanical properties of the hybrid layer of the two groups.

4 CONCLUSION

The settings here adopted allow one to homogenize the contribution of the collagen fibers and the cement trying to understand if the application of EDC for a period of 30 s modifies the mechanical response of the hybrid layer as whole. The slight reduction of the nanoindentation modulus of the hybrid layer observed for the EDC treated sample must be ascribed to the microstructure of the dentin, that is characterized by a lower mean value of nanoindentation modulus compared to that of the specimens of group B. Indeed, it can be speculated that the cross-linker application of 30 s increases the links between the collagen fibers strengthening the demineralized dentin and the resin-dentin bonding.

REFERENCES

[1] Vitale, M.C., Chiesa, M., Coltellaro, F., Bignardi, C., Celozzi, M. & Poggio, C., FEM analysis of different dental root canal-post systems in young permanent teeth. *European Journal of Paediatric Dentistry*, **9**(3), pp. 111–117, 2008.

[2] Calì, M. et al., Influence of thread shape and inclination on the biomechanical behaviour of plateau implant systems. *Dental Materials*, **34**(3), pp. 460–469, 2018.

[3] Menicucci, G., Ceruti, P., Barabino, E., Screti, A., Bignardi, C. & Preti, G., A preliminary in vivo trial of load transfer in mandibular implant-retained overdentures anchored in 2 different ways: allowing and counteracting free rotation. *The International Journal of Prosthodontics*, **19**(6), pp. 574–576, 2006.

[4] Boero Baroncelli, A., Reif, U., Bignardi, C. & Peirone, B., Effect of screw insertion torque on push-out and cantilever bending properties of five different angle-stable systems. *Veterinary Surgery*, **42**(3), pp. 308–315, 2013.

[5] Manzella, C., Burello, V., Bignardi, C., Carossa, S. & Schierano, G., A Method to improve passive fit of frameworks on implant-supported prostheses: An in vivo study. *The International Journal of Prosthodontics*, **26**(6), pp. 577–579, 2014.

[6] Bignardi, C., Zanetti, E.M., Terzini, M., Ciccola, A.R., Schierano, G. & Audenino, A.L., Reliability, learnability and efficiency of two tools for cement crowns retrieval in dentistry. *The Open Biomedical Engineering Journal*, **12**(1), pp. 27–35, 2018.

[7] Zanetti, E.M., Pascoletti, G., Calì, M., Bignardi, C. & Franceschini, G., Clinical assessment of dental implant stability during follow-up: What is actually measured, and perspectives. *Biosensors*, **8**(3), p. 68, 2018.

[8] Bresciano, M., Schierano, G., Manzella, C., Preti, G., Screti, A. & Bignardi, C., Retention of luting agents on implant abutments of different height and taper. *Clinical Oral Implants Research*, **16**, pp. 594–598, 2005.

[9] Bignardi, C., Petraroli, M. & Pugno, N., Nanoindentations on conch shells of gastropoda and bivalvia molluscs reveal anisotropic evolution against external attacks. *Journal of Nanoscience and Nanotechnology*, **10**(10), pp. 6453–6460, 2010.

[10] Peluccio, M.S., Bignardi, C., Lombardo, S., Montevecchi, F.M. & Carossa, S., Comparative study of nanomechanical properties of cements used in teeth restoration. *Journal of Physics: Condensed Matter*, **19**(39), p. 395003, 2007.

[11] Buonocore, M.G., A simple method of increasing the adhesion of acrylic filling materials to enamel surfaces. *Journal of Dental Research*, **34**, pp. 849–853, 1955.

[12] Gwinnett, A.J. & Matsui, A. A study of enamel adhesives. The physical relationship between enamel and adhesive. *Archives of Oral Biology*, **12**(12), pp. 1615–1620, 1967.

[13] Nakabayashi, N., Kojima, K. & Matsuhara, E., Promotion of adhesion by infiltration monomers into tooth substrates. *Journal of Biomedical Materials Research*, **16**, pp. 265–73, 1982.

[14] Inokoshi, S., Hosoda, H., Harnirattisai, C. & Tatsumi, T., A study on the resin-impregnated layer of dentin Part I. A comparative study on the decalcified and undecalcified sections and the application of argon ion beam etching to disclose the resin-impregnated layer of dentin. *Japanese Journal of Conservative Dentistry*, **33**, pp. 427–442, 1990.

[15] Jain, A. & Bahuguna, R., Role of matrix metalloproteinases in dental caries, pulp and periapical inflammation: An overview. *Journal of Oral Biology and Craniofacial Research*, **5**(3), pp. 212–218, 2015.

[16] Van Meerbeek, B. et al., Buonocore memorial lecture. Adhesion to enamel and dentin: current status future challanges. *Operative Dentistry*, **23**, pp. 215–235, 1998.

[17] Bhoopathi, P.G., Kittappa, K.K., Sanjeev, K. & Sekar, M., Effect of chlorhexidine and cystatin incorporated adhesives on MMPs and cysteine cathepsin-an in-vitro zymographic analysis. *Journal of Clinical and Diagnostic Research*, **12**(7), pp. ZC01-ZC05, 2018.

[18] Hashimoto, M., Ohno, H., Sano, H., Kaga, M. & Oguchi, H., Degradation patterns of different adhesives and bonding procedures. *Journal of Biomedical Materials Research - Part B Applied Biomaterials*, **66**(1), pp. 324–330, 2003.

[19] Kern, M., Pashley, D., Adelung, R., Ludwig, K., Bosmann, K. & Yang, B., Effect of structural change of collagen fibrils on the durability of dentin bonding. *Biomaterials*, **26**(24), pp. 5021–5031, 2005.

[20] Hashimoto, M., Oguchi, H., Sano, H., Endo, K., Ohno, H. & Kaga, M., In vivo degradation of resin-dentin bonds in humans over 1 to 3 years. *Journal of Dental Research*, **79**(6), pp. 1385–1391, 2000.

[21] Mazzoni, A. et al., Carbodiimide inactivation of MMPs and effect on dentin bonding. *Journal of Dental Research*, **93**(3), pp. 263–268, 2014.

[22] Montagner, A.F., Sarkis-Onofre, R., Pereira-Cenci, T. & Cenci, M.S., MMP inhibitors on dentin stability. *Journal of Dental Research*, **93**(8), pp. 733–743, 2014.

[23] Tjäderhane, L., Dentin bonding: can we make it last? *Operative Dentistry*, **40**(1), pp. 4–18, 2015.

[24] Tjäderhane, L. et al., Optimizing dentin bond durability: Control of collagen degradation by matrix metalloproteinases and cysteine cathepsins. *Dental Materials*, **29**(1), pp. 116–135, 2013.
[25] Bedran-Russo, A.K.B., Pashley, D.H., Agee, K., Drummond, J.L. & Miescke, K.J., Changes in stiffness of demineralized dentin following application of collagen crosslinkers. *Journal of Biomedical Materials Research Part B: Applied Biomaterials*, **86B**(2), pp. 330–334, 2007.
[26] Zhang, Z. et al., Effect of carbodiimide on the fatigue crack growth resistance of resin–dentin bonds. *Dental Materials*, **32**(2), pp. 211–222, 2016.
[27] Castellan, C.S., Pereira, P.N.R., Viana, G., Chen, S.N., Pauli, G.F. & Bedran-Russo, A.K., Solubility study of phytochemical cross-linking agents on dentin stiffness. *Journal of Dentistry*, **38**(5), pp. 431–436, 2010.
[28] Mazzoni, A. et al., Effect of carbodiimide (EDC) on the bond stability of etch-and-rinse adhesive systems. *Dental Materials*, **29**(10), pp. 1040–1047, 2013.
[29] Roos, J.R. et al., Assessment by nano-indentation of the hardness and elasticity of the resin-dentin bonding area. *Journal of Dental Research*, **72**(10), pp. 1434–1442, 1993.
[30] Celis, J.P. et al., Chemical characterization of the resin-dentin interface by micro-raman spectroscopy. *Journal of Dental Research*, **72**(10), pp. 1423–1428, 1993.
[31] Tezvergil-Mutluay, A., Zhang, Z., Arola, D., Pashley, D.H., Mutluay, M. & Tay, F.R., Effects of EDC crosslinking on the stiffness of dentin hybrid layers evaluated by nanoDMA over time. *Dental Materials*, **33**(8), pp. 904–914, 2017.
[32] Ryou, H. et al., Effect of biomimetic remineralization on the dynamic nanomechanical properties of dentin hybrid layers. *Journal of Dental Research*, **90**(9), pp. 1122–1128, 2011.
[33] Mutluay, M.M., Yahyazadehfar, M., Ryou, H., Majd, H., Do, D. & Arola, D., Fatigue of the resin-dentin interface: A new approach for evaluating the durability of dentin bonds. *Dental Materials*, **29**(4), pp. 437–449, 2013.
[34] Ricci, H.A., Sanabe, M.E., De Souza Costa, C.A., Hebling, J. & Pashley, D.H., Chlorhexidine increases the longevity of in vivo resin-dentin bonds. *European Journal of Oral Sciences*, **118**(4), pp. 411–416, 2010.
[35] Habelitz, S., Marshall, S.J., Marshall, G.W. Jr & Balooch, M., Mechanical properties of human dental enamel on the nanometre scale. *Archives of Oral Biology*, **46**, pp. 173–183, 2001.
[36] Brauer, D.S., Hilton, J.F., Marshall, G.W. & Marshall, S.J., Nano- and micromechanical properties of dentine. *Journal of Biomechanics*, **44**(8), pp. 1626–1629, 2012.
[37] Kinney, J.H., Balooch, M., Marshall, S.J., Marshall, G.W. & Weihs, T.P., Hardness and young's modulus of human peritubular and intertubular dentine. *Archives of Oral Biology*, **41**(1), pp. 9–13, 1996.
[38] Miller, M., Bobko, C., Vandamme, M. & Ulm, F.J., Surface roughness criteria for cement paste nanoindentation. *Cement and Concrete Research*, **38**(4), pp. 467–476, 2008.
[39] Oliver, W.C. & Pharr, G.M., An improved technique for determining hardness and elastic modulus using load and displacement sensing indentation experiments. *Journal of Materials Research*, **7**(6), pp. 1564–1583, 1992.
[40] VanLandingham, M.R., Review of instrumented indentation. *Journal of Research of the National Institute of Standards and Technology*, **108**(4), p. 249, 2012.
[41] Del Gaudio, C. et al., Natural polymeric microspheres for modulated drug delivery. *Materials Science and Engineering: C*, **75**, pp. 408–417, 2017.

[42] Hochstetter, G., Jimenez, A. & Loubet, J.L., Strain-rate effects on hardness of glassy olymers in the nanoscale range. Comparison between quasi-static and continuous stiffness measurements. *Journal of Macromolecular Science: Part B*, **38**(5–6), pp. 681–692, 1999.

[43] Mazeran, P.E., Beyaoui, M., Bigerelle, M. & Guigon, M., Determination of mechanical properties by nanoindentation in the case of viscous materials. *International Journal of Materials Research*, **103**(6), pp. 715–722, 2012.

[44] Romberg, E. et al., Indentation damage and mechanical properties of human enamel and dentin. *Journal of Dental Research*, **77**(3), pp. 472–480, 1998.

[45] Kinney, J.H., Balooch, M., Marshall, G.W. & Marshall, S.J., A micromechanics model of the elastic properties of human dentine. *Archives of Oral Biology*, **44**(10), pp. 813–822, 1999.

[46] Yoshikawa, T. et al., Long-term durability of dentin bonds made with a self-etching primer, in vivo. *Journal of Dental Research*, **78**(4), pp. 906–911, 1999.

[47] Comba, A. et al., Carbodiimide effect on MMPs and hybrid-layer micro-hardness in radicular dentine. *Abstracts of the Academy of Dental Materials Annual Meeting*, pp. e29–30, 2018.

SECTION 8
PERFORMANCE BASED DESIGN MATERIALS

NUMERICAL MODELLING AND SIMULATION OF VEHICULAR IMPACTS ON TIRE-MOUNTED CABLE BARRIERS

HOWIE FANG* & AMEY BADAR
The University of North Carolina at Charlotte, USA

ABSTRACT

This research focused on evaluating retrofit designs for cable barriers using reclaimed tires so as to reduce the chance of vehicle under-riding for small passenger cars. Finite element (FE) analysis was used to conduct simulations of full-scale vehicular crashes into these retrofit cable barriers and the barriers' performance was evaluated. Four designs of the tire-mounted cable barriers were modelled and evaluated under impacts by a 1996 Dodge Neon, which was in compliance with the 1100C test vehicle specified by the Manual for Assessing Safety Hardware (MASH). Ten different material models were used in the FE models, including the piecewise linear plasticity model for steel components, elastic model for the tires and a few other components, viscous damping model for the shock absorbers, low-density foam model for the radiator core, spot-weld model for sheet metal connections, Blatz-Ko rubber model for rubber cushions, rigid model for most mounting hardware, and null material model defined for contact purposes. The simulations of vehicular crashes were conducted on both a flat terrain and a sloped median. The impact conditions were based on the MASH Test Level 3 (TL-3) requirements, i.e., at an impact speed of 100 km/h (62.1 mph) and a 25° impact angle. The simulation results showed that the tire-mounted cable barrier could improve the safety performance of the current cable barrier system (i.e., without tires) under impacts of small passenger cars.

Keywords: finite element, modelling, simulation, impact, contact, cable barrier, tire mounted.

1 INTRODUCTION

Roadside barriers are used on highways to prevent errant vehicles from entering dangerous areas such as steep slopes, rocks, or oncoming traffic. Roadside barriers can be categorized based on their overall rigidity as flexible barriers (such as cable barriers), semi-rigid barriers (such as W-beam and Thrie-beam guardrails), and rigid barriers (such as concrete barriers). Cable barriers are typically installed on highway medians and generally cause less severe damage and injuries to the striking vehicles and occupants than W-beam guardrails and concrete barriers. However, cable barriers require the median to have sufficient width to accommodate cable deflections in the transverse direction.

In 1967, Graham et al. developed four mathematical models for predicting the trajectory of a vehicle in collision with cable median barriers (CMBs) and W-beam guardrails [1]. The research included vehicle speeds up to 97 km/h (60 mph) and impact angles up to 35 degrees to represent the worst-case scenario of highway accidents. Based on suggestions from this research, New York State adopted a standard barrier height of 0.762 m (30 inches) with a 0.076 m (3 inch) cable spacing and a 4.88 m (16 foot) post spacing. In 1990, Phillips et al. studied the performance of terminal ends of cable barriers in crash events [2]. Twelve full-scale tests were conducted in accordance with the NCHRP report 230 to evaluate three distinct terminal designs. It was concluded that the terminal used in the standard CMB could snag and overturn the vehicle in an impact and a new terminal design demonstrated acceptable characteristics was developed. In 1993, Laker et al. studied a new cable barrier design to overcome the deficiencies of an earlier two-cable system [3]. The new design,

ORCID: http://orcid.org/0000-0002-3483-0374

WIT Transactions on Engineering Sciences, Vol 124, © 2019 WIT Press
www.witpress.com, ISSN 1743-3533 (on-line)
doi:10.2495/MC190201

which included a second pair of ropes added to the slots on the 6 mm gauge steel posts, were crash tested and found to meet the standards by the UK Department for Highways. Also in 1993, Yang et al. investigated the causes of tension loss in cable barriers and developed corrective measures for the same [4]. This study used laboratory tests on normal and pre-stressed cables to determine if there were any significant differences in cable strain due to long term loading. In 1997, Hunter et al. evaluated the in-service performance of three-strand CMBs on interstate highways [5]. Based on crash data collected from 1990 to 1997, several regression models were developed and used to estimate the effects of CMB installation on crash rates. The study showed that, while the number of crashes was increased due to CMB installations, the number of fatal crashes and head-on collisions and the overall severity indices were greatly reduced.

In 1998, Sposito and Johnston studied the effectiveness of the three-strand CMBs in preventing crossover accidents on interstate and evaluated the maintenance and repair costs [6]. They found out that CMBs were cost effective compared to the concrete barriers and that CMBs worked the best in medians with a minimum of 7 m width. In the work by Cooner et al., they evaluated the in-service performance of various proprietary CMBs in Texas [7]. It was found from the study that the overall fatalities caused by cross-over crashes were significantly reduced since the installation of the CMBs. In 2013, Stolle and Sicking investigated CMB containment failures and crash severity and found that low-tension CMBs had lower, crash severity rate but higher penetration and rollover rates than high-tension CMBs [8]. In 2014, Stolle studied the contributing factors to CMB penetrations [9]. They found that high-tension CMBs experienced overriding due to cable entrapment by the posts and that low-tension CMBs experienced under-riding penetrations due to the low strength of connections to attached bottom and middle cables to the post. The study by Burns and Bell on CMB performance showed that while crash occurrence was increased following the installation of CMBs, the number of fatal crashes and head-on collisions were reduced by a significant amount [10].

Besides evaluation of in-service CMBs, modelling and simulation work were also conducted on CMB performance evaluations. In 2005, Hiser and Reid developed improved methods for modelling slip base structures in impact analysis [11]. In 2007, Marzougui et al. conducted a study using finite element (FE) and vehicle dynamics analysis on the effect of sloped terrain to CMB performance [12]. The FE models were validated with full-scale crash testing data and the simulation results showed that, when installed on flat terrain, the CMB studied met the safety requirement of NCHRP Report 350. On a 6:1 sloped median, the simulation results showed that the CMB could redirect a pickup truck with the CMB at an offset of 1.22 m (4 ft) from the centre of the median. The simulation results also confirmed the observation from full-scale crash test that the CMB might not redirect mid-size and small vehicles if the CMB was placed more than 0.3 m (1 ft) from centre of the median. In 2010, Mohan et al. developed and validate the FE model of a 3-strand CMB that included details of the soil, posts, cables, and hook bolts [13]. Fang et al. developed full-scale FE models of North Carolina CMBs on a sloped median and evaluated their performance at various impact speeds and angles [14]. A new retrofit design was developed in the study and was found to have improved performance of the current CMB design. Subsequently, Fang et al. studied the performance of CMBs on sloped medians with horizontal curvatures [15]. In the work of Stolle and Reid, they developed a wire rope FE model for cable barriers and performed a mesh sensitivity analysis to determine the optimum mesh size with good accuracy and computational efficiency [16]. Stolle et al. used computer simulations to evaluate a new cable to post attachment for high tension cable barriers [17]. In 2013, Wang et al. studied the modelling of slender members in CMBs, such as hook bolts and cables [18]. It was found that the beam-element model outperformed those using solid and shell elements, in both accuracy and computational efficiency.

Since the 1970's, tires have been suggested for use in barriers to improve energy absorption in vehicular crashes, as seen in a number of patents [19]–[28]. However, few studies exist in literature on the performance of barrier systems designed or retrofitted with reclaimed tires. Although CMBs have been shown to be generally effective in preventing cross-median crashes, there still exists vehicle under-riding when impacted by some small passenger cars. In this study, nonlinear FE simulations were utilized to evaluate the performance of CMBs mounted with reclaimed tires. The goal of this research was to develop a retrofit CMB design with reclaimed tires that could prevent small passenger cars from under-riding. In the remaining portion of the paper, the FE models of a test vehicle and CMBs are first presented. The impact conditions based on the Manual for Assessing Safety Hardware (MASH) [29] were then introduced. Finally, the simulation results are analysed and discussed in detail, followed by some concluding remarks and recommendations for future work.

2 MODELLING OF TEST VEHICLE AND CABLE BARRIERS

2.1 FE modelling of test vehicle

The test vehicle was a 1996 Dodge Neon with a curb weight of 2,414 lb (1,095 kg). The FE model of this vehicle was originally developed for LS-DYNA [30] and validated using full-scale crash tests at the National Crash Analysis Centre, had a total of 339 parts composed of 283,683 nodes and 270,953 elements (2,852 solid, 92 beam, 267,775 shell, and 234 other elements). Ten different material models were used in the FE models, including the piecewise linear plasticity model for most steel components, elastic model for the tires and a few other components, viscous damping model for the shock absorbers, low-density foam model for the radiator core, spot-weld model for sheet metal connections, Blatz-Ko rubber model for rubber cushions, rigid model for most mounting hardware, and null material model defined for contact purposes. Hourglass control was used on various components that could potentially experience large deformations. The FE model of the 1996 Dodge Neon is shown in Fig. 1.

Figure 1: FE model of a 1996 Dodge Neon.

2.2 Modelling of tire-mounted CMBs (TMCMBs)

The FE model of the tire-mounted CMB (TMCMB) was created based on the NCDOT low-tension, three-cable CMB, as shown in Fig. 2 along with a real NCDOT CMB.

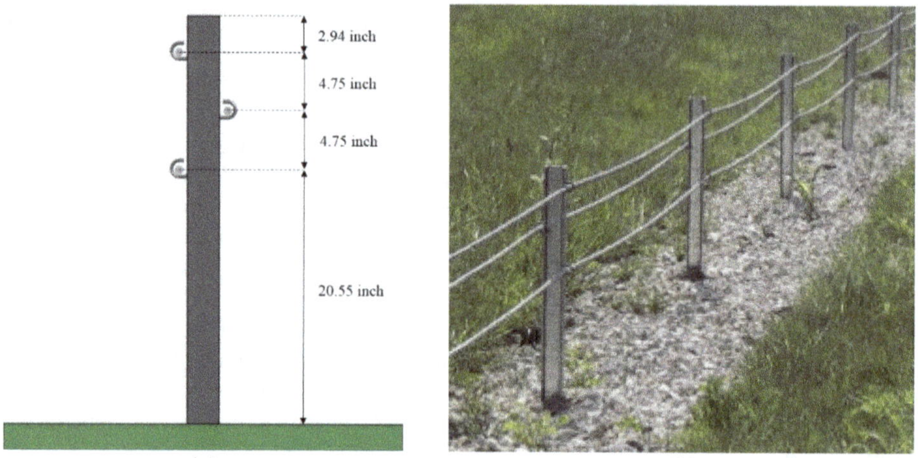

(a) (b)

Figure 2: A cable median barrier. (a) FE model of the current NCDOT CMB; and (b) An NCDOT low-tension CMB.

The tire used to mount onto the CMB was a P245/70R17 tire from a 2007 Chevrolet Silverado. The FE model of the tire, as shown in Fig. 3(a), was composed of 4,380 shell elements with mesh sizes from 12–15 mm for both the tread and sidewall. The tire was modelled as elastic material with Young's modulus of 29.4 MPa and Poisson's ratio of 0.45. The thickness of the shell elements was kept uniform at 7.5 mm so that the total mass of the tire was 35 lb (15.9 kg). The tire was mounted to the two cables on the same side of the CMB, as schematically shown in Fig. 3(b).

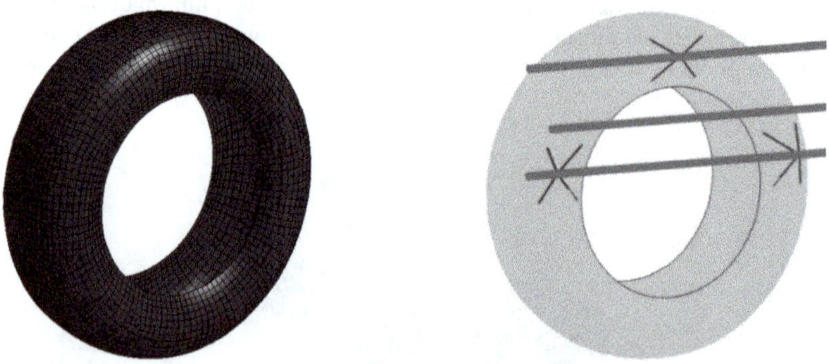

(a) (b)

Figure 3: FE model of tire and tire mounting on CMBs. (a) FE model of a P245/70R17 tire; and (b) Tire mounting on CMB cables.

In this study, four TMCMB designs were evaluated on their performance in vehicle redirection and/or preventing vehicle under-riding. The four designs, as shown in Fig. 4, are named based on cable heights and hook types: original cable heights with J-hooks (TMCMB-O-J), original cable heights with U-hooks (TMCMB-O-U), new cable heights witth J-hooks

(TMCMB-N-J), and new cable heights witth U-hooks (TMCMB-N-U). It should be noted that the J-hooks, upon yielding, allow for cables to be released from the hooks, while the U-hooks will not allow the cables to be detached unless ultimate failure occurs on the U-hooks.

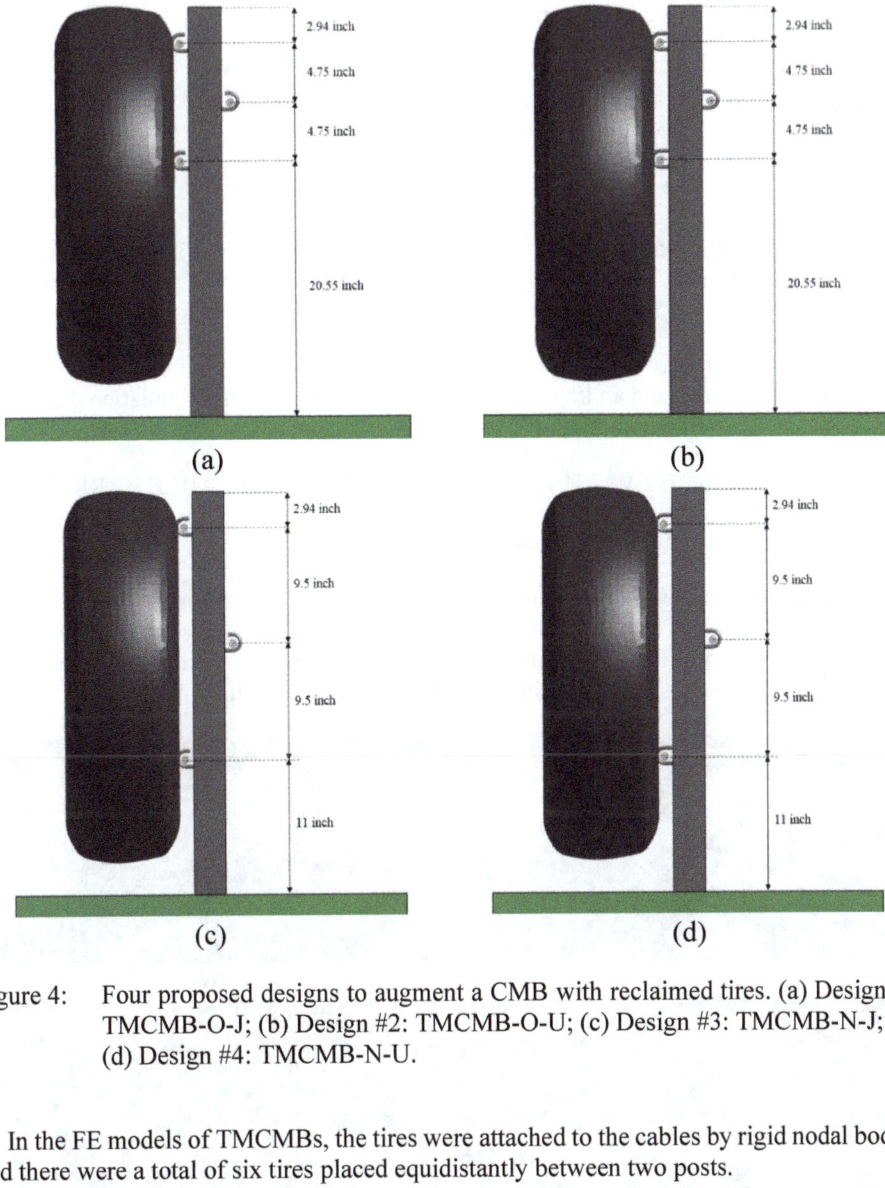

Figure 4: Four proposed designs to augment a CMB with reclaimed tires. (a) Design #1: TMCMB-O-J; (b) Design #2: TMCMB-O-U; (c) Design #3: TMCMB-N-J; and (d) Design #4: TMCMB-N-U.

In the FE models of TMCMBs, the tires were attached to the cables by rigid nodal bodies, and there were a total of six tires placed equidistantly between two posts.

2.3 Simulation setup

In this study, the four TMCMB designs were first evaluated using a 400 ft (120 m) section placed on a flat terrain and impacted by the vehicle from the side with the mounted tires. The best of the four TMCMB designs was then evaluated on a 6H:1V sloped median with two TMCMB placements (see Fig. 5), one on the slope where the vehicle would impact the tires on the barrier before reaching to the bottom of the ditch and the other on the slope where the vehicle would reach the bottom of ditch before impacting the tires on the barrier. For both TMCMB placements, the vehicle impacted the barrier from both front-side (i.e., the vehicle impacted the barrier before reaching the bottom of ditch) and backside (i.e., the vehicle reached the bottom of ditch before hitting the barrier).

Figure 5: TMCMB placement on a 6H:1V sloped median. (a) Placement 1; and (b) Placement 2.

Crash simulations were performed under MASH TL-3 impact conditions, i.e., at a speed of 62.1 mph (100 km/h) and an impact angle of 25 degrees. In all the simulations, the initial impact points were at the mid-point of two posts at mid-span of the TMCMB.

3 PERFORMANCE EVALUATION OF TIRE-MOUNTED CABLE BARRIERS

3.1 Vehicular impacts on TMCMB placed on a flat terrain

Numerical simulations were first conducted for vehicular impacts on the four TMCMBs placed on a flat terrain. The FE simulation results of vehicle trajectories' after impacting the TMCMBs are shown in Fig. 6 in which the barriers are shown in their undeformed shapes and the rectangular boxes in dashed lines give the exit boxes specified in MASH.

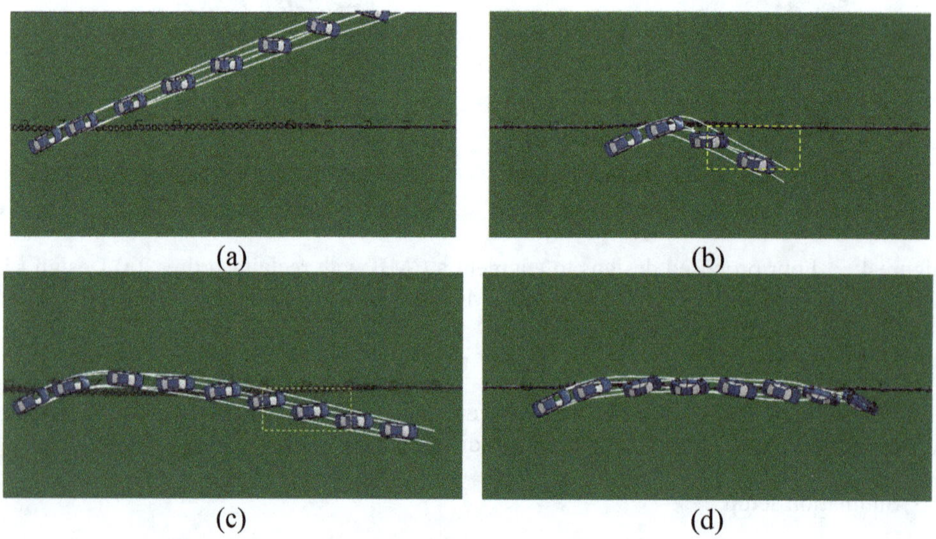

Figure 6: Vehicle trajectories after impacting the TMCMBs on a flat terrain. (a) Design #1: TMCMB-O-J; (b) Design #2: TMCMB-O-U; (c) Design #1: TMCMB-N-J; and (d) Design #2: TMCMB-N-U.

In the case of TMCMB-O-J, the vehicle under-rode and penetrated the barrier since the cables were released from the J-hooks. This was considered a failed case because the vehicle entered into the oncoming travel lane. For the TMCMB-O-U design, the cables were not released due to the use of U-hooks and thus retained and redirected the vehicle. However, due to the rigidity of the TMCMB, the vehicle was bounced back and went out of the bottom side of the exit box, which was considered as a failed case by MASH. In the case of TMCMB_N_J, which used J-hooks but had lowered cables, the vehicle was successfully redirected and passed the MASH exit box criterion, i.e., leaving the exit box from the right side. Finally, in the case of TMCMB_N_U, the vehicle was redirected but rolled over towards the impacting side, which was also considered a failed case as per MASH. It should be noted that since the vehicle stayed in contact with TMCMB_N_U, the exit box criterion was not applicable according to MASH. Based on the above analysis, the TMCMB_N_J design was determined to outperform the other three proposed designs and was further evaluated on a sloped median, as presented next. Fig. 7 shows the front views of vehicle-barrier interactions for the case of TMCMB-N-J.

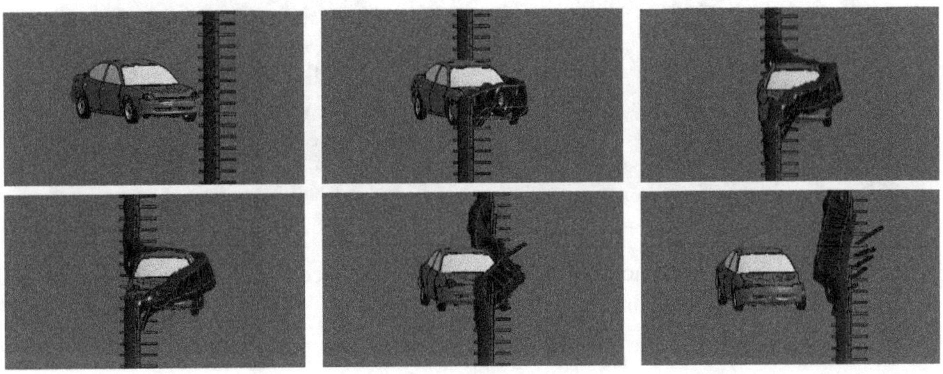

Figure 7: Vehicle-interactions for the TMCMB-N-J on a flat terrain.

3.2 Vehicular impacts on TMCMB-N-J placed on a 6H:1V sloped median

The TMCMB-N-J design was shown to outperform other designs on a flat terrain, which was the standard test site required by MASH. To evaluate its performance under in-service conditions, which is also recommended by MASH, the TMCMB-N-J design was tested on a 6H:1V sloped median with two TMCMB placements shown in Fig. 5. Full-scale FE simulations were performed for vehicular impacts from both front-side and backside of the TMCMB-N-J design and the simulation results are summarized in Table 1.

Fig. 8 shows the top view of vehicle trajectories for both front-side and backside impacts on the TMCMB-N-J for Placement 1. The exit boxes defined by MASH are shown by the

Table 1: Simulation results of the TMCMB-N-J design place on a 6H:1V sloped median.

TMCMB placement	Impact side	Vehicular responses
Placement 1	Front	Redirected (exit box criterion satisfied)
	Back	Redirected (exit box criterion satisfied)
Placement 2	Front	Redirected (exit box criterion satisfied)
	Back	Redirected (exit box criterion not applicable)

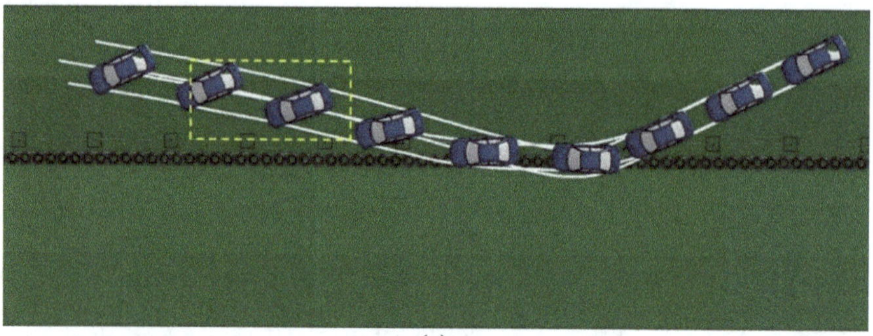

(a)

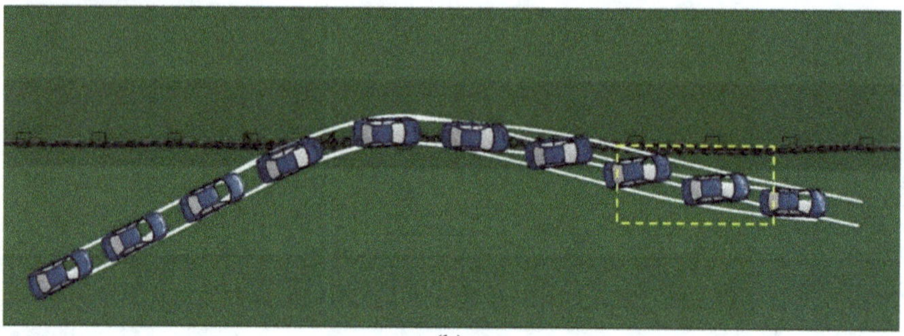

(b)

Figure 8: Dodge Neon impacting the TMCMB-N-J for Placement 1. (a) Front-side impact; and (b) Backside impact.

rectangles in dashed lines. It can be seen that the vehicle was redirected in both cases and the MASH exit box criterion was satisfied, i.e., the vehicle's tire marks were within the exit box before leaving the exit box from the exit side.

Fig. 9 shows the yaw, roll, and pitch angles of the Dodge Neon impacting the TMCMB-N-J for Placement 1. For both front-side and backside impacts, the vehicle's roll and pitch

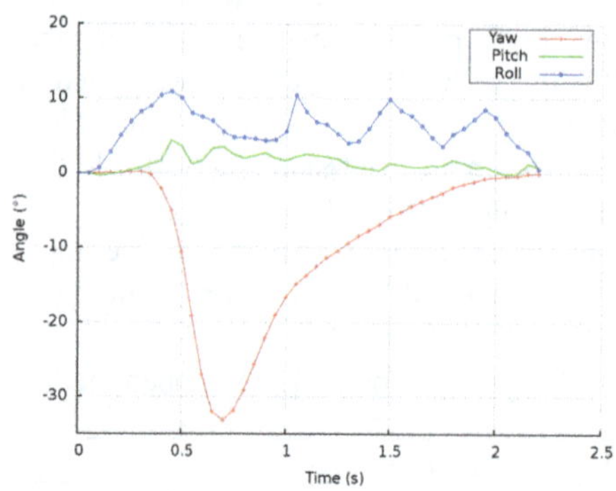

(a)

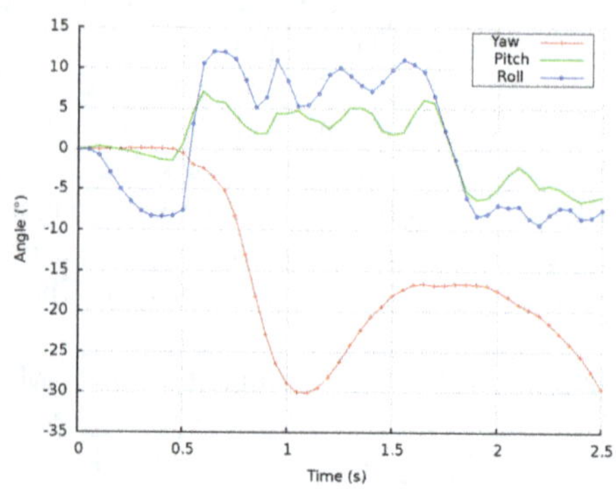

(b)

Figure 9: Yaw, pitch and roll angles of the Dodge Neon impacting the TMCMB-N-J for Placement 1. (a) Front-side impact; and (b) Backside impact.

angles were less than 15° in both positive and negative directions, and thus passed the MASH Criterion *F*, which specified a maximum of 75° of roll or pitch angle. Fig. 10 shows the vehicle's transverse velocities towards the travel lane in the two impact cases shown in Fig. 8. The vehicle's transverse velocities were approximately 10 km/h after redirected, indicating a small likelihood of getting involved in a secondary collision.

Fig. 11 shows the top view of vehicle trajectories for both front-side and backside impacts on the TMCMB-N-J for Placement 2. It can be seen that the vehicle was redirected in both

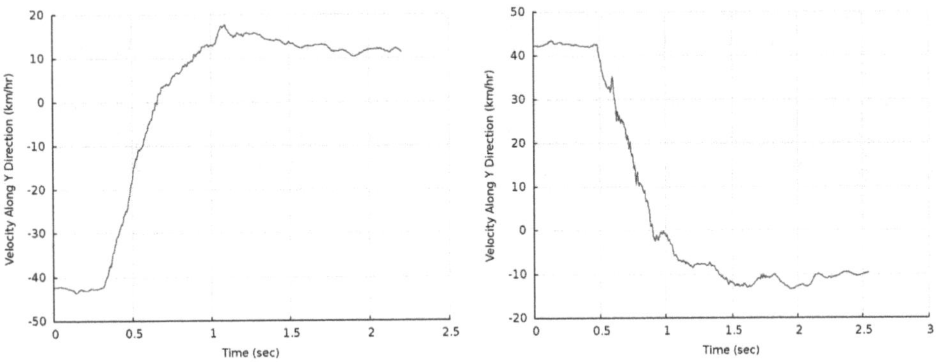

(a) (b)

Figure 10: Transverse velocities of the Dodge Neon impacting the TMCMB-N-J for Placement 1. (a) Front-side impact; and (b) Backside impact.

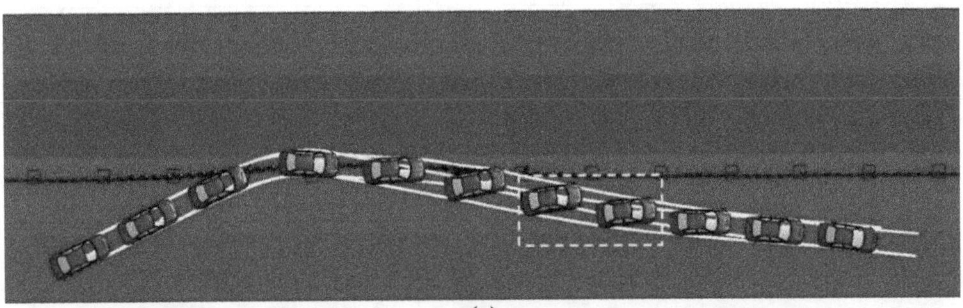

(a)

(b)

Figure 11: Dodge Neon impacting the TMCMB-N-J for Placement 2. (a) Front-side impact; and (b) Backside impact.

cases and the MASH exit box criterion was satisfied for the case of front-side impact. For backside impact, the vehicle stayed in contact with the TMCMB after being redirected; therefore, the exit box criterion was not applicable according to MASH.

Fig. 12 shows the yaw, roll, and pitch angles of the Dodge Neon impacting the TMCMB-N-J for Placement 2. For both front-side and backside impacts, the vehicle's roll and pitch angles were less than 15° in both positive and negative directions, thus passing the MASH Criterion *F*. Fig. 13 shows the vehicle's transverse velocities towards the travel lane for both cases. The vehicle's transverse velocities were less than 10 km/h after redirected, indicating a small likelihood of getting involved in a secondary collision.

The simulation results shown in Figs 8–13 indicated that the TMCMB-N-J design worked well when placed on a 6H:1V sloped median for both placements. The use of reclaimed tires to retrofit the CMB was shown to be effective in eliminating vehicle under-riding by small passenger cars such as the Dodge Neon, which was seen on the currently used CMB systems. It should be noted that the TMCMB-N-J design should also be tested under impacts by a

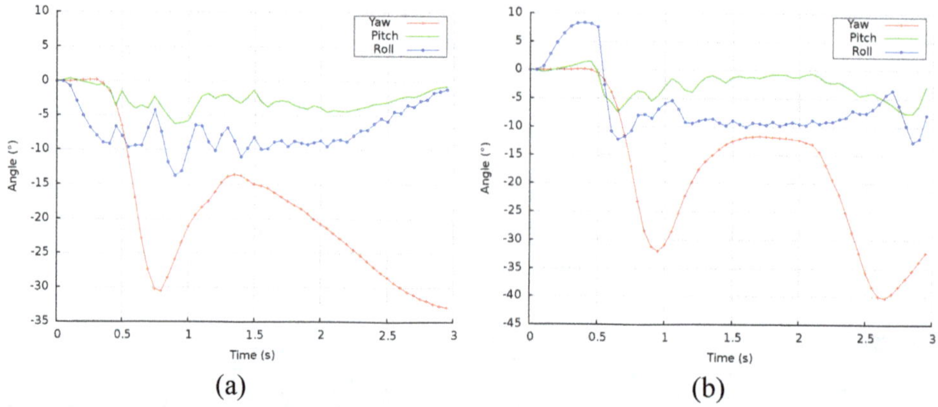

Figure 12: Yaw, pitch and roll angles of the Dodge Neon impacting the TMCMB-N-J for Placement 2. (a) Front-side impact; and (b) Backside impact.

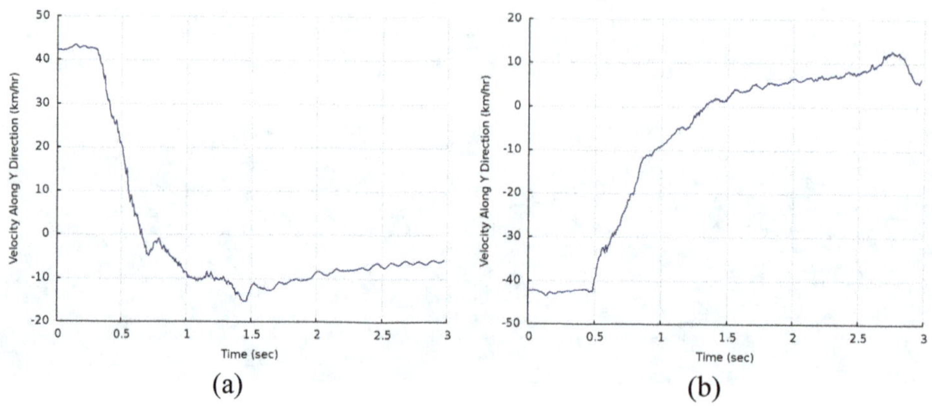

Figure 13: Transverse velocities of the Dodge Neon impacting the TMCMB-N-J for Placement 2. (a) Front-side impact; and (b) Backside impact.

pickup truck as specified by MASH. Since the focus of this study was to solve the issue of vehicle under-riding, the evaluation of TMCMB-N-J under impacts by a pickup truck was not considered in the current work but should be conducted in the future.

4 CONCLUSIONS

In this study, finite element (FE) modelling and simulations were used to evaluate tire-mounted cable median barriers (TMCMBs) to solve the issue of vehicle under-riding that was seen on the current cable barriers. Ten different material models were used in the FE models, including the piecewise linear plasticity model for most steel components, elastic model for the tires and a few other components, viscous damping model for the shock absorbers, low-density foam model for the radiator core, spot-weld model for sheet metal connections, Blatz-Ko rubber model for rubber cushions, rigid model for most mounting hardware, and null material model defined for contact purposes. Four different TMCMB designs were first evaluated when placed on a flat terrain and impacted by a small passenger car, i.e, a 1996 Dodge Neon. The best of the four designs evaluated, namely the TMCMB-N-J design, was then further evaluated on a 6H:1V sloped median, which was the most commonly used for cable barriers. The simulation results showed that the TMCMB design was effective for both front-side and backside impacts for two placements, one on the front slope and the other on the back slope. In the four cases evaluated, the TMCMB-N-J design was capable of redirecting the Dodge Neon and satisfy the requirements by Manual for Assessing Safety Hardware (MASH). This research provided valuable insights regarding the use of reclaimed tires on the current cable median barriers. For future investigations, the TMCMB-N-J design can be evaluated under impacts by a pickup truck based on MASH specifications to ensure its effectiveness before it can be used as a retrofit option for the current cable barrier systems.

REFERENCES

[1] Graham, M.D., Burnett, W.C., Gibson, J.L. & Freer, R.H., *New Highway Barriers: The Practical Application of Theoretical Design*, Highway Research Record No. 174, 1967.

[2] Phillips, R., Tyrell, A., Bryden, J. & Fortuniewicz, J., Cable guiderail breakaway terminal ends. Report FHWA/NY/RR-90/148, New York State Department of Transportation, Albany, NY, 1990.

[3] Laker, I.B. & Naylor, A.W., Development and proving tests of a four-rope safety fence. *Transportation Research Record*, **1419**, pp. 1–8, 1993.

[4] Yang, W., Bendana, L.J. & Bruno, N.J., Performance of cable guiderail in New York. *Transportation Research Record*, **1419**, pp. 9–20, 1993.

[5] Hunter, W.W., Stewart, J.R., Eccles, K.A., Huang, H.F., Council, F.M. & Harkey, D.L., Three-strand cable median barrier in North Carolina in-service evaluation. *Transportation Research Record*, **1743**(1), pp. 97–103, 1997.

[6] Sposito, B. & Johnston, S., Three-cable median barrier. Report No. OR-RD-99-03, Oregon Department of Transportation, 1998.

[7] Cooner, S.A., Rathod, Y.K., Alberson, D.C., Bligh, R.P., Ranft, S.E. & Sun, D., Performance evaluation of cable median barrier systems in Texas. Report No. FHWA/TX-09/0-5609-1, US Department of Transportation, 2009.

[8] Stolle, C.S. & Sicking, D.L., Impact conditions associated with cable median barrier failure. *93rd Annual Meeting of the Transportation Research Board*, Paper No. 13-1564, 2013.

[9] Stolle, C.S., Factors contributing to cable median barrier penetrations. *93rd Annual Meeting of the Transportation Research Board*, Paper No. 14-4624, 2014.

[10] Burns, K. & Bell, K., Performance evaluation of a cable median barrier system on an Oregon highway with a narrow median. *Transportation Research Record*, **2588**, pp. 137–144, 2016.

[11] Hiser, N.R. & Reid, J.D., Modeling slip base mechanisms. *International Journal of Crashworthiness*, **10**(5), pp. 463–472, 2005.

[12] Marzougui, D., Mohan, P., Kan, C. & Opiela, K., Performance evaluation of low-tension three-strand cable median barriers. *Transportation Research Record*, **2025**, pp. 34–44, 2007.

[13] Mohan, P., Marzougui, D., Meczkowski, L. & Bedewi, N., Finite element modeling and validation of a 3-strand cable guardrail system. *International Journal of Crashworthiness*, **10**(3), pp. 267–273, 2005.

[14] Fang, H., Wang, Q. & Weggel, D.C., Crash analysis and evaluation of cable median barriers on sloped medians using an efficient finite element model. *Advances in Engineering Software*, **82**, pp. 1–13, 2015.

[15] Fang, H., Li, N., DiSogra, M., Gutowski, M. & Weggel, D.C., Recommendations for placement of cable median barriers on 6:1 and 4:1 sloped medians with horizontal curvatures. Report No. FHWA/NC/2011-09, North Carolina Department of Transportation, 2012.

[16] Stolle, C.S. & Reid, J.D., Development of a wire rope model for cable guardrail simulation. *International Journal of Crashworthiness*, **16**(3), pp. 331–341, 2011.

[17] Stolle, C.S., Reid, J.D. & Bielenberg, R.W., Improved models of cable-to-post attachments for high-tension cable barriers. Report No. TRP-03-267-1212, Nebraska Department of Transportation, 2012.

[18] Wang, Q., Fang, H., Li, N., Weggel, D.C. & Wen, G., An efficient FE model of slender members for crash analysis of cable barriers. *Engineering Structures*, **52**, pp. 240–256, 2013.

[19] Walker, B., US Patent No. 3,661,359. US Patent and Trademark Office: Washington, DC, 1972.

[20] Way, M. & Pas, J., US Patent No. 3,848,853. US Patent and Trademark Office: Washington, DC, 1974.

[21] Vincent, C.K., US Patent No. 4,090,694. US Patent and Trademark Office: Washington, DC, 1978.

[22] Yoho, L.E., US Patent No. 4,066,244. US Patent and Trademark Office: Washington, DC, 1978.

[23] Moon, D.G., US Patent No. 5,238,228. US Patent and Trademark Office: Washington, DC, 1993.

[24] Bernaquez, N. & Sabourin, I., US Patent No. 5,480,255. US Patent and Trademark Office: Washington, DC, 1996.

[25] Yunick, H., US Patent No. 5,645,368. US Patent and Trademark Office: Washington, DC, 1997.

[26] Kramer, R., US Patent No. 20,030,081,997. US Patent and Trademark Office: Washington, DC, 2003.

[27] Diem, R., US Patent No. 6,505,993. US Patent and Trademark Office: Washington, DC, 2003.

[28] Talbott, A.F., US Patent No. 7,387,295. US Patent and Trademark Office: Washington, DC, 2008.

[29] MASH, *Manual for Assessing Safety Hardware (MASH)*, American Association of State Highway and Transportation Officials (AASHTO): Washington, DC, 2009.

[30] LSTC, *LS-DYNA Keyword User's Manual – Version 971*, Livermore Software Technology Corporation (LSTC): Livermore, CA, 2007.

EXPERIMENTAL–NUMERICAL CALIBRATION OF THE FRACTURE LOCUS OF WEATHERING STEEL

FRANCO CONCLI & LORENZO MACCIONI
Faculty of Science and Technology, Free University of Bolzano-Bozen, Italy

ABSTRACT

Weathering steel, specifically Cor-Ten (or CORTEN) steel is a material particularly exploited in the last century for various outdoor applications, e.g. bridges, building facades, artworks etc. In addition to a tensile strength comparable with other construction steels, the natural oxide of this material, that is common rust, has the same specific volume as the metallic core. This ensures the adhesion of the oxidized protective layer as for aluminium. Therefore, the stable rust layer protects the raw material from further corrosion. This characteristic overcomes the need for painting and maintenance. These properties boost the exploitation of Cor-Ten in several civil applications, also where safety is a fundamental requirement, e.g. guard rails used, for example, in the South-Tyrolean region along the highways. With the aim of verifying or optimizing such safety applications, it is important to know the ductile behavior of the material. Indeed, during an impact, the main purpose of the structure is to absorb energy and this implies that large deformations will take place. Experimental quasi-static tests were performed on several sample geometries made of Cor-Ten. The same tests were also numerically reproduced, to retrieve the actual stress state, quantify the plastic strain at failure and calibrate a ductile damage model. The material model is based on both the classical incremental model of plastic response with isotropic hardening and the phenomenological concept of damage in continuum mechanics.

Keywords: ductile fracture, Cor-Ten, fracture locus, experimental, FEM.

1 INTRODUCTION

In 1933, US Steels launched a new self-passivating steel (weathering steel) with Cu, Cr and P in the alloy [1]. To underline its remarkable resistance to CORrosion and to TENsile stress, the new material was called Cor-Ten and its composition underwent some adjustments in the following years [2]. However, its strength properties were not the promoters of its exploitation and interest in research. Indeed, the anticorrosive and passivating characteristics of this material inspired engineers, architects and artists to use this material. Cor-Ten was initially exploited in the railway sector [3] in order to reduce the corrosion of wagons. It was also used in the construction of bridges [4] with the aim of reducing the need of painting and maintenance. Its application extends to building facades [5] for aesthetic reasons and in many outdoor artworks e.g. [6].

Researchers focused on its corrosion behavior under different environmental conditions [7]–[9]. Indeed, Cor-Ten requires special conditions to create a protective layer; these are wet/dry cycles [10]. This phenomenon is negatively affected by the harmful action of Cl^-, as may occur close to the marine environment [11] and by SO_2, associated with high humidity especially in polluting environment [12]. Moreover, the corrosion behavior was studied with varying alloy elements [13], metal microstructures [14] and surfaces finish [15].

Nowadays, the safety barriers (guardrail) of the Brenner Highway [16] and in most of the South-Tyrolean region are made in Cor-Ten. Firstly, its natural color presents a low visual impact (with respect to the galvanized alternative) in a UNESCO world heritage such as this alpine region. Secondly, the maintenance costs are expected to be low.

Despite experimental tests being conducted to homologate the Cor-Ten guardrail installations [17], in some specific areas this application is not exempt from corrosion [18].

WIT Transactions on Engineering Sciences, Vol 124, © 2019 WIT Press
www.witpress.com, ISSN 1743-3533 (on-line)
doi:10.2495/MC190211

Therefore, it is possible that the installation has undergone geometric changes capable of compromising the safety of the structure, especially during a crash impact.

In the light of the above considerations, it is essential to fine-tune an FE modeling of the crash behavior of the Cor-Ten barriers to assert their safety performance [19] both in their initial conditions and after their eventual modifications undergone by maintenance and/or corrosion. Furthermore, the model could be exploited for traffic accident reconstruction. In addition, since the functional goal of safety barriers is "to absorb as much impact energy as possible through their deformation and at the same time maintain their integrity" [20] having a characterization of the Cor-Ten plastic and fracture behavior is fundamental to developing these models.

The work presented in this paper represents the first step towards fine-tuning an FE model able to describe the crash behavior of traffic barriers made of Cor-Ten. Therefore, the present paper aims to characterize the ductile behavior of the Cor-Ten through experimental quasi-static tests with different geometries thus different level of triaxiality. The same tests were also numerically reproduced, to retrieve the actual stress state, quantify the plastic strain at failure and calibrate a ductile damage model.

2 MATERIALS AND METHODS

The material of the study is a Cor-Ten Weathering steel. In order to characterize the behavior of the material, dedicated tensile tests were performed on a MTS Criterion 45 testing machine. The machine is capable to apply loads up to 100 kN.

In order to determine the relevant constitutive law, 4 mono-axial tests were performed on smooth samples (Fig. 1). Applied force, crosshead displacement as well as deformation, measured with an extensometer, were acquired.

While for the elastic region the $\sigma - \varepsilon$ relation can be derived directly from the measurements, in the plastic (hardening) region, an inverse numerical procedure is required. Several simulations were performed and the *F-s* (Force–crosshead displacement) curve (Fig. 2) obtained is numerically compared with the measured one. The constitutive relation can be described by the Voce law [21], [22].

$$\sigma = C_1 + C_2(1 - e^{-C_3\varepsilon_p}) + C_4\varepsilon_p,$$

where constants C_i (i = 1,4,2,3) are the yielding stress, the linear coefficient, the exponential coefficient and the so-called exponential saturation parameter, respectively, and ε_p is the accumulated plastic strain (Table 1).

In this way, with an iterative procedure, it is possible to calibrate the constitutive law.

Several ductile damage models have been developed over the years. One of the first works was carried out by Mirza et al. [23]. They studied the influence of the stress triaxiality using copper specimens concluding that the strain triaxiality plays a fundamental role in the failure of metals. Other models were proposed by Rice and Tracey [24], Hancock and Mackenzie [25] and Johnson and Cook [26]. All these models propose a relation between the maximum plastic strain at fracture ε_{peeq} and the triaxiality η. Triaxiality is a parameter that well represents the state of stress and is defined as:

$$\eta = \frac{\Sigma\frac{\sigma_i}{3}}{\sqrt{\frac{1}{2}\left[\Sigma(\sigma_i-\sigma_j)^2\right]}}.$$

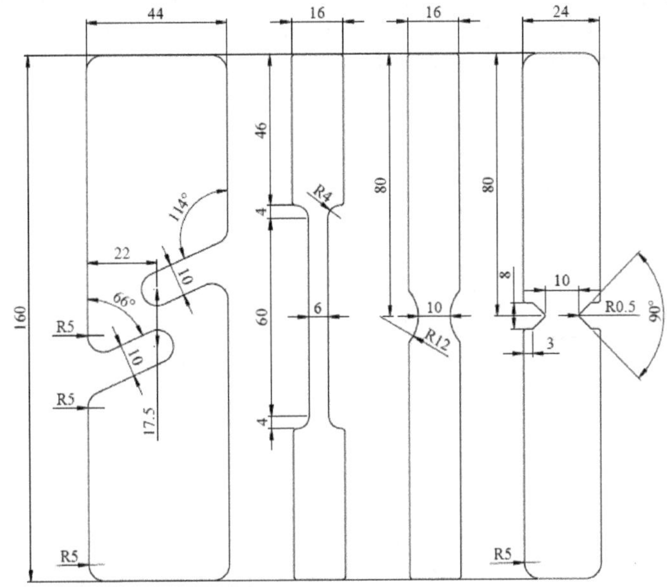

Figure 1: Geometries of the samples.

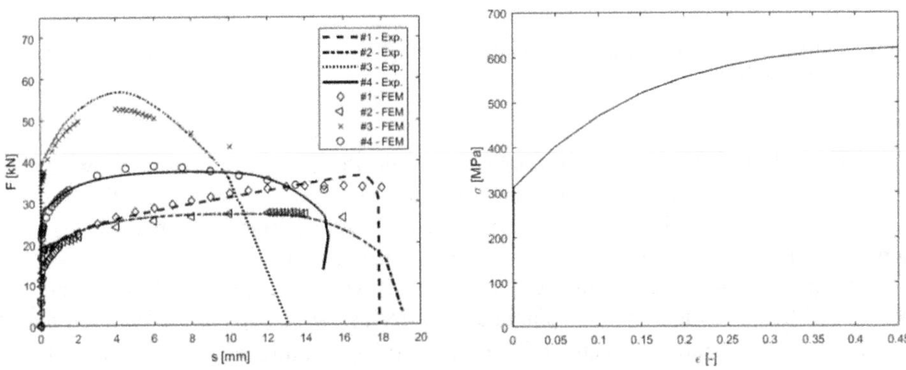

(a) (b)

Figure 2: (a) *F*-*s* relation – comparison between numerical and experimental curves; and (b) True stress–true strain curve.

Table 1: Voce hardening parameters for the Cor-Ten steel.

C_1 (MPa)	C_2 (MPa)	C_3 (-)	C_4 (MPa)
320	380	5	500

The relation proposed by Rice and Tracey [24] is an exponential decreasing function. Starting from this work, Hancock and Mackenzie [25] and, successively, Johnson and Cook

[26] proposed new relations able to take into consideration the effects of temperature and strain-rate. Further studies by Bao [27]–[30], performed on aluminium samples, have shown that the effect of the triaxiality cannot be modelled with a monotonic function. Different triaxiality levels are associated with different failure mechanisms (Fig. 3).

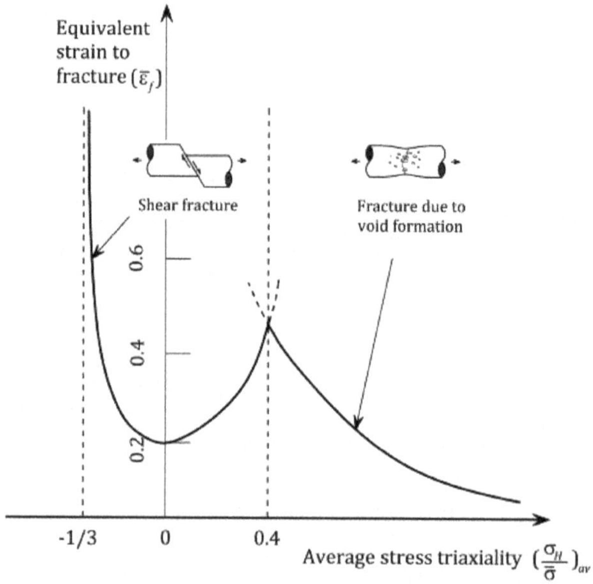

Figure 3: Influence of stress triaxiality on the equivalent strain to fracture. Adapted from [31].

It is possible to subdivide the triaxiality axis into three main regions. At low levels of triaxiality, the failure is mainly due to shearing; for high levels of triaxiality, the fracture is mainly due to the nucleation-growth-coalescence mechanism. In the intermediate region, the fracture originates due to a combination of the two mechanisms.

In this preliminary stage of the research, low triaxiality range of a Cor-Ten steel was studied. Further investigations are planned in order assess the $\eta - \varepsilon_{peeq}$ relation also for higher values of triaxiality. Effects of strain rate and temperature were also neglected.

The sample geometries used to calibrate the ductile fracture model are shown in Fig. 1. The experimental tests were performed up to the rupture of the sample itself. The same geometries were reproduced numerically [32]–[33] by means of the open-source free solver Code-Aster. The choice of an open-source tool was made because it allows more flexibility with respect to close-source commercial software, making it possible to customise the code with the implementation of specific models for the analysis of the physical problem of interest. Simulations were interrupted in correspondence to the imposed displacement that has been shown experimentally to cause failure (Figs. 4–7). From the numerical results, the value of equivalent plastic strain at fracture ε_{peeq} and the triaxiality η at the same point were obtained (Fig. 8). The values are shown in Table 2.

The equation describing the fracture locus (Fig. 9) results in:

$$\varepsilon_{peeq} = 38.592 \cdot \eta^2 - 20.611 \cdot \eta + 3.2838.$$

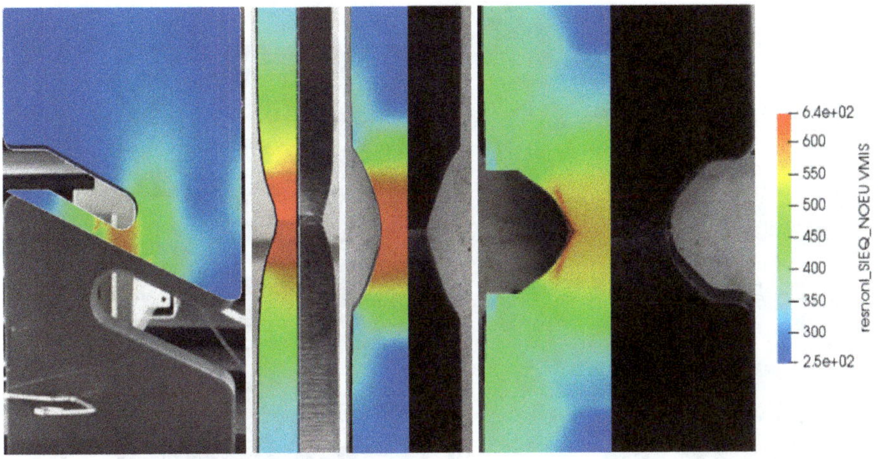

Figure 4: Numerical models of the test to retrieve plastic strain at fracture; from left geometries #1, #2, #3 and #4 (von Mises stress).

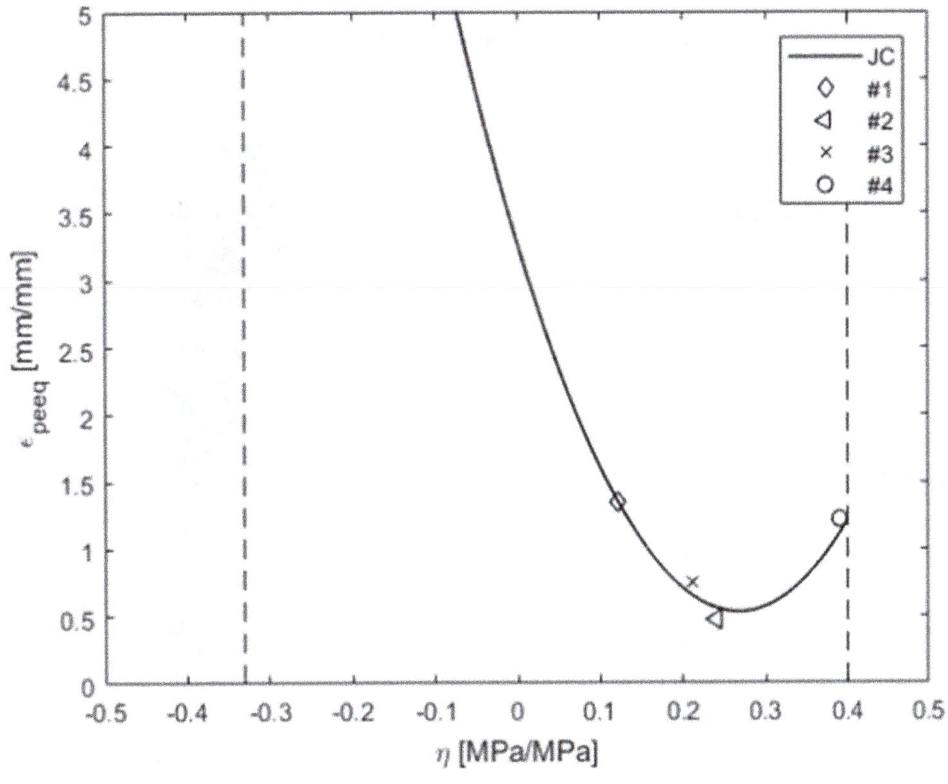

Figure 5: Fracture locus.

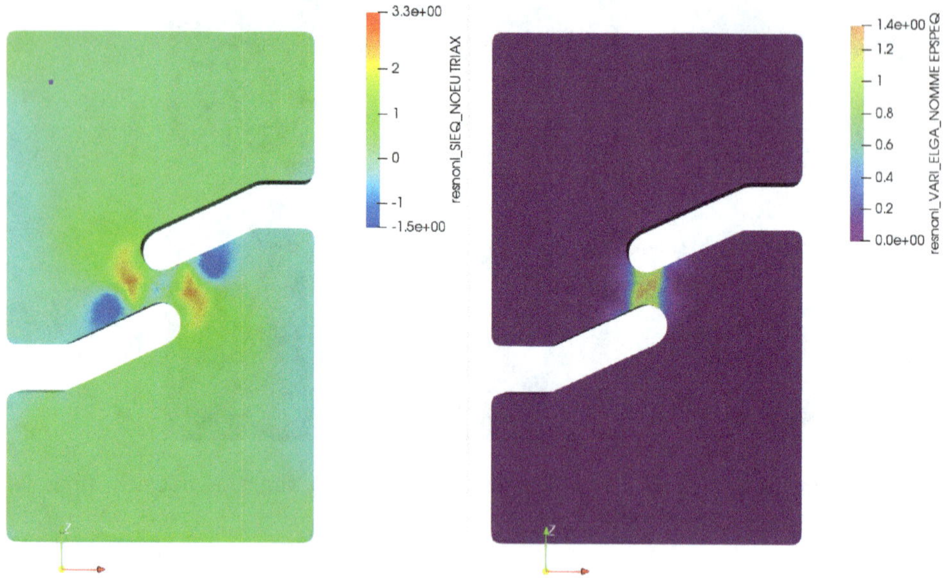

Figure 6: Triaxiality and equivalent plastic strain at fracture, sample #1.

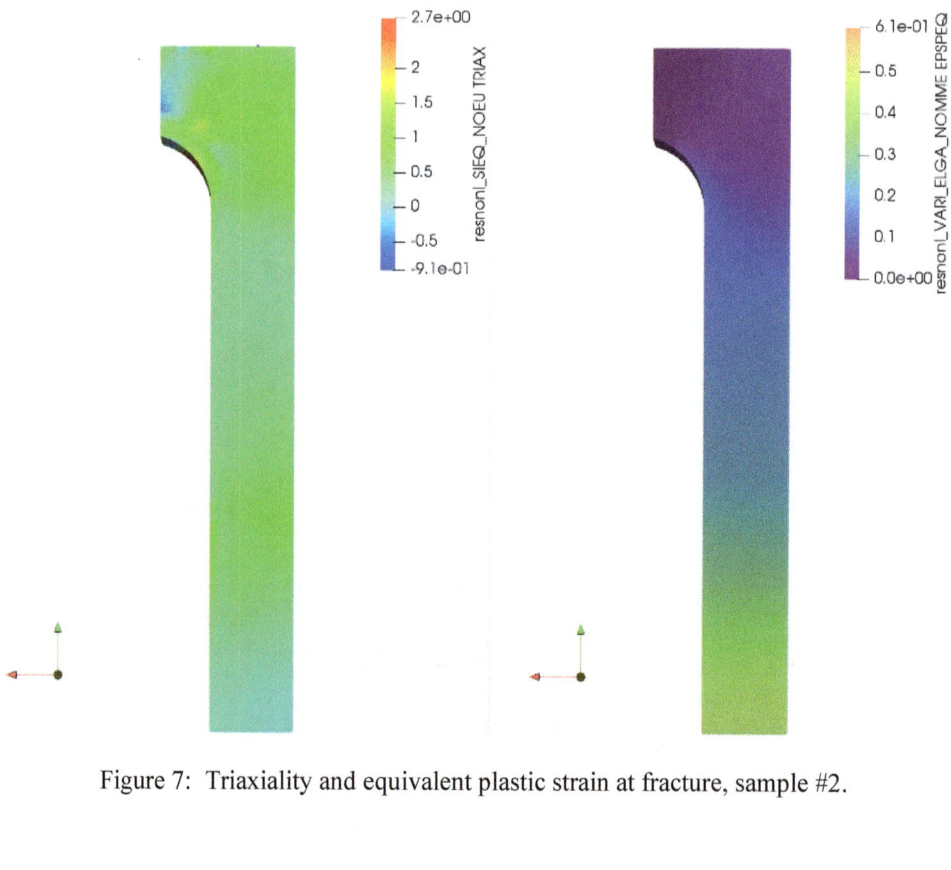

Figure 7: Triaxiality and equivalent plastic strain at fracture, sample #2.

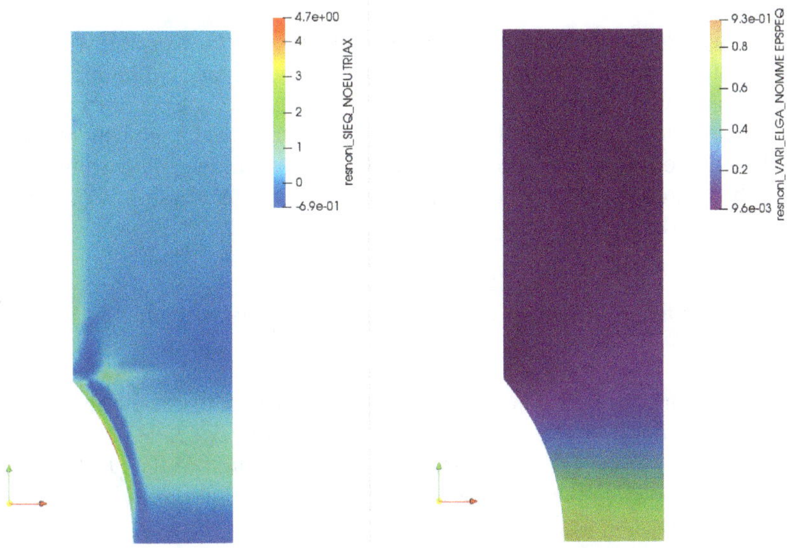

Figure 8: Triaxiality and equivalent plastic strain at fracture, sample #3.

Table 2: Results of the calibration procedure: triaxiality η and equivalent plastic strain at fracture ε_{peeq}.

	Sample geometry			
	#1	#2	#3	#4
η	0.25	0.23	0.21	0.39
ε_{peeq}	1.35	0.475	0.753	1.22

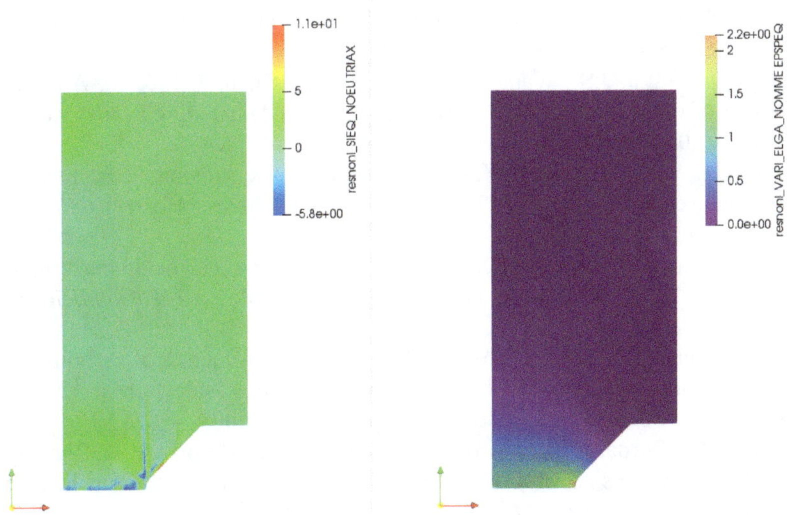

Figure 9: Triaxiality and equivalent plastic strain at fracture, sample #4.

3 CONCLUSIONS

In this research, a ductile fracture model has been calibrated for Cor-Ten weathering steel. The effect of strain rate and temperature has been neglected but will be included in future studies. The calibration of the fracture locus was performed by means of experimental tests on 4 different sample geometries. This characterization can thus be implemented through an FE model in order to study the plastic behavior of Cor-Ten during different situations. Future studies will be addressed to the study of the behavior of a guard rail in Cor-Ten, new and with some defects, during impact.

ACKNOWLEDGEMENT

The authors would like to thank A.ERRE [34] for the financial support.

References

[1] Decker, P., Brüggerhoff, S. & Eggert, G., To coat or not to coat? The maintenance of Cor-Ten® sculptures. *Materials and Corrosion.*, **59**(3), pp. 239–247, 2008.
[2] Revie, R.W., *Uhlig's Corrosion Handbook: Third Edition*, Wiley: Hoboken, 2011.
[3] Dunkley, F.G., Painting of railway rolling stock. *Journal of the Institution of Locomotive Engineers*, **57**(319), pp. 509–553, 1967.
[4] Fischer, M., Weathering steel in bridges. *Structural Engineering International*, **5**(1), pp. 51–54, 1995.
[5] Mostafavi, D. & Leatherbarrow, M., *On Weathering: The Life of Buildings in Time*, MIT: Cambridge, MA, 1993.
[6] Corten + Art, Pinterest. www.pinterest.it/trackdesign9867/corten+art/?lp=true.
[7] Kamimura, T., Hara, S., Miyuki, H., Yamashita, M. & Uchida, H., Composition and protective ability of rust layer formed on weathering steel exposed to various environments. *Corrosion Science*, **48**(9), pp. 2799–2812, 2006.
[8] Morcillo, M., Chico, B., Díaz, I., Cano, H. & de la Fuente, D., Atmospheric corrosion data of weathering steels. A review. *Corrosion Science*, **77**, pp. 6–24, 2013.
[9] Concli, F. et al., Load independent power losses of ordinary gears: Numerical and experimental analysis. *5th World Tribology Congress, WTC 2013*, **2**, 2013.
[10] S. RJ and G. WP, For architectural applications-unpainted high strength low alloy steel. *Matls Prot.*, **8**(12), pp. 70–77, 1969.
[11] Zhang, Q.C., Wu, J.S., Wang, J.J., Zheng, W.L., Chen, J.G. & Li, A.B., Corrosion behavior of weathering steel in marine atmosphere. *Materials Chemistry and Physics*, **77**(2), pp. 603–608, 2003.
[12] Wang, J.H., Wei, F.I., Chang, Y.S. & Shih, H.C., The corrosion mechanisms of carbon steel and weathering steel in SO2 polluted atmospheres. *Materials Chemistry and Physics*, **47**(1), pp. 1–8, 1997.
[13] Wang, R., Luo, S., Liu, M. & Xue, Y., Electrochemical corrosion performance of Cr and Al alloy steels using a J55 carbon steel as base alloy. *Corrosion Science*, **85**, pp. 270–279, 2014.
[14] Guo, J., Yang, S., Shang, C., Wang, Y. & He, X., Influence of carbon content and microstructure on corrosion behaviour of low alloy steels in a Cl^- containing environment. *Corrosion Science*, **51**(2), pp. 242–251, 2009.
[15] Chiavari, C., Bernardi, E., Martini, C., Passarini, F., Motori, A. & Bignozzi, M.C., Atmospheric corrosion of Cor-Ten steel with different surface finish: Accelerated ageing and metal release. *Materials Chemistry and Physics*, **136**(2–3), pp. 477–486, 2012.
[16] Autostrada del Brennero SpA Brennerautobahn AG.

www.autobrennero.it/documenti/4Areatecnica/Pubblicazioni/2010/barrieresicurezza_A22.pdf.

[17] Autostrada del Brennero SpA Brennerautobahn AG. Design and Testing of Safety Barriers of their own Design for the Benefit of User. www.autobrennero.it/it/la-rete-autostradale/la-sicurezza/barriere-sicurezza/.

[18] Deflorian, F. & Rossi, S., Premature corrosion failure of structural highway components made from weathering steel. *Engineering Failure Analysis*, **9**(5), pp. 541–551, 2002.

[19] Whitworth, H.A., Bendidi, R., Marzougui, D. & Reiss, R., Finite element modeling of the crash performance of roadside barriers. *International Journal of Crashworthiness*, **9**(1), pp. 35–43, 2004.

[20] Ren, Z. & Vesenjak, M., Computational and experimental crash analysis of the road safety barrier. *Engineering Failure Analysis*, **12**(6), pp. 963–973, 2005.

[21] Voce, E., A practical strain-hardening function. *Metallurgia*, **51**, pp. 219–226, 1955.

[22] Voce, E., The relationship between stress and strain for homogeneous deformation. *Journal of the Institute of Metals*, **74**, pp. 537–562, 1948.

[23] Mirza, M.S., Barton, D.C, Church, P. & Sturges, J.L., Ductile fracture of pure copper: An experimental and numerical study. *Journal De Physique. IV (Proceedings)*, **7**(3), pp. C3-891-C3-896, 1997.

[24] Rice, J.R. & Tracey, D.M., On the ductile enlargement of voids in triaxial stress fields. *Journal of the Mechanics and Physics of Solids*, **17**(3), pp. 201–217, 1969.

[25] Hancock, J.W. & Mackenzie, A.C., On the mechanisms of ductile failure in high-strength steels subjected to multi-axial stress-states. *Journal of the Mechanics and Physics of Solids*, **24**(2–3), pp. 147–160, 1976.

[26] Johnson, G.R. & Cook, W.H., A constitutive model and data for metals subjected to large strains, high strain rates and high temperatures. *Proceedings of the 7th International Symposium on Ballistics*, pp. 541–547, 1983.

[27] Bao, Y. & Wierzbicki, T., On fracture locus in the equivalent strain and stress triaxiality space. *International Journal of Mechanical Sciences*, **46**(1), pp. 81–98, 2004.

[28] Bao, Y., *Prediction of Ductile Crack Formation in Uncracked Bodies*, Massachusetts Institute of Technology: Cambridge, MA, 2003.

[29] Bao, Y., Dependence of ductile crack formation in tensile tests on stress triaxiality, stress and strain ratios. *Engineering Fracture Mechanics*, **72**(4), pp. 505–522, 2005.

[30] Gilioli, A., Manes, A., Giglio, M. & Wierzbicki, T., Predicting ballistic impact failure of aluminium 6061-T6 with the rate-independent Bao-Wierzbicki fracture model. *International Journal of Impact Engineering*, **76**, pp. 207–220, 2015.

[31] Suárez, F., Gálvez, J.C., Cendón, D.A. & Atienza, J.M., Distinct fracture patterns in construction steels for reinforced concrete under quasistatic loading—A review. *Metals*, **8**(3), pp. 171, 2018.

[32] Concli, F. & Gilioli, A., Numerical and experimental assessment of the static behavior of 3D printed reticular Al structures produced by Selective Laser Melting: progressive damage and failure. *Procedia Structural Integrity*, **12**, pp. 204–212, 2018.

[33] Concli, F., Gilioli, A. & Nalli, F., Experimental–numerical assessment of ductile failure of Additive Manufacturing selective laser melting reticular structures made of Al A357. *Proceedings of the Institution of Mechanical Engineers, Part C: Journal of Mechanical Engineering Science*, 2019.

[34] A.ERRE Srl. di Serravalle Pistoiese (PT), Italy. www.aerrecarpenteria.it/.

MECHANICAL PROFILING OF COMMERCIAL GRADE BIODEGRADABLE PLASTIC FILMS POST EXPOSURE TO NATURAL AND ACCELERATED WEATHERING

SULTAN MAJED AL-SALEM[1], AFAF YA'QOUB AL-NASSER[1], MAJED HAMEED AL-WADI[1], HANAN HMOUD SULTAN[1], HAJAR JAWAD KARAM[1], AYYAD TOMAN AL-DHAFEERI[2], MOHAMMAD AL-FOUDAREE[1] & ZAKIA RASHEED[1]

[1]Environment and Life Sciences Research Centre, Kuwait Institute for Scientific Research, Kuwait

[2]Water Research Centre, Kuwait Institute for Scientific Research, Kuwait

ABSTRACT

Market standards for plastic products are typically governed by mechanical properties that manufacturers and converters adhere to. However, in the case of biodegradable plastic film products, namely bags used on the market, the standards must convey more environmentally related factors. In this paper, mechanical profiling of three commercial grade market products were tested. The plastic films were deemed biodegradable of oxo-biodegradable nature containing calcium carbonate (CaCO3) amongst other prodegradant chemical agents. The evaluation of mechanical properties was conducted in laboratory conditions to determine the impact of exposing the samples to various degradation media, in accordance to ASTM D 882. Experiments were conducted using a universal testing machine (50 kN load cell) set to a 500 mm min-1 test speed. The measured properties included Young's modulus (MPa), ultimate stress (MPa), ultimate strain (%), stress at break (MPa), strain at break (%) and force at break (N). The width and thickness of the film specimens were measured using an internally calibrated Vernier callipers to the nearest 0.1 mm. Processing and cutting variation was minimal between the width and the thickness over the length of some samples accounting for less than 1% of total number of samples measured. The determination of the mechanical properties was conducted post accelerated and natural weathering as per the standard test methods of ASTM D 4329 and ASTM D 5272, respectively. Statistical analysis utilizing the means of one-way analysis of variance (ANOVA) was also conducted among the samples exposed to natural weathering with respect to variables of meteorological conditions and primary/secondary airborne pollutants. The change of the mechanical properties has led to the understanding of the response of polymeric films deemed as biodegradable in response to exposure to natural environment, where waste minimization and standard development can be achieved for future mitigation.

Keywords: strain, oxo-biodegradable, polymer, stress, biodegradable plastic.

1 INTRODUCTION

In the current climate, there is a surge of controversy surrounding the topic of plastics. Undeniably and unavoidably, plastics have become a crucial part of this generation. However, the inappropriate disposal of plastic waste, as well as, the increase in consumption provokes serious environmental issues. The rising environmental awareness on a global scale has imposed considerable pressure for associated authorities to take actions. This includes the research into alternative options to conventional plastics which is latched on to a negative reputation. Nevertheless, the *greenness* of these so-called environmentally friendly alternatives is often questionable. This is especially true when the routes of biodegradation are in question, i.e. oxo-biodegradation and hydro-biodegradation. Since April 2015, the European Parliament and the European Council established Directive EU 2015/720 [1]; which is an amendment of Directive 94/62/EC [2] as regards to reducing the consumption of lightweight plastic carrier bags. Subsequently, the directive aims to reduce the level of littering of these bags which accumulates in the environment where more plastics related problems (i.e. marine pollution) aggravates.

WIT Transactions on Engineering Sciences, Vol 124, © 2019 WIT Press
www.witpress.com, ISSN 1743-3533 (on-line)
doi:10.2495/MC190221

The official (bio)degradation mechanism for oxo-degradable plastic is defined by the British Standard as a "degradation mechanism resulting from oxidative and cell-mediated phenomena, either simultaneously or successively" [3]. This definition is also referenced in CEN/TR 15351: 2006. However, it must be emphasised that only the process is defined, but the terminology "oxo-degradation" is not standardised. In the current market, there are conventional plastics enriched with additives which aim to aid in the degradation process when initiated by the presence of oxygen and accelerated by light and/or heat. This type of plastics can be divided into two groups. The first group is accountable for the majority of these additivated plastics which is called oxo-degradable plastics. The behaviour of biodegradable polymers (in general) and oxo-biodegradable plastics (in particular) under weathering tests and exposure to climatic conditions is a scant topic in literature.

Materials exposed to certain conditions in the outdoors, will deteriorate with time. These materials include synthetic polymers, including common polyolefin (PO) plastics such as polyethylene (PE) and polypropylene (PP), as well as elastomers, wood, hair, wool, skin protein and cellulose [4]. Majority of PO polymers are used in the packaging industry among other applications. In Europe, out of the 58 million tonnes produced in the year 2015, about 40% went to satisfy the packaging sector's demand [5]. Other applications of such materials will typically include indoor products oriented for customer use such as carrier bags and outdoor products such as mulches, plastics covers and greenhouses. These materials will degrade over time slowly due to exposure to solar radiation in a photodegradation induced mechanism, losing their integrity and attributes. Other influencing parameters have a major contribution on the rate of photodegradation were temperature, humidity and other meteorological conditions contribute immensely to it. However, it was deemed that ultra violet (UV) levels and exposure is the most significant factor resulting in breaking the polymer chains, producing radicals and reducing the molecular weight of the polymer. When these polymers are exposed to the natural weather and start to degrade due to a negative influence of environment, the process is called weathering [6]. A detailed discussion on the influence of UV and climatic conditions impact on polymeric materials can be found elsewhere [7]–[9].

The intensity of light due to solar radiation that reaches to the surface of the earth is associated with the breakage of these chemical bonds that tie polymer chains together. The solar radiation that reaches the surface of the earth is based on wave lengths between 295 and 2500 nm. Solar radiation between 280 and 315 nm is termed UVB and has an energy of 426–380 kJ mol^{-1}. UVA has an energy of 389 and 300 kJ mol^{-1} between the wave lengths of 315 and 400 nm [10]. Only 4–6% of the solar radiation that reaches the earth's surface is capable of breaking chemical bonds (290–400 nm) [11]. Based on the aforementioned principles, weathering tests have been established and standardised to determine the loss of products integrity after exposure.

In this work, and following the current needs of both governmental and private entities alike within the state of Kuwait; three commercial grade oxo-biodegradable film products are studied within the scope of this communication. The products are claimed to be both biodegradable under climatic conditions of Kuwait, and of oxo-biodegradable origin. The products were assessed for their biodegradability and their properties under climatic conditions are reported with respect to accelerated and natural weathering tests. The mechanical profile of the specimens was investigated as an indicator to the degradation profile of the materials.

2 MATERIALS AND METHODS

2.1 Samples acquirement and assessment

Three types of plastic bags used in the Kuwaiti market were secured from local dealers with their main characteristics depicted in Table 1. The samples acquired were presumed to be environmentally friendly and biodegradable, which were secured from various sources within the state. The samples were cut into standard testing dimension specimens accepted by international protocols for thin plastic film characterization as described previously in [12]–[14].

Table 1: Types and description of plastic bags considered for testing [12]–[14].

Biodegradable bag type	Application	Average thickness per test specimen
Type 1 – Yellow in colour	Trash bag	70 μm
Type 2 – Pigmented with blue	Trash bag	40 μm
Type 2 – Pigmented with white	Carrier bag	30 μm

2.2 Fourier transformed infrared spectrometry (FTIR)

An IS10 Thermo-Nicolet FTIR unit equipped with an attenuated reflectance (diamond) attachment was used for the analysis and the identification of material's chemical fingerprint. The test was conducted in accordance with ISO 10640. Sample were tested using 32 scans for the background and each individual spectrum in a spectra range of 4000–400 cm^{-1} following the methodology depicted previously in Al-Salem et al. [14] and Al-Salem [15].

2.3 Natural weathering of film samples

In this work, the films intended for characterization post natural weathering were arrayed over non-expanded metal trays of non-corroding aluminium mesh with openings covering ≈70% of its surface area in accordance with ASTM D 5272-08 [16] using design for exposure rack "A" with an approximate individual diameter of openings equal to 0.5 cm. Each rack had seven trays, each divided into two segments allowing 60 specimens to be mounted on. All racks were placed at a 45° angle as per ASTM D 1435-05 [17] and ASTM D 5272-08 [16]. The first tray was positioned at 90 cm from ground level, and each tray was of 15 cm in width and 177 cm in length. Two sites were chosen to simulate the impact of weather in Kuwait. The first was located in an arid area close to the bay (Shwiakh area, latitude N 29°13'49"; longitude 47°58'31" E) with a mean annual ambient temperature of 32°C and relative humidity of 30%. The second, Salmyia area (latitude N 29°33'33"; longitude 48°8'33" E), is a seafront area with a mean annual ambient temperature of 30°C and relative humidity of 60%. The start of the natural weathering loading was on the 26th December 2017 and were kept with the intention of exposing the samples for a continuous period of one year. The film samples were mounted on the trays with a minimum spacing of 11 mm between each individual specimen. The racks were facing southward direction in accordance with ASTM 1435-05. A minimum of ten specimens of each type were unloaded of each tray at each site after 1, 3, 6, 9 and 12 months for further testing and characterization. The unloaded film samples were carefully placed in plastic bags and rinsed for 3–5 s before drying and conducting the various experiments. Readers are referred to Al-Salem et al. [7] and Al-Salem [9] for more details on the methodology.

2.4 Accelerated ageing of test specimens

Accelerated (artificial) indoor weathering (ageing) tests are commonly used for studying materials integrity and were conducted in this work to correlate the results obtained with natural weathering ones. Films of the study samples were exposed to accelerated weathering (AW) in accordance with ASTM D 4329. Samples were mounted on the racks facing the UV lamps with no empty spaces in the panels. This is in order to maintain a uniform repeatable test conditions. Cycle A procedure was used for general applications durability testing, i.e. 8 h of UV exposure at 60°C followed by 4 h of condensation at 50°C. At the end of each continuous weathering test, the chamber was cooled to room temperature and the trays were set to rest on a flat surface for a minimum of 24 h. Samples were laid to rest for a minimum of 72 h before characterization following internationally recognised methodologies and laboratory testing protocols of weathering. A minimum of four replicates were exposed to the different exposure durations in the QUV machine chamber. UV lamps irradiance was also selected according to ASTM D 4329 [18], and the lamp type was set to be 0.68 W m^{-2} (irradiance) for normal lamp operation, which was maintained for almost 5000 h of operation. The irradiance sensor was calibrated every 400 h of lamp operation during the UV cycle under normal test temperature. The equipment used was cleaned every 800 h to remove scale deposits resulting from water evaporation during the condensation cycles. Readers are referred to Al-Salem et al. [7], [8], [14] for a detailed depiction on the accelerated weathering procedure.

2.5 Mechanical profiling

The evaluation of mechanical properties was conducted at room temperature to determine the impact of exposing the samples to various degradation media in following ASTM D 882 [19]. Experiments were conducted using a Tinuis Olsen (UK) 50 ST Model – universal testing machine (50 kN load cell) set to a 500 mm min^{-1} test speed. The measured properties included Young's modulus (MPa), Ultimate Stress (MPa), Ultimate Strain (%), the stress at break (MPa), the strain at break (%) and the force at break (N). The width and thickness of the film specimens were measured using an internally calibrated Vernier callipers to the nearest 0.1 mm. Processing and cutting variation was minimal between the width and the thickness over the length of some samples accounting for less than 1% of total number of samples measured. The variation not being more than ±0.02 mm for width and ±0.01 mm for thickness with a constant width of 10 mm and thickness as depicted previously. The maximum extension available was 1065 mm equivalent to a strain of 900%. Strain at break (elongation at break) was measured based on cross head displacement and a gauge of 100 mm. For yield strength, it was observed that some specimens showed multiple yield points and some had a distinct yield. To standardize the results for easy comparison between all specimens, it was decided to consider the first yield point as the yield strength of the specimen. The measurement values obtained came from the average of the replicate specimens stored directly from the machine's software (Horizon 10.2.5.0). The tests were conducted with a gauge grip separation of 100 mm as suggested by the experimental protocol followed and was also set to terminate at 75% of ultimate force. The threshold of detection was input in the test program at 3 N. To assure accurate readings that reflects the true modulus of elasticity determined from the stress and strain curves, the grip separation was reduced to 25 mm for the accelerated weathering specimens at threshold limit of weathering exposure. This was done due to the fact that samples were shorter due to exposure to UV for prolonged periods of time in the weathering chamber.

2.6 Statistical analysis

Analysis of variance (ANOVA) testing (F-test) for the comparison of the significant variance between two or more groups was conducted to analyse the interaction of independent variables considered in this study. These included physical properties of the different plastic types (haze (%), light transmission (%) and ΔE), as well as the mechanical properties. Independent one-way ANOVA test was conducted amongst the studied groups with respect to the different natural weathering durations considered in this work. F values as a ratio of mean sum of squares due to treatment (MST) to mean sum of squares due to error (MSE), and p values as the total number of population, were both examined to determine the dependency of the variables with respect to group. *F* critical taken as upper limits of the normal distribution was considered as the limit of acceptance/rejection of null hypothesis. The analysis covered both sites considered in this work. Relevant pollutant levels and meteorological conditions were used to check dependency of variables on each other with respect to the dependent variables. The variables considered were as per the following to study the extent of pollutant and meteorological conditions impact of exposed samples: Particulate matter (PM_{10}, mg m^{-3}), total hydrocarbon (THC, ppm), methane (CH_4, ppm), non-methane hydrocarbons (NMHC, ppm), ozone (O_3, ppb), sulphur dioxide (SO_2, ppb), hydrogen sulphide (H_2S, ppb), nitrogen oxides (NOx, ppb), ammonia (NH_3, ppb), benzene (C_6H_6, ppb), carbon monoxide (CO, ppm), carbon dioxide (CO_2, ppm), temperature (°C), wind speed (m s^{-1}), wind direction (°) and relative humidity (%). Readers are referred to Al-Salem [9] for more details on the pollutants levels data collection. The analysis was conducted utilizing the means of Microsoft Office Excel 2007 – Data Analysis tool as a built-in function. The methodology has been previously reported in Al-Salem [9].

3 RESULTS AND DISCUSSION

3.1 Degressive behaviour and physical deterioration

The natural weathering test conducted in this project was achieved successfully for nine continuous months. All samples exposed to the action of weathering in this work appeared consistent and comparable by touch. Physical observations were noted as a first stage and a means of initial assessment for impact of accelerated and natural weathering on the exposed samples. Signs observed *visually* can lead to the early detection of deterioration and can indicate the severity of the weathering tests conducted [7]. The aim of the test as per the proposed workplan of this research, is to test the specimens after one year of continuous natural weathering at both sites. The work is ongoing to date.

The samples had a stable and gradual degradation with respect to exposure time judging by the visual appearance of the plastic films of all three types. The winter season faced in Kuwait last year was observed to be somewhat cloudy at the start of the experimentation stage with an average direct sunlight exposure of 5 h. It should also be noted that signs of crazing and haziness were visible from the first unloading session (i.e. 30 days) on the 25th January 2018. The notes were taken after rinsing and examining the specimens in the laboratory. More pictorial analysis is shown elsewhere [14]. It was also apparent that the photo-degradation of the PO polymers serving as substrates to the biodegradable blends, was clearly related to the meteorological conditions which accelerated the degradation process with the changes in seasons of exposure. Increase in blistering and crazing was noted with both site loadings as well, as a function of weathering time. Understanding the mechanism of photodegradation is essential to study the phenomena itself and analyse the status of the samples after various exposure durations. There exists various factors affect the photodegradation of polymers in the outdoors. It was apparent that chain propagation affects the samples in different rates, with respect to the sites used in this work. The combination of UV and thermal degradation at site no. 1 was apparently more severe than site no. 2. This could be attributed to the arid nature of the first location [9]. Yousif and Haddad [10] reported that the combination of UV and ambient temperatures influence the samples to a great extent with respect to natural weathering. This goes hand in hand with the findings in this research based on the total colour change (ΔE) estimated and shown previously in Al-Salem [20].

It is well established that the exposure to UV is the most influential of all these factors [10]. It causes plastic products to lose integrity, dye and pigments where a clear discolouration is typically noted with the ageing [21]. Typically stabilisers are added to these materials to overcome such phenomena and enhance various mechanical properties. It was also determined that such stabilizers are added to the samples studied in this work due to the fact that samples are biodegradable in nature (as chemical analysis will reveal). The samples of Type 3 started to be more brittle after exposure to 5 months in site no. 1. This supports the claim that harsh weather with less humidity has more acute effect on the samples than humidity in site no. 2.

3.2 Plastic films assessment and biodegradable type investigation

A sample of the FTIR spectra obtained from scanning the bag samples are shown in Fig. 1. The main peaks of the spectra are characteristic peaks for a PE grade plastic material. Additional peak areas that are associated with plastic additives are also visible with the IR spectroscopy. These additional areas are noted within the range of 850 cm^{-1} and 1450 cm^{-1} of the IR spectrum [14]. Type 2 showed a spectrum that is different than the other two products. No additional peak was evident at the 1450 cm^{-1} region and the peak at 850 cm^{-1} was significantly less intense when compared with the other sample bags. This is an

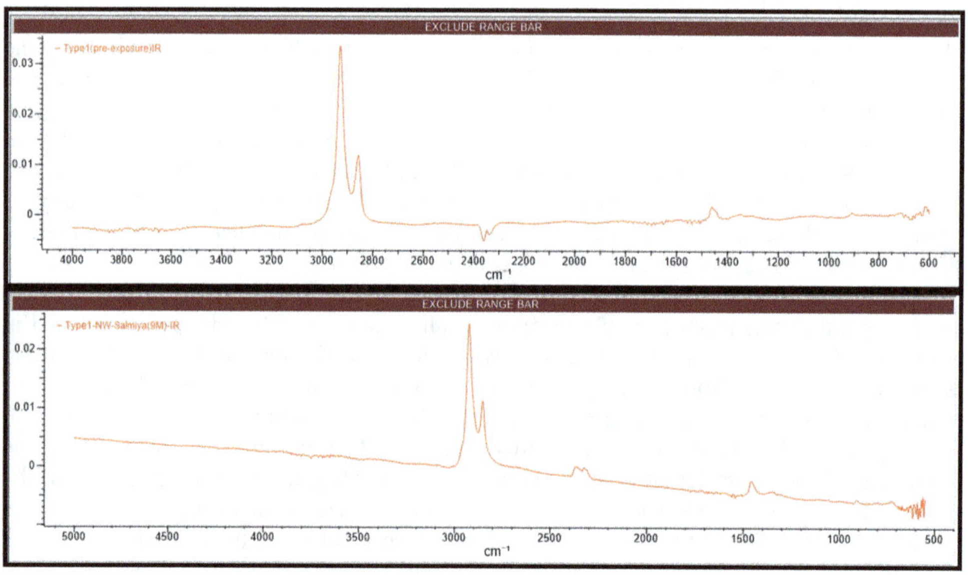

Figure 1: Sample of peak identification from FTIR analysis of Type 1 sample (pre-exposure (top) and after exposure to natural weathering by 9 months (bottom).

indication that the formulation of these bags is different. From the assessment aforementioned, and depicted in our past work [12]–[14], these bags are characterized as oxo-biodegradable bags with high content of calcium carbonate ($CaCO_3$). Readers can refer to Al-Salem et al. [14] for the thermal stability experimental results that confirm the aforementioned analysis.

Past studies have implemented similar techniques to identify the products evolved from TG/IR. The main idea of the study reported in Ewing and Kazarian [22] was the use of supercritical carbon dioxide ($scCO_2$) for polymeric processing which is an increasing area in industrial processes. FTIR spectroscopy provided information of the interactions between CO_2 and polymers under high-pressure and supercritical conditions. It was used to study the intermolecular interactions between CO_2 and polymers functionalized with different functional groups such as, a polyketone, polybutadiene, polyalcohol and polyoxime. Morphological changes were studied for polyethylene terephthalate (PET), a semi-crystalline polymer after exposing it to supercritical ($scCO_2$) for a range of exposure times in the work of Fleming et al. [23]. FTIR determined the $scCO_2$-induced morphological gradient. The range of the spectra between 1320–1360 cm^{-1} region revealed that the CO_2-induced influenced the morphological reorganization as a function of depth for about 30, 40 and 50 μm for the 15, 20 and 30 min samples, respectively. Other studies in the investigation of the degradation profile focused on polyoxymethylene (POM)/thermoplastic polyesterurethane (TPU) by thermogravimetric analysis coupled with FTIR spectroscopy [24]. The work showed four different types of samples (i.e. POM, TPU, POM/TBU 90:10 blend and POM/TBU 80:20 blend). The work was carried out at operating conditions of heating rate equal to 20 K min^{-1}, sample mass = 3 mg, nitrogen flow rate = 50 cm^3 min^{-1}). The FTIR spectra of gases evolved during thermal degradation of POM show peaks at 1745 cm^{-1} carbonyl group, and at 2800 cm^{-1} aldehyde group where TPU show butyl ether and formaldehyde same as the POM/TPU blends [24]. By surveying the technical literature related to the topic, it can be noted that the samples under investigation are of polyolefin nature based on the CO_2 signal being detected by the IR signal. The evolution of the additives in the signal post exposure to weathering, shows that the samples are biodegradable and of commercial grade nature. The signal was compatible was past work related to plastic grades used on the market [15]. The shift in the signal for the carbon based compounds, and the loss of additives due to weathering (excluding evolved compounds from UV exposure – site no. 2) can be seen in Fig. 1 for Type 1 plastic sample.

3.3 Mechanical profile determination

In this section, the mechanical properties measured are reported with respect to each degradation media and exposure duration. The film samples of the various types were exposed to different durations of soil burial, accelerated and natural weathering, and the properties were measured thereafter. The accelerated weathering data are discussed at first. The degradation threshold limits determined from the accelerated weathering tests were reported in our previous work [14]. Tables 2 and 3 show a sample the mechanical properties evaluated post accelerated and natural weathering in this work. Figs 2 and 3 show a sample of the mechanical properties depicted as a function of accelerated weathering exposure. The threshold limit (i.e. point of total deformation) were determined for Types 1, 2, and 3, as 11, 13 and 19 days of continuous aging [14].

It was noted that for Types 1 and 3, the Young's modulus has increased gradually until the threshold limit of the deterioration point was reached by 7.5% and 2.8% for Types 1 and 3, respectively. Type 2 showed a slight decrease in the modulus after 312 h of UV exposure

Table 2: Mechanical properties evaluated post natural weathering (site no. 2) for Type 1.

Exposure duration (h)	Young's modulus (MPa)	std.	Ultimate stress (MPa)	std.	Ultimate strain (%)	std.	Break stress (MPa)	std.
0	320.00	63.29	14.00	1.83	387.00	23.53	14.00	2.00
30	281.40	57.27	13.56	1.75	356.40	60.90	13.08	1.71
90	293.80	45.93	10.95	1.68	65.10	40.57	10.534	1.70
180	489.60	43.00	15.76	1.06	12.47	2.83	1.01	0.61
270	332.5	30.5	4.375	0.645	9.635	7.765	0.0343	0.0083

Table 3: Mechanical properties evaluated post accelerated weathering for Type 1.

Exposure duration (h)	Young's modulus (MPa)	std.	Ultimate stress (MPa)	std.	Ultimate strain (%)	std.	Break stress (MPa)	std.
0	320.00	6.00	14.00	1.83	387.00	23.53	14.00	2.00
88	384.00	4.00	12.49	1.61	56.46	6.00	6.32	0.50
176	283.71	5.00	11.47	3.83	78.38	24.00	2.07	0.09
246	344.25	15.00	16.75	2.74	384.78	18.00	14.93	3.21

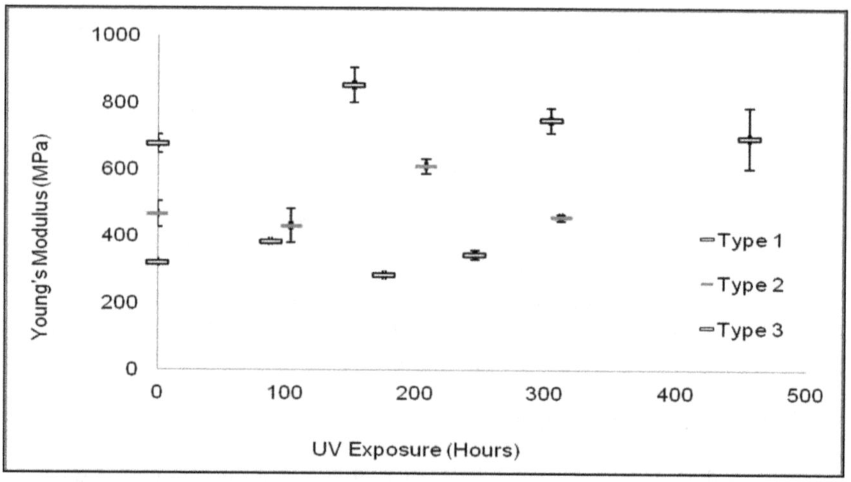

Figure 2: Young's modulus (MPa) with respect to accelerated weathering duration for the studied specimens.

by 2.5% (Fig. 2). Previous reports show that the modulus is governed by the crystallinity of polymers and the chemical changes with continuous exposure [25]–[27]. PE has a reported crystallinity of 40–60% that is highly susceptible to degradation with UV exposure and its molar masses were reported to photodegrade even with the presence of secondary antioxidants in composites [28]. Hence, an alteration to the Young's modulus was expected in this study due to crystallinity changes as a consequence of UV exposure. Parts of this discussion was shown previously in Al-Salem [9] and Al-Salem et al. [14] for various blends of polyolefin materials exposed to weathering conditions. It was suspected that the

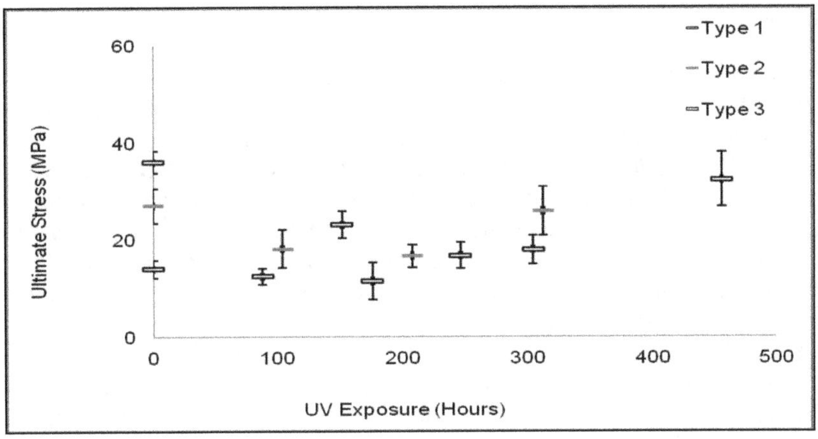

Figure 3: Ultimate stress (MPa) with respect to accelerated weathering duration for the studied specimens.

amorphous regions in the PE had undergone degradation unlike the crystalline ones, which has remained unaffected without the formation of new crystalline regions in the polymer's matrix with weathering time [29]. However, there is a possibility of forming new crystalline regions in the matrix with PE polymers that results in an increased total crystallinity, which is attributed to the chain scission reaction of the degradation mechanism according to Sam et al. [29]. It was reported by Roy et al. [30] that natural weathering and, consequently, by mimicking its effect, accelerated weathering caused polymers to become more brittle, which was noted in this work with samples exposed to natural weathering. The embrittlement of the samples which was noted in this work with samples exposed to accelerated weathering is in agreement with the alteration in Young's modulus due to the occurrence of crack initiation from chain scission that gave rise to sufficient chain mobility, producing secondary crystallization [30].

Alexander [31] and Gulmine et al. [32] indicated that HDPE degraded very rapidly and was more prone than other PO polymers to weathering. On the other hand, Winslow [33] and Iring et al. [34] indicated that LLDPE degraded faster than other polyolefins. Hence, it was hard to pinpoint the exact polymer that could degrade faster in such products causing the collapse in the polymer's matrix which has to be judges on each case individually. PE and PP are susceptible to UV irradiation, and the mechanism of degradation has been reported previously in detail by Philip et al. [11]. This effect lead to the loss of the amorphous region in the polymers matrix due to cross-linking reactions, which lead to a change in the modulus measured at various intervals of accelerated weathering indicated by the change in the stiffness/brittleness of the samples (Figs 2 and 3).

Stress at break (MPa) is reported with respect to the duration of accelerated weathering exposure in Fig. 3. A decreasing trend till the point of 2/3 of the total duration of the accelerated weathering test. was noted with the studied samples which was expected due to cross-linking effect with exposure to UV. However, all types studied showed an increased strain value at break at threshold limit. This behaviour was related to chain scission in semi-crystalline polymers, which lead to a decrease in the molecular weight of the material and consequently to a loss in strain and elongation at break with weathering [35]. The natural weathering results are shown below (Tables 2 and 3).

3.4 Statistical analysis of variance

The result obtained from one way ANNOVA that revealed dependency and independency of the variables for the different properties at two sites tested three different types of polymers. With regards to site no. 1 (Shwaikh) which is noted to be arid in climate, Type 1 of the studied samples showed great dependency of haze on ED, PM_{10}, NO, NO_2, NO_X, O_X, NH_3 and RH. This was also noted for samples of Type 2. The duration of the exposure leads the materials for prolonged duration of exposure under various climatic conditions that might affect the surface of the plastics exposed. This is measured with haze (%) and was noted clearly in the analysis conducted on site no. 1. Therefore, it can be confirmed that the materials are degrading with such exposure. The thickness of the materials plays a major role in the fact that haze (%) Type 3 was dependent on ED, PM_{10} and NO_X only. The level of oxygen in the atmosphere is suspected to be irrelevant at thicknesses below 40 μm. Light transmission on the other hand was noted to be dependent on ED, PM_{10}, NO_X, O_X, NH_3 and RH, for Type 1 of the studied samples. It was also dependent on ED, PM_{10}, NO_X, O_X and NH_3 for Type 2. Light transmission depends on ED, PM_{10}, NO_X and NH_3 for Type 3 of the studied samples at site no. 1. It can be deduced that the light transmission is highly sensitive to days of exposure and levels of particulate matter in the air. Both Types 1 and 2 were pigmented plastic articles. ΔE was dependent on ED, O_3, toluene, temperature and TRS for Type 1 (site no. 1) and ED, THC, CH_4, NMHC, toluene, benzene, o-xylene, ethylbenzene, CO, mp-xylene, temp. and WS for Type 2 (site no. 1). Samples of Type 3 were dependent on Δ ED, THC, CH_4, NMHC, O_3, SO_2, H_2S, NO_2, toluene, benzene, o-xylene, ethylbenzene, CO, mp-xylene, temperature, WS and TRS. On a whole, ΔE at site no. 1 was dependent very strongly on the exposure temperature and duration. This can be noted as a very sensible conclusion since the exposure to prolonged weathering duration will lead to loss of pigmentation.

Samples exposed to site no. 2 were also analysed for physical properties dependence on climatic conditions. Haze (%) was noted to be very dependent on ED, PM_{10}, NO_2, NO_X, O_X, NH_3 and RH for samples of Type 1 (site no. 2). For Types 2 and 3, haze (%) was dependent on ED, PM_{10}, NO_X, O_X and NH_3 (Type 2) and ED, PM_{10}, NO_X and NH_3 (Type 3). Types 2 and 3 showed that Young's modulus was highly dependent on CO_2, whilst Type 1 was dependent on WD (site no. 1). However, it was dependent on WD and CO_2 concentration for Types 1 and 2, respectively at site no. 2, and independent of any variable for Type 3. Ultimate stress was strongly dependent on ED and TRS for all types in both locations. The continuation of the work will lead to better understanding to dependency of variables on climatic conditions.

4 CONCLUSION

Three types of biodegradable plastic bags used in the form of standard films as per international protocols for plastic characterization, were examined and studied in this work. The work aims at determining the degradation profile with respect to natural and accelerated weathering exposure. In addition, the determination of the nature and type of the bags was conducted. The following summarises the main technical findings of this work. The results can also lead to the determination of the optimal engineering standards based on the characterization campaign conducted in this research:

- Natural weathering as per internationally recognised protocols was achieved in this work. The samples of the three types were exposed to two sites. The first was noted to be of a more arid and harsh climatic conditions, where the second has a more humid climate.

- It was noted that samples at site no. 1 were more severely crazed with respect to weathering exposure time. The samples with the lowest thickness (i.e. Type 3) were noted to deteriorate more severely especially after 5 months of continuous exposure. This was attributed to the thickness of the samples and UV penetration to the surface of the polymeric material.
- UV radiation is noted to be the most influential parameters that degrades the plastic films which leads to their photodegradation in the environment. The fact that samples exposed to arid environments have deteriorated more rapidly than the ones exposed to the humid site, indicates that starch component is not active in the formulated blends. This supports the conclusion that the samples studied contain prodegradants that are different in nature than hydro-based compounds.
- The infrared spectra identified the materials to be of polyethylene in nature containing other additives between the 850 cm^{-1} and 1450 cm^{-1} regions of the infrared spectrum. The polyethylene grade was LLDPE, LDPE and HDPE depending on the formulation of the polymeric film and the type of the material tested. Calcium carbonate ($CaCO_3$) was noted to be present in the plastic bags as a main oxo-biodegradable additive.
- Weatherometer results indicated that the samples are degraded under UV attack which is also consistent with oxo-biodegradable polymers. FTIR analysis for Type 1 and Type 3 after weathering revealed the presence of carbonyls, which is a direct indication of degradation.
- Type 1 of the samples had the largest change in accelerated weathering with respect to haze (%) indicating that it was the sample with the highest change in surface chemistry.
- Type 3 of the studied samples showed the highest change in colour where ΔE was estimated to be 23.001 after 456 h of continuous exposure to accelerated weathering [14]. This is in agreement with natural weathering tests conducted at both sites and compliments the work conducted in this research.
- The statistical analysis showed that the dependency of the haze, light transmission, ΔE and mechanical properties with the specific parameters as shown in the one way ANOVA test. The significant variables on the samples tested at site no. 2 (Samlyia) on Type 1 were determined as exposure duration, ozone and total reduced sulphur. The significant variables on the samples tested at Salmiya site on Type 2 were exposure duration and total reduced sulphur. The significant variables on the sample tested at Salmiya site on Type 3 were exposure duration, ammonia, sulphur dioxide and total reduced sulphur.
- Focusing on the significant variables on the samples tested at site no. 1 (Shwaikh site) on Type 1, these were exposure duration, wind direction and total reduced sulphur. The significant variables on the sample tested at Shwaikh on Type 2 were exposure duration and total reduced sulphur. Finally, the significant variables on the sample tested at Shwaikh on Type 3 were exposure duration, sulphur dioxide and total reduced sulphur. The regression coefficient determined for both sites showed that the dependency is very weak on the aforementioned variables.

The work in this research can also direct various agencies in Kuwait and the Gulf Council Countries to develop standards that can govern the use of biodegradable plastics under climatic conditions. It can also lead to the development of various products that can be compounded with the specific needs of localities in both Kuwait and surrounding countries that suffer from harsh climatic conditions. Results so far indicate that Type 1 is an appropriate blend to be used for oxo-biodegradables in Kuwait, where the deterioration based on their mechanical profile was more rapid. Type 1 also showed signs of crazed cracks on the surfaces, however, during the exposure time-spans to the natural climate, the samples showed sustained modulus values.

ACKNOWLEDGEMENTS

The Project Leader would like to thank the Kuwait Institute for Scientific Research and Kuwait Municipality for funding and supporting this research project under the grant for project EM097C. The continuous support and encouragement of the management of the Environment Affairs Department of Kuwait Municipality is highly appreciated, whom have granted their approval to publish this work. The contribution of the Kuwait Foundation for the Advancement of Sciences for funding the research mission is gratefully acknowledged.

REFERENCES

[1] Directive EU 2015/720. OJ L 115 of 6/5/2015, p. 11, 2015.

[2] Directive 94/62/EC of 20/12/1994 on packaging and packaging waste. OJ L 365 of 31/12/1994, p. 10, 1994.

[3] BS 8472, Methods for the assessment of the oxo-biodegradation of plastics and of the phyto-toxicity of the residues in controlled laboratory conditions, 2011.

[4] Sionkowska, A., Modification of collagen films by ultraviolet irradiation. *Polymer Degradation Stability,* **68**, pp. 147–151, 2000.

[5] Plastics Europe, *Plastics: The Facts, an Analysis of European Plastics Production, Demand and Waste Data*, 2016.

[6] Wypych, G., *Handbook of Material Weathering*, 4th ed., Chemtec Publishing: Toronto, 2008.

[7] Al-Salem, S.M., Abraham, G., Al-Qabandi, O.A. & Dashti, A.M., Investigating the effect of accelerated weathering on the mechanical and physical properties of high content plastic solid waste (PSW) blends with virgin linear low density polyethylene (LLDPE). *Polymer Testing*, **46**, pp. 116–121, 2015.

[8] Al-Salem, S.M. et al., Effect of die head temperature (DHT) at compounding stage on the degradation of linear low density polyethylene (LLDPE)/plastic film waste blends post accelerated weathering. *International Journal of Polymer Science,* 5147209, 2016.

[9] Al-Salem, S.M., Influential parameters on natural weathering under harsh climatic conditions of mechanically recycled plastic film specimens. *Journal of Environmental Management*, **230**, pp. 355–365, 2019.

[10] Yousif, E. & Haddad, R., Photodegradation and photostabilization of polymers, especially polystyrene: review. *SpringerPlus,* **2**, 398, 2013.

[11] Philip, M., Attwood, J., Hulme, A., Williams, G. & Shipton, P., Evaluation of weathering in mixed polyethylene and polypropylene products. The Waste and Resources Action Programme (WRAP) Report (UK), ISBN: 1-84405-092-0, 2004.

[12] Al-Salem, S.M. et al., Biodegradable films physical properties change with respect to degradation exposure to various media. *Proceedings of the 7th International Conference on Engineering for Waste and Biomass Valorisation*, Prague, Czech Republic, pp. 272–282, 2018.

[13] Al-Salem, S.M., Al-Nasser, S.H. & Karam, H., In situ aerobic biodegradation in soil contact using the standard test method of ASTM D 5988-12. *Proceedings of the 3rd International Conference on Energy and Environmental Science,* Seoul, South Korea, 2019.

[14] Al-Salem, S. et al., Investigating biodegradable bags standards and properties under Kuwait's environmental conditions. Annual Progress Report, KISR No. 15164, Project Code: EM097C, 2018

[15] Al-Salem, S.M., The first chemical identification of polyolefin (PO) waste blends using infrared spectroscopy. *Proceedings of the 9th International Conference on Waste Management and the Environment*, Seville, Spain, pp. 2–18.

[16] American Society for Testing and Materials (ASTM), Standard practice for outdoor exposure testing of photodegradable plastics, ASTM D 5272-08, Philadelphia, PA, 2013.

[17] American Society for Testing and Materials (ASTM), Standard practice for outdoor weathering of plastics, ASTM D 1435-05, Philadelphia, PA, 2005.

[18] American Society for Testing and Materials (ASTM), Standard practice for fluorescent UV exposure of plastics, ASTM D 4329, Philadelphia, PA, 2005.

[19] American Society for Testing and Materials (ASTM), Standard test method for tensile properties of thin plastic sheeting, ASTM D 882, Philadelphia, PA, 2010.

[20] Al-Salem, S.M., Waste not, want not: Standardising biodegradable plastics as a waste management mitigation strategy. *Proceedings of the International Conference on Engineering Challenges in Gulf Countries*, Kuwait City, Kuwait, 2019.

[21] Goldshtein, J. & Margel, S., Synthesis and characterization of polystyrene/2- (5-chloro 2Hbenzotriazole-2-yl)-6-(1,1-dimethylethyl)-4-methyl-phenol composite micro-spheres of narrow size distribution for UV irradiation protection. *Colloid Polymer Science,* **289**, pp. 1863–1874, 2011.

[22] Ewing, A.V. & Kazarian, S.G., Current trends and opportunities for the applications of in situ vibrational spectroscopy to investigate the supercritical fluid processing of polymers. *Journal of Supercritical Fluids*, **134**, pp. 88–95, 2018.

[23] Fleming, O.S., Chan, K.L.A. & Kazarian, S.G., FT-IR imaging and Raman microscopic study of poly(ethylene terephthalate) film processed with supercritical CO_2. *Vibrational Spectroscopy*, **35**, pp. 3–7, 2004.

[24] Pielichowski, K. & Leszczyńska, A., TG-FTIR study of the thermal degradation of polyoxymethylene (POM)/thermoplastic polyurethane (TPU) blends. *Journal of Thermal Analysis and Calorimetry*, **78**, 2004.

[25] Lundin, T., Cramer, S.M., Falk, R.H. & Felton, C., Accelerated weathering of natural fiber-filled polyethylene composites. *Journal of Materials in Civil Engineering*, **16**, pp. 547–555, 2004.

[26] Mendes, L.C., Rufino, E.S., de Paula, F.O.C. & Torres, A.C. Jr., Mechanical, thermal and microstructure evaluation of HDPE after weathering in Rio de Janeiro city. *Polymer Degradation and Stability,* **79**, pp. 371–383, 2003.

[27] Mourad, A.I., Thermo-mechanical characteristics of thermally aged polyethylene/ polypropylene blends. *Materials and Design,* **31**, pp. 918–929, 2010.

[28] Ojeda, T. et al., Degradability of linear polyolefins under natural weathering. *Polymer Degradation and Stability*, **96**, pp. 703–707, 2011.

[29] Sam, S.T., Ismail, H. & Ahmad, Z., Effect of cobalt stearate on natural weathering of LLDPE/soya powder blends. *Polymer-Plastics Technology and Engineering,* **50**(9), pp. 957–968, 2011.

[30] Roy, P.K., Surekha, P., Rajagopal, C., Chatterjee, S.N. & Choudhary, V., Studies on the photo-oxidative degradation of LDPE films in the presence of oxidised polyethylene. *Polymer Degradation and Stability,* **92**(6), pp. 1151–1160, 2007.

[31] Alexander, M., Biodegradation: problems of molecular recalcitrance and microbial fallibility. *Advance Applied Microbiology,* **7**, pp. 35–80, 1965.

[32] Gulmine, J.V., Janissek, P.R., Heise, H.M. & Akcelrud, L., Degradation profile of polyethylene after artificial accelerated weathering. *Polymer Degradation and Stability*, **79**, pp. 385–397, 2003.

[33] Winslow, F.H., Photooxidation of high polymers. *Pure and Applied Chemistry,* **49**, pp. 495–502, 1977.

[34] Iring, M., Foldes, E., Barabas, K., Kelen, T. & Tudos, F., Thermal oxidation of linear low density polyethylene. *Polymer Degradation and Stability,* **14**, pp. 319–332, 1986.

[35] Rajakumar, K., Sarasvathy, V., Thamarai Chelvan, A., Chitra, R. & Vijayakumar, C.T., Natural weathering studies of polypropylene. *Journal of Polymers and the Environment,* **17**, pp. 191–202, 2009.

SECTION 9
NUMERICAL MODELS

MODELLING OF VOID NUCLEATION AND APPLICATION TO THE SIMULATION OF SPALL EXPERIMENTS

FENG-GUO ZHANG
Institute of Applied Physics and Computational Mathematics, Beijing, China

ABSTRACT

Spallation in ductile metals is often dominated by nucleation, growth and coalescence of voids. Previous research clearly shows that the volume density of nucleated voids is closely related to many factors such as grain size, loading condition and the like. In addition, it is confirmed that the nucleation of voids is inhibited by the existing voids with the plastic region around them. Therefore, a new model of dynamic damage is proposed, where the void nucleation equation is coupled with loading condition, porosity and the void growth is described by using a hollow sphere model with micro-inertia effects. The model is used to simulate spall experiments, and simulation results agree well with experimental data.

Keywords: spallation, void nucleation, ductile metal.

1 INTRODUCTION

Damage in materials subjected to dynamic tension is of great importance in many scientific fields. When a target plate is impacted by a flyer plate, compressive shock waves are generated in the two plates. As they encounter the free surface of the target and of the flyer, they reflect back as rarefaction waves. Tensile stress is generated wherever the rarefaction waves interact. As for ductile materials, damage (in the form of voids) will nucleate, grow and coalesce under the higher tensile stress, which can ultimately lead to the formation of a spall layer. Now, many theoretical damage models, both empirical and micro-physical in nature, have been developed to describe and predict phenomenon of spall fracture [1]–[6]. Some of these models are dependent on stress threshold, while others are dependent on strain threshold. Usually the damage level is quantified according to the void volume in the region of fracture for all these models. Up to now, there is no general and satisfactory spall model, and current models are confined in a narrow applicative range and should be further improved [7].

In this paper we focus our attention only on void formation. Recent research on the mechanics of void growth clearly shows that void nucleation may be represented as a bifurcation phenomenon, where a void forms spontaneously followed by highly localized plastic flow around the new void and the growing voids in turn induce relaxation zones in which local stresses decrease, thus decreasing the probability of nucleating voids in these zones [8]. In existing models, nucleation and coalescence are generally dealt with in an empirical fashion [3], [9]–[11]. In recent years, however, renewed attention has been paid to damage processes. In the present work, a new model of dynamic damage is proposed, where the void nucleation equation is coupled with loading condition, porosity and the void growth is described by using a hollow sphere model with micro-inertia effects.

2 VOID GROWTH

As suggested by Czarnota et al. [4] and Jacques et al. [12], the porous ductile material is modelled as an assemblage of hollow spheres where the pairs (a, b) and (A, B) are the current and initial values of the inner and outer radii respectively, and the ratio of b to a is assumed

WIT Transactions on Engineering Sciences, Vol 124, © 2019 WIT Press
www.witpress.com, ISSN 1743-3533 (on-line)
doi:10.2495/MC190231

to be identical for all hollow spheres. So, the macroscopic damage can be described by the porosity $\phi = a^3 / b^3$. A single hollow sphere is subjected to the external stress p, where p is a positive hydrostatic tension.

For void early growth ($b < c \leq a$), radius c is the elastic–plastic interface), the evolution law of hollow sphere can be written [13] as

$$\begin{aligned} &\rho(1 - \phi^{1/3})a\ddot{a} + \rho(1.5 - 2\phi^{1/3} + 0.5\phi^{4/3})\dot{a}^2 \\ &= p/(1 - \phi) - \frac{2}{3}Y_0(1 - \ln(Y_0 / 2G)) - \frac{4}{3}G \cdot \phi, \end{aligned} \tag{1}$$

where Y_0 and G are the yield strength and the shear modulus of the solid respectively. From eqn (1), for an elastic perfectly plastic material, the critical pressure at void nucleation is given by

$$p_c = \frac{2}{3}Y_0(1 + \ln(2G / Y_0)), \tag{2}$$

and radius c is

$$c^3 = (a^3 - A^3) \cdot 2G / Y_0. \tag{3}$$

While the entire sphere has yielded, the evolution law of hollow sphere can be written [2] as

$$\begin{aligned} &\rho(1 - \phi^{1/3})a\ddot{a} + \rho(1.5 - 2\phi^{1/3} + 0.5\phi^{4/3})\dot{a}^2 \\ &= p/(1 - \phi) - \frac{2}{3}Y_0 \ln\frac{1}{\phi} - 2\eta\frac{a\dot{a}}{a_0^2}. \end{aligned} \tag{4}$$

From eqns (1), (3) and (4), we can determine the parameter a_0

$$a_0^3 = a^3(1 - \frac{Y_0}{2G\phi}). \tag{5}$$

3 VOID NUCLEATION

In spallation conditions, where the negative pressures are large, void nucleation tends to be determined more by stress than strain. The volume density of nucleated voids increases with shock pressure and action time of negative pressures increases. At the same time, considering nucleation inhibition, as induced by the growth of already nucleated voids, we propose a new void nucleation rate function as

$$\dot{N} = k \cdot \frac{1}{t_c} \cdot \frac{p - p_c}{p_c}(1 - \phi \cdot 2G / Y_0), \tag{6}$$

where p_c is the critical pressure at void nucleation (see eqn (2)), t_c is the time when the tensile stress p equals to p_c.

4 COMPARISON OF CALCULATION WITH EXPERIMENT

Based on the proposed model, finite element simulations of plate impact tests are carried out to compare with the experiments on OFHC copper. There, the targets (4 mm thick) with the average grain sizes 30, 60, 100, and 200 μm were impacted by the quartz impactor (2 mm) of ~131m/s velocity. Results from the experiments show that void coalescence is observed to dominate the void growth behavior in the 30 and 200 μm samples, whereas void growth is dominated by the growth of isolated voids in 60 and 100 μm samples [14]. Because the present model does not include the process of void coalescence, we choice experiment data of simple with the average grain sizes 60 μm to simulate. Both the calculated results and the experimental data for the free surface velocity profiles are shown in Fig. 1 and damage statistics are listed in Table 1, where the ΔFSV is the free surface velocity difference between the first peak plateau and the first valley, the porosity is the void volume fraction. Comparison between the calculated and experimental results shows that they are in good agreement.

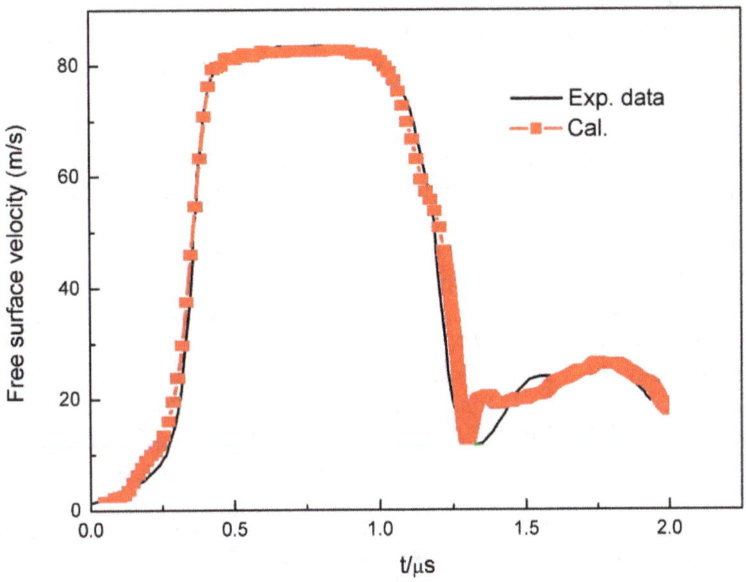

Figure 1: Free surface velocity histories.

Table 1: Damage statistics from optical analysis and numerical results.

	Number of voids (10^6/cm^3)	ΔFSV (m/s)	Porosity	Average void diameter (μm)
Exp.	27.15	71	0.166	11.3
Cal.	24.35	68.92	0.107	10.17

5 CONCLUSION

A new void nucleation rate function considering the effects of shock pressure and nucleation inhibition by the growth of already nucleated voids was proposed. The present approach, which includes a new void nucleation rate function and void growth rate in hollow sphere, is used to characterize spall damage in ductile metals subjected to dynamic loading conditions. For simulated plate impact experiment, calculated free surface velocity, number of voids, porosity and average void diameter agree well with experiment data, demonstrating the applicability of the newly developed model.

ACKNOWLEDGEMENTS

This work was supported by the Science Challenge Project (Grant Nos. TZ2018001 and TZ2016001) and the National Natural Science Foundation of China (Grant No. U1530261).

REFERENCES

[1] Gurson, A.L., Continuum theory of ductile rupture by void nucleation and growth. yield criteria and flow rules for porous ductile media. *J. Eng. Mater. Technol.*, **99**(2), 1977.

[2] Johnson, J.N., Dynamic fracture and spallation in ductile solids. *J. Appl. Phys.*, **52**(4) p. 2812, 1981.

[3] Tonks, D.L., Zurek, A.K., Thissell, W.R. & Schwartz, D.S., Modeling incipient copper damage from the tensile Hopkinson bar and gas gun. *Proceedings of the Conference of the American Physical Society Topical Group on Shock Compression of Condensed Matter*, vol. 706, p. 507, 2003.

[4] Czarnota, C., Jacques, N., Mercier, S. & Molinari, A., Modelling of dynamic ductile fracture and application to the simulation of plate impact tests on tantalum. *J. Mech. Phys. Solids*, **56**(4), p. 1624, 2008.

[5] Seaman, L., Curran, D. & Shockey, D., Computational models for ductile and brittle fracture. *J. Appl. Phys.*, **47**(11), p. 4814, 1976.

[6] Wilkerson, W. & Ramesh, K.T., Unraveling the anomalous grain size dependence of cavitation. *Phys. Rev. Lett.*, **117**(21), p. 215503, 2016.

[7] Chen, X., Asay, J.R., Dwivedi, S.K. & Field, D.P., Spall behavior of aluminum with varying microstructures. *J. Appl. Phys.*, **99**(2), p. 023528, 2006.

[8] Trumel, H., Hild, F., Roy, G., Pellegrini, Y. & Denoual, C., On probabilistic aspects in the dynamic degradation of ductile materials. *J. Mech. Phys. Solids*, **57**(12), p. 1980, 2009.

[9] Bontaz-Carion, J. & Pellegrini, Y.P., X-ray microtomography analysis of dynamic damage in tantalum. *Adv. Engng. Mat.*, **8**(6), p. 480, 2006.

[10] Thomason, P.F., Ductile spallation fracture and the mechanics of void growth and coalescence under shock-loading conditions. *Acta. Mater.*, **47**(13), p. 3633, 1999.

[11] Tonks, D.L., Zurek, A.K. & Thissell, W.R., Coalescence rate model for ductile damage in metales. *J. Phys. IV France*, **110**, p. 893, 2003.

[12] Jacques, N., Mercier, S. & Molinari, A., Void coalescence in a porous solid under dynamic loading conditions. *Int. J. Fract.*, **173**(2), p. 203, 2012.

[13] Carroll, M.M. & Holt, A.C., Static and dynamic pore-collapse relations for ductile porous materials. *J. Appl. Phys.*, **43**(4), p. 1626, 1972.

[14] Escobedo, J.P. et al., Effects of grain size and boundary structure on the dynamic tensile response of copper. *J. Appl. Phys.*, **110**(3), p. 033513, 2011.

ATOMISTIC SIMULATION OF EJECTA PRODUCTION FROM NANOCRYSTALLINE CU WITH A GROOVED SURFACE UNDER SHOCKWAVE LOADING

AN-MIN HE
Institute of Applied Physics and Computational Mathematics, Beijing, China

ABSTRACT

Large-scale molecular dynamics simulations are carried out to investigate the ejecta production from nanocrystalline Cu with a grooved surface under various shock intensities. The microjetting processes from the sample before and after shock-reduced melting are obtained, and the effects of polycrystalline as well as solid–liquid transition on ejecta properties are analyzed. It is found that polycrystalline structure has a significant influence on microjetting properties, including the critical condition of microjet formation and the total mass of ejecta, before shock-induced melting. After shock-induced melting, the microjetting from the nanocrystalline system tends to be consistent with the monocrystalline system.

Keywords: ejecta, molecular dynamics, nanocrystal.

1 INTRODUCTION

Periodic micron-scale surface perturbations are usually generated on the surface of metal during machining. When a shock wave generated by explosive or high energy laser reflects from the surface, microjet usually forms, which expands continuously and finally fragments into a great number of small particles called ejecta [1]. Due to its great importance in many fields such as inertial confinement fusion (ICF) and implosion dynamics, extensively experimental and theoretical studies have been conducted on this phenomenon [2]–[4]. Although a great progress in theoretical modelling of ejecta is achieved in recent years [5], [6], some underlying mechanisms related to microjetting process still remain unrevealed.

In the area of ejecta research, molecular dynamics (MD) simulation, which has been proven a powerful tool to explore the dynamic responses of material under extreme conditions, has been widely adopted to study the microscopic mechanism of ejecta production. However, previous works mainly focus on single crystals [7], and ejecta production from a more realistic polycrystalline remains unexplored. Here, we perform large-scale MD simulations to investigate the ejecta production from nanocrystalline Cu with a grooved surface under various shock loading conditions.

2 MODEL AND SIMULATION METHODS

Classical MD simulations are performed to study ejecta production from nanocrystalline Cu. The initial specimen is an idealized hexagonal columnar nanocrystal with a wedged surface groove as shown in Fig. 1. The atomic configuration of the columnar nanocrystal, which is composed of three types of idealized hexagonal columnar grains, is the same as the one in our previous work [8]. Type 1 grain is the reference grain with [100], [010], and [001] crystallographic directions along the x-, y-, and z-axes. Type 2 and 3 grains are rotated around the columnar axis by 60° and 30° relative to type 1 grain, respectively. The depth and vertex angle of the groove are 10 nm and 120°, respectively. For comparison, a monocrystalline Cu with the same surface perturbation is also constructed. The widely used Embedded Atom Model (EAM) potential developed by Mishin et al. [9] is adopted. Prior to shock wave loading, the crystal is equilibrated at 300 K and zero pressure with the constant–pressure–temperature

(NPT) ensemble and three-dimensional (3D) periodic boundary conditions. Then a shock wave is generated using the momentum mirror method, where the sample impacts a wall of infinite mass at a given velocity of $-u_p$ along the x-axis, and a shock wave is induced in the opposite direction. The shock intensity can be varied by changing u_p.

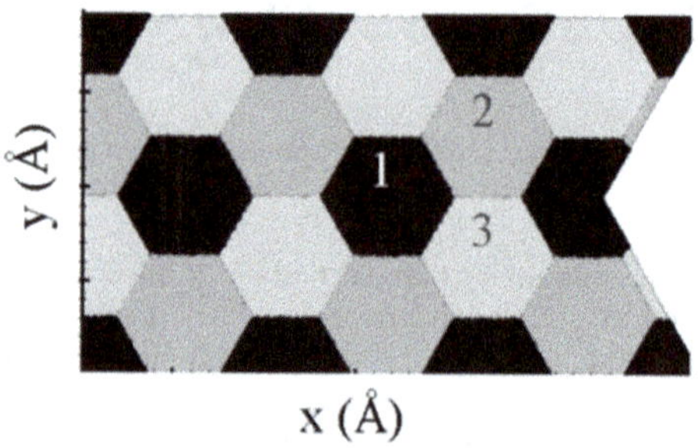

Figure 1: Initial configuration of the nanocrystalline Cu with a grooved surface.

3 RESULTS AND DISCUSSION

The microjetting processes of the nano- and monocrystalline Cu under different shock intensities are studied. Fig. 2 shows representative pictures of microjet formation for $u_p = 1.5\ \mathrm{km/s}$, where the specimens remain solid during shock compression and subsequent unloading. For the monocrystal (Fig. 2(b)), it can be seen that a typical microjet forms due to the interaction of shock wave and surface groove, which expands under velocity gradient and finally breaks up into ejecta particles. However, for the case of nanocrystal (Fig. 2(a)), although a microjet is produced at the earlier times (40 ps), it stops expanding and shrinks to the metal surface. This phenomenon is attributed to the different shockwave profiles in the two samples. A shockwave with a ramp front travels inside the nanocrystal. When it reaches the surface and interacts with the groove, a microjet with lower tip velocity compared to the monocrystal case forms. As a result, the kinetic energy of the microjet is not big enough to overcome the effect of surface tension and the microjet finally becomes a liquid surface bump.

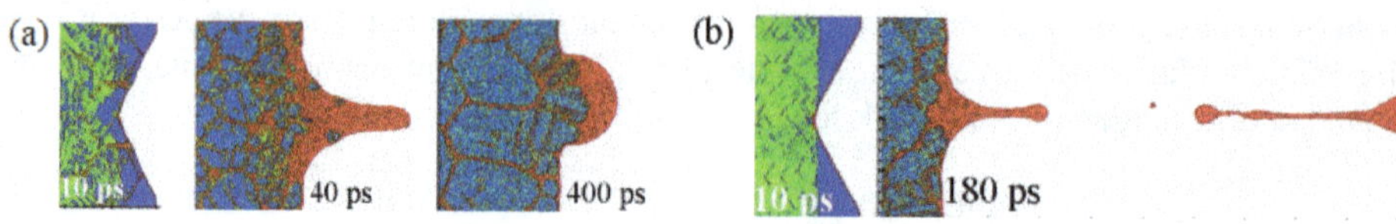

Figure 2: Morphology evolution of microjet for (a) nano- and (b) monocrystalline Cu shocked at $u_p = 1.5\ \mathrm{km/s}$.

As the shock intensity increases to $u_p = 2.0\ \mathrm{km/s}$ shown in Fig. 3, partial melting in the nanocrystal occurs during shock unloading, resulting in a solid-liquid mixing state in the bulk (20 ps in Fig. 3(a)). However, for the monocrystal, only a limited surface region transforms into liquid state (20 ps in Fig. 3(b)). As a consequence, the total mass of microjet from the nanocrystal is higher than that from the monocrystal, especially in the jet root region.

As the shock intensity increases further to $u_p = 3.5$ km/s as shown in Fig. 4, shock induced melting takes place both in the nano- and monocrystals. Therefore, the two microjets form and grow in liquid states, resulting in approximately identical jetting states.

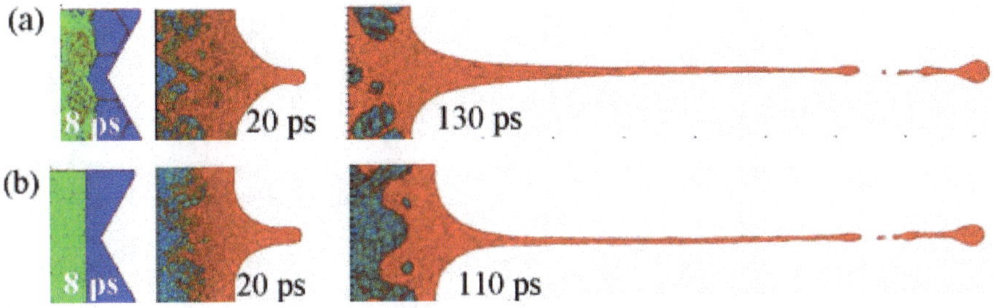

Figure 3: Morphology evolution of microjet for (a) nano- and (b) monocrystalline Cu shocked at $u_p = 2.0$ km/s.

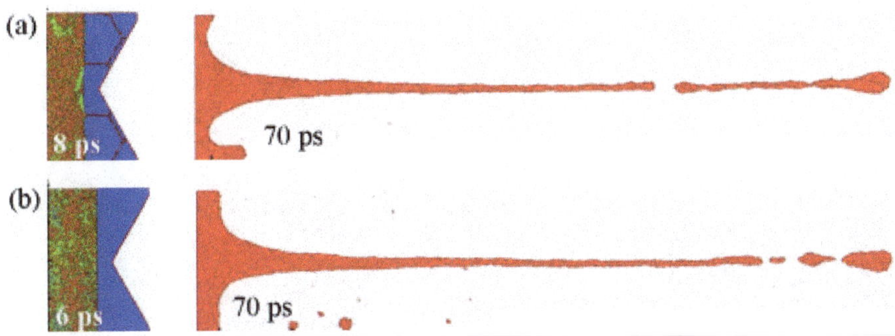

Figure 4: Morphology evolution of microjet for (a) nano- and (b) monocrystalline Cu shocked at $u_p = 3.5$ km/s.

The variation of jetting factor with u_p is displayed in Fig. 5. The jetting factor R_ej is defined as

$$R_\mathrm{ej} = \frac{m_\mathrm{ej}}{\rho V_\mathrm{def}}, \tag{1}$$

where m_ej is the total mass of microjet above the free surface, ρ is the material density, and V_def is the initial volume of the surface groove. It can be clearly seen in Fig. 5 that the jetting factors are nearly identical in the shock induced melting region ($u_p \geq 3.5$ km/s), consistent with the jetting states given in Fig. 4. In the solid and mixing region (2.0 km/s $\leq u_p \leq$ 3.0 km/s), the jetting factor for the nanocrystal is obviously higher than that of the monocrystal due to the different melting dynamics as explained above.

The variation of microjet tip velocity with post-shock particle velocity for the two systems is presented in Fig. 6. It can be seen that the tip velocity of microjet in the nanocrystal is almost the same as in the monocrystal case, and increases linearly with post-shock particle velocity.

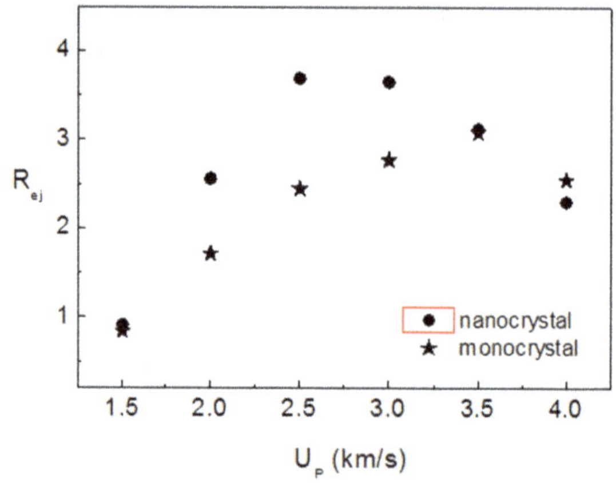

Figure 5: Variation of jetting factor with post-shock particle velocity.

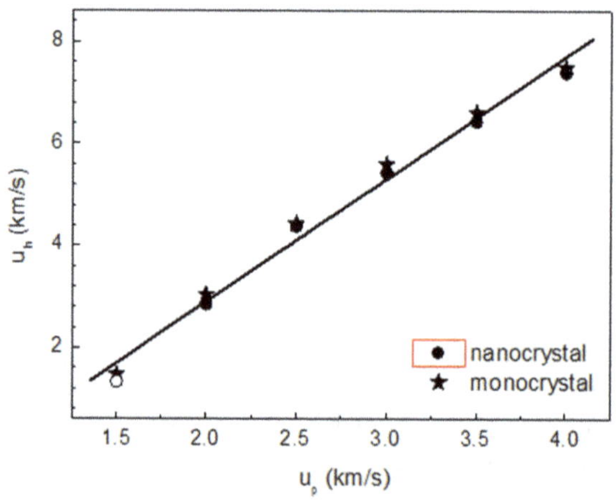

Figure 6: Variation of microjet tip velocity with post-shock particle velocity.

4 CONCLUSION

We performed molecular dynamics simulations to investigate the ejecta production from nanocrystal Cu with a grooved surface under various shock loading conditions. It is found that the microjetting morphology and the total mass of ejecta are different from the monocrystal in the shock pressure region before shock induced melting, caused by the special features of shock wave propagation and the melting dynamics in the nanocrystal. However, the tip velocity of microjet is independent on the micro-structure of the system.

ACKNOWLEDGEMENTS

This work was supported by the National Natural Science Foundation of China (Grant No. U1530261), and the Science Challenge Project (Grant Nos. TZ2016001 and TZ2018001).

REFERENCES

[1] Asay, J.R., Mix, L.P. & Perry, F.C., Ejection of material from shocked surfaces. *Appl. Phys. Lett.*, **29**, p. 284, 1976.

[2] Sorenson, D.S., Minich, R.W., Romero, J.L., Tunnel, T.W. & Malone, R.M., Ejecta particle size distributions for shock loaded Sn and Al metals. *J. Appl. Phys.*, **92**, p. 5830, 2002.

[3] Buttler, W.T. et al., Dynamic comparisons of piezoelectric ejecta diagnostics. *J. Appl. Phys.*, **101**, p. 063547, 2007.

[4] Zellner, M.B. et al., Probing the underlying physics of ejecta production from shocked Sn samples. *J. Appl. Phys.*, **103**, p. 123502, 2008.

[5] Cherne, F.J., Hammerberg, J.E., Andrews, M.J., Karkhanis, V. & Ramaprabhu, P., On shock driven jetting of liquid from non-sinusoidal surfaces into a vacuum. *J. Appl. Phys.*, **118**(18), p. 185901, 2015.

[6] He, A.-M., Liu, J., Liu, C. & Wang, P., Numerical and theoretical investigation of jet formation in elastic-plastic solids. *J. Appl. Phys.*, **124**, p. 185902, 2018.

[7] Shao, J.L., Wang, P., He, A.M., Duan, S.Q. & Qin, C.S., Atomistic simulations of shock-induced microjet from a grooved aluminium surface. *J. Appl. Phys.*, **113**, p. 153501, 2013.

[8] He, A.M., Duan, S.Q., Shao, J.L., Wang, P. & Luo, S.N., Local and bulk melting of shocked columnar nanocrystalline Cu: Dynamics, anisotropy, premelting, superheating, supercooling, and re-crystallization. *J. Chem. Phys.*, **139**, p. 074502, 2013.

[9] Mishin, Y., Mehl, M.J., Papaconstantopoulos, D.A., Voter, A.F. & Kress, J.D., Structural stability and lattice defects in copper: Ab initio, tight-binding, and embedded-atom calculations. *Phys. Rev. B*, **63**(22), p. 224106, 2001.

INDIRECT TENSILE STRENGTH METHOD FOR DEFINING A PROPER PRACTICE OF ASPHALT MIXTURES DESIGN

ROSA VEROPALUMBO, NUNZIO VISCIONE & FRANCESCA RUSSO
Department of Civil, Construction and Environmental Engineering, University of Naples Federico II, Italy

ABSTRACT

This paper shows the results of a laboratory investigation of indirect tensile strength test effects on bituminous mixtures characterization. In particular, the research focusses, in the first analysis, on the procedure of Marshall mix design, where 4 different mixtures of hot mix asphalt (HMA) were prepared at different bitumen contents between 3.5% and 4.25% by total aggregate weight at 0.25% increments. In the second phase, bituminous specimens made-up by gyratory compactor in compliance with Marshall optimization were tested to evaluate their indirect tensile strength. It was ascertained how the air voids percentage and indirect tensile strength can change by varying the number of revolutions during gyratory compaction and the height of specimens. A one-way ANOVA test was finally carried out to verify whether statistically significant differences exist in terms of mechanical properties when the number of revolutions and the specimen size change. This research may help to provide administrations with fast laboratory procedures to use in situ for all activities related to acceptance of mixtures.

Keywords: specimens size, gyratory method, number of revolutions, indirect tensile strength.

1 INTRODUCTION

The bituminous mixtures adopted in road construction are subject to repetitive loads corresponding to the continuous passage of vehicles; such tons, if sufficiently high, cause the loss of stiffness of the materials constituting the package and can lead to cracking and breakage, through the accumulation of stresses in the long term. In general, the superstructure performs three different tasks:

- It reduces the stresses transmitted to the soil to a level compatible with its bearing capacity so that the ground does not suffer deformation.
- It forms a structure that is stable over time and not very deformable, able to withstand repeated loads applied by vehicles and such as to ensure adequate driving comfort.
- It guarantees traffic safety in relation to problems of tire-pavement adherence in the presence of weather agents (rain, mud, snow, ice, rubber deposits).

In reference to the second point listed above, in a flexible pavement an important role is played by a base layer to ensure a suitable structure of resistance. The base layer of flexible pavements is constituted of bituminous mixture with nominal aggregate size typically less than those of a subgrade layer. This layer aims to resist fatigue phenomena because it is an inflected layer, so it is mainly subjected to tensile stresses.
For this reason, it is fundamental to choose the right materials and make a good mix design of asphalt mixture to improve the performance of the pavement and for the increment of pavement useful life.

Some studies [1] have shown that the principal factor influencing the strength of bituminous mixture is the filler and in particular the type of filler adopted. Researchers have been studying the reemployment of second raw materials [2]–[4] such as the digested spent bentonite filler. They have shown that the use of this alternative material as filler in hot mix

WIT Transactions on Engineering Sciences, Vol 124, © 2019 WIT Press
www.witpress.com, ISSN 1743-3533 (on-line)
doi:10.2495/MC190251

asphalt (HMA) mixture increases by about 7% the indirect tensile strength (ITS) value if compared to the traditional mixture. Also, the use of waste brick powder [5] as mineral filler in asphalt mixture has been studied. The results have shown that the recycled brick fillers (at 40°C) increase the stiffness modulus of asphalt binder which may improve the rutting resistance of asphalt mixture. Other types of filler such as steel slag, ladle furnace slag, waste foundry sand, glass wastes and coal ash [6] have been reemployed in cement concrete mixtures for base and sub-base layers in road pavement using the ITS value to complete the mix design of the mixtures.

In reference to the potential of fine aggregates recycled in powder form as fillers in the bituminous mixture, other researchers [7] have focused their studies on the use of such materials for the base layers and foundation. The traditional Marshall procedure was used for the mix design of mixtures and by the analysis of results it was concluded that mixtures containing recycled fine aggregates in powder have a higher indirect tensile strength and a higher fatigue strength than the traditional mixtures. An effective method is also provided to better understand the internal structure shape after fatigue damage [8] and the influence of microstructure characteristics on fatigue performance. The analysis of results has shown that the air–void ratio increases after fatigue damage at different test temperatures. Therefore, the rising temperature is harmful to the fatigue performance over low temperature range from -10°C to 10°C. As has been seen, the indirect tensile strength test is the primary test carried out to characterize the stiffness of a bituminous mixture. Most frequently, the problem is to find if the specimens compaction type used for ITS evaluation during the mix design phase is also valid during the control phase of the mixture. The purpose of this research was to offer claims and references for mixture packaging process with 100% limestone aggregates for a base layer, evaluating the behaviour and the changes of some mechanical properties such as Marshall Stability and Indirect Tensile Strength. In particular the research aims to evaluate ITS on mixtures optimized with the right percentage of bitumen according to Technical Rules of the Contract not only compacted with impact compaction but also replicated using another type of compaction method such as gyratory compaction in an effort to obtain the same value.

2 MATERIALS AND METHOD

2.1 Aggregates

In this study traditional limestone aggregates were used for the mix production. The limestone material is the product of limestone crushing [9] and it is configured as a different size aggregate white in colour and characterized by an amorphous structure. The limestone aggregates have been selected because it is one of the most suitable materials for the HMA production for a base layer in Italy. The chemical properties have been evaluated according to EPA 7000B-2007 (Table 1) and results shown that this type of material is constituted (main properties) by calcium (66%), steel (4.2%), silicon (26%) and magnesium (2.7%).

Table 1: Chemical properties of limestone aggregates.

Parameters	Unit	Value
Calcium	mg/kg	190000
Steel	mg/kg	12000
Silicon	mg/kg	75000
Magnesium	mg/kg	7800

The physical properties of limestone aggregates samples were evaluated, and the results are reported in Table 2.

Table 2: Aggregate physical properties.

Aggregate test	Unit	Value	Standard
Los Angeles abrasion value	%	20.4	UNI EN 1097-2
Shape index	-	70.7	UNI EN 933-4
Flakiness index	-	89.0	UNI EN 933-3

2.2 Bitumen

A standard 50/70 penetration grade bitumen was used as binder, most commonly used in Italy for traditional HMA mixtures. Preliminary evaluation of the collected asphalt samples was conducted to verify that the significant properties are compliant with the specifications of the National Italian Autonomous Roads Corporation for a fresh binder (Table 3).

Table 3: Properties of asphalt 50/70 pen binder.

Test	Unit	Value	Standard
Penetration @ 25°C	dmm	61	UNI EN 1426
Soft. point	°C	48.8	UNI EN 1427
Dynamic Visc. @ 160°C	Pa s	0.22	UNI EN 13702
Frass	°C	-10	UNI EN 12593

An additional test was conducted to evaluate the viscoelastic properties of the bitumen, in particular the shear modulus G*, with a Dynamic Shear Rheometer (DSR) according to EN 14770 Standard. The bitumen samples were tested at 5 different temperatures (10°C, 20°C, 30°C, 40°C, 50°C) and in a frequency range of 0.1–10 Hz. The test consists to bring the test specimen to the selected gap setting plus from 0.025 mm to 0.050 mm for parallel plate geometries and applied a constant oscillating stress.

To represent the behaviour of the binder in function of the temperature and frequency, the G* obtained are reported in terms of Master curve (Fig. 1), developed on the bases of Williams–Landel–Ferry (WLF) law:

$$\log\frac{a(T)}{a(T_0)} = \frac{-C_1\cdot(T-T_0)}{C_2+T-T_0},$$

where a(T) and a(T_0) are the shift factors at temperature T and T_0; T is the temperature to shift; T_0 is the temperature of reference for the shift; and C_1 and C_2 are constants that depend by nature of the material.

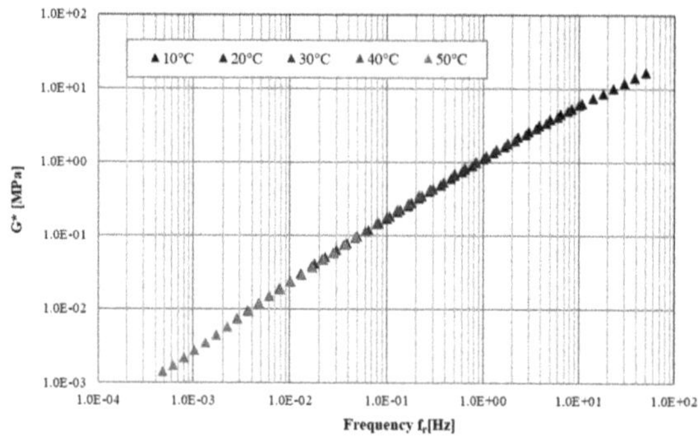

Figure 1: Master curve.

2.3 Mix design

2.3.1 Grading curve

The particle size distribution of aggregates is one of the most important characteristics to determine how it will perform as a pavement material. In this research the grading curve study was established in accordance with several local and international requirements as follows:

- According to EN13108-1 (see Table 1 in EN 13108-Overall limits of target composition – basic sieve set plus set 2).
- According to Technical Specifications on road maintenance of the National Italian Autonomous Roads Corporation that reflects EN 13108-1 but it is more restrictive regarding the calculation of the percentage of material passing through some specific sieves.
- According to maximum density gradation obtained using n=0.45 in the equation developed by Fuller and Thompson in 1907.

The aggregates mixture was produced in a laboratory using different sizes of limestone – adequately proportioned – and the optimal distribution obtained is reported in Table 4 and Fig. 2.

Table 4: Aggregate size proportion.

Sieve size (mm)	**Passing (%)**
0.063	7.2
0.25	10.6
0.5	12.5
2	21.7
6.3	38.0
10	56.9
14	70.7
20	89.0
31.5	100.0

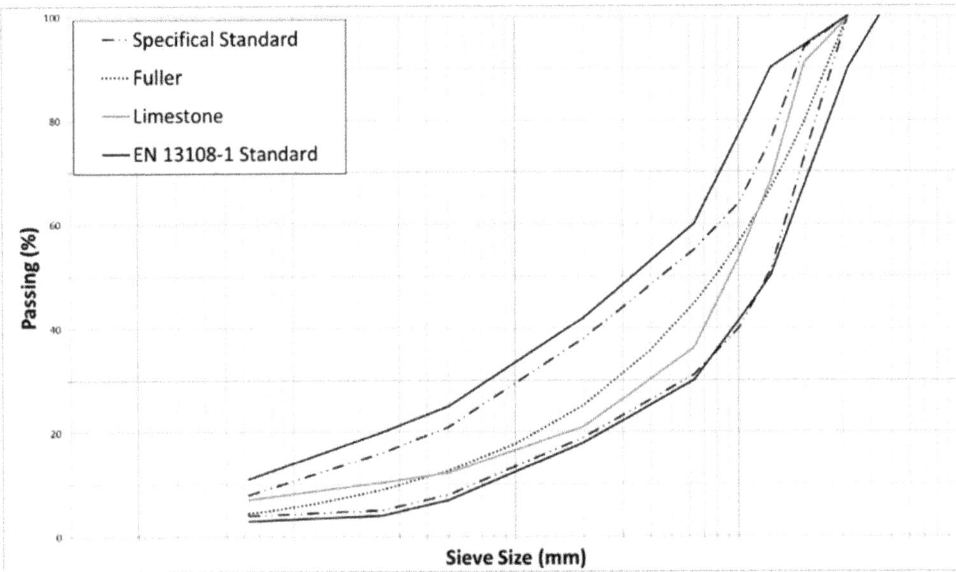

Figure 2: Grading curve of limestone aggregates.

2.3.2 Optimal bitumen content

The optimal bitumen content has been evaluated according to Marshall method. This method is based on the volumetric approach of design mix materials. Though, all the volumetric parameters are important and required to be within the allowable limit to have better performance of the HMA, the bitumen content and percentage air voids plays a dynamic role in the durability of the HMA. The Marshall method seeks to select the asphalt binder content at a desired density that satisfies minimum stability and range of flow values required by technical standards. Different 4 HMA mixtures were prepared at 4 different bitumen contents between 3.5% and 4.5% by total aggregate weight at 0.25% increments for the bituminous mixture design solutions according to Technical Specifications on road maintenance of the National Italian Autonomous Roads. Sixteen cylindrical specimens (4 for each HMA mixture) with a diameter of 101.6 ± 0.1 mm and a height of 63.05 ± 2.5 mm were prepared according to UNI EN 12697-30 using a sliding mass of 7850 ± 50 g falling from a height of 457 ± 5 mm with 75 blows on each side. For every specimen packaged, volumetric properties [10] have been evaluated according to UNI EN 12697-8 (dimensional and SSD procedure) and at the same time mechanical performances in terms of Marshall Stability and Flow have been evaluated according to UNI EN 12697-34 to determine the optimal bitumen content.
The optimum binder content has been selected based on the combined results of Marshall stability and flow and voids analysis according to requirements of the National Italian Autonomous Roads Corporation specification.

The optimal bitumen content (Fig. 3) has been reached in correspondence of 4.00% on total weight of aggregates to which corresponds a stability value of 818 kg; flow value of 2.65 mm and a percentage of SSD and dimensional voids respectively of 4.4% and 11.1%.

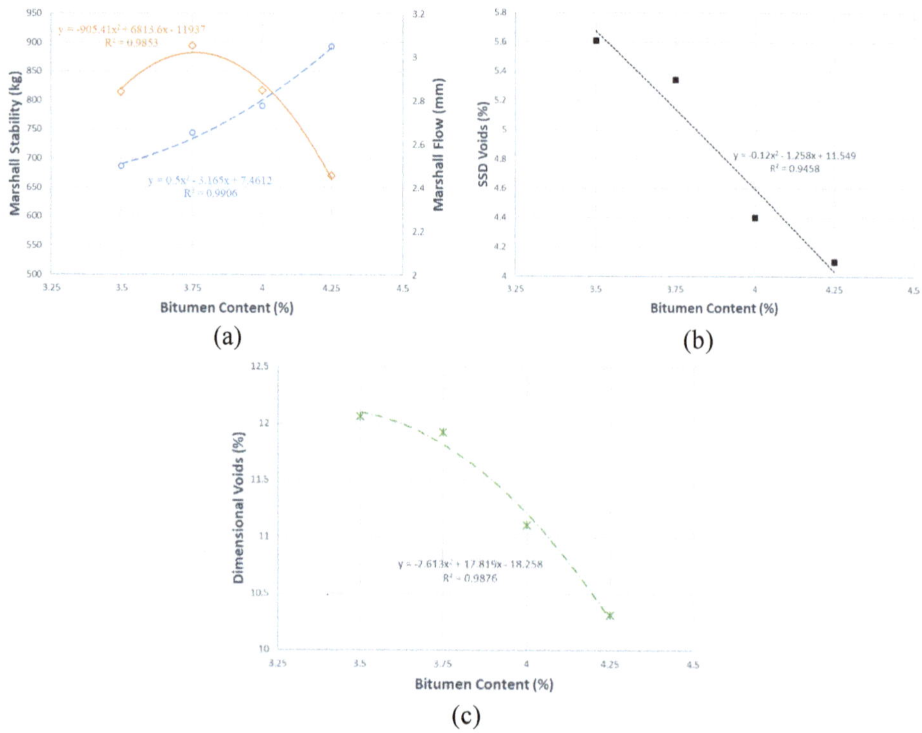

Figure 3: Bitumen content. (a). Marshall stability and Marshall flow; (b) SSD voids; (c) Dimensional voids.

2.4 ITS evaluation

The effect of traffic load in a long period could be affect the strength of an asphalt mixture showing fatigue cracks and/or rutting. To determine this damage is commonly used the Indirect Tensile Strength (ITS). The ITS test was performed according to UNI EN 12697-23. To evaluate the ITS, the cylindrical specimen must be placed in the compression testing machine between the loading strips and be loaded diametrically along the direction of the cylinder axis with a constant speed of displacement until it breaks. The indirect tensile strength is the maximum tensile stress calculated from the peak load applied at break and the dimensions of the specimen according the following eqn:

$$ITS = \frac{2P}{\pi DH},$$

where *ITS* is the indirect tensile strength, expressed in gigapascals (GPa); *P* is the peak load, expressed in kilonewtons (kN); *D* is the diameter of the specimen, expressed in millimetres (mm); and *H* is the height of the specimen, expressed in millimetres (mm).

Besides the strength of mixture is also compromised by water effect thus to evaluate its behaviour and durability the ITS value was determinate before and after a soaked in a water bath for 72h at 40°C. The relation of the strength values before and after water storage is called

Indirect Tensile Strength Ratio according to UNI EN 12697-12 standard. Therefore, the following eqn has been adopted:

$$ITSR = \frac{ITS_{wet}}{ITS_{dry}} \cdot 100,$$

where $ITSR$ is the indirect tensile strength ratio (%); ITS_{wet} is the average indirect tensile strength of the wet group (kPa); and ITS_{dry} is the average indirect tensile strength of the dry group (kPa).

3 RESULTS AND DISCUSSION

One of the most difficult problem that incur in the pavement world, in particular during the control phase of mixture is choosing the correct size of the specimen for ITS evaluation. For this reason, the research in exam deals with specimen compacted with Marshall compactor and also with specimens obtained with the gyratory compactor according to UNI EN 12697-31 at different size because the gyratory standard prescribe a range of height different from those required by ITS standard which requires that the specimens shall be compacted in accordance to UNI EN 12697-31.
Therefore, in this research the ITS parameter was evaluated on:

- Specimens packaged according to UNI EN 12697-30 with a diameter (101.6±0.1) mm and height (63.5±2.5) mm.
- Specimens packaged according to UNI EN 12697-31 with a diameter of 150 mm (maximum aggregate size is more than 16 mm) using a height falling within the range defined by ratio between h_{min} and the diameter D of 0.66 to 1.05.
- Specimens packaged according to UNI EN 12697-23 with a diameter of 150 mm (maximum aggregate size not exceed 40 mm) and a height of between 35 mm and 75 mm.

The evaluation of ITS, at the optimum bitumen content obtained using Marshall mix design method, has been evaluated on specimens compacted using the gyratory compactor. Furthermore, by means of the gyratory compactor, the asphalt mixtures have been designed at a specific level of compactive effort, that is a function of the number of gyrations N_{design}. Two different numbers of gyrations have been chosen (100 and 180) to have an air voids correlation with Marshall specimen. To comply with the requirements prescribed by UNI EN 12697-31, 4500 gr of HMA have been introduced into the mould during specimen preparation; however, to comply with the requirements fixed by UNI EN 12697-23, 3000 gr of HMA was used.

3.1 Laboratory phase results

The ITS value has been evaluated on 7 specimens for each compaction process described above. The size of the specimen and different number of revolutions as reported in Table 5. Each specimen, before to test, was kept in an air chamber at 25°C.

Fig. 4 reports the mean value of the ITS of each abovementioned combination. The figure shows that the ITS value is the same for G100B and G180B while there is a difference of 4% between G100A and G180A. Besides, after being conditioned in a water bath for 72 h at 40°C, the specimen G180B seems not to affect any reduction in terms of ITS; while the other specimens show a reduction of 14, 20, 22 and 30% respectively for Marshall, G100A, G180B and G180A.

Table 5: Sample obtained from Marshall compaction and gyratory compaction.

Label	Compaction Type	Height	N_{design}	N° specimens
MT	Marshall	h∈(63,5 ± 2,5)	-	7
G100A	Gyratory	hmin/D∈(0.66;1.05)	100	7
G180A	Gyratory	hmin/D∈(0.66;1.05)	180	7
G100B	Gyratory	h∈(35mm;75mm)	100	7
G180B	Gyratory	h∈(35mm;75mm)	180	7

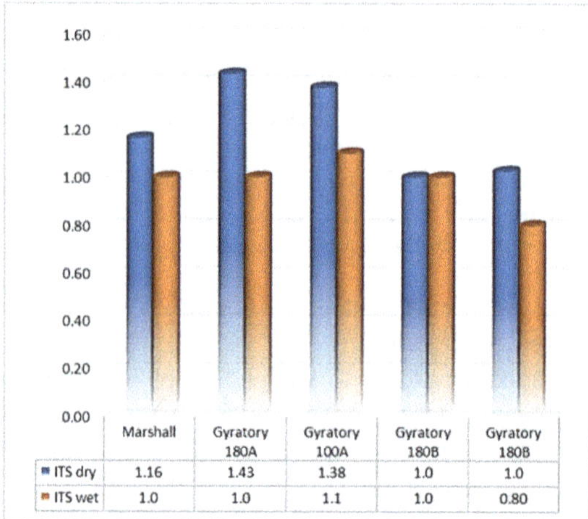

Figure 4: ITS wet and dry.

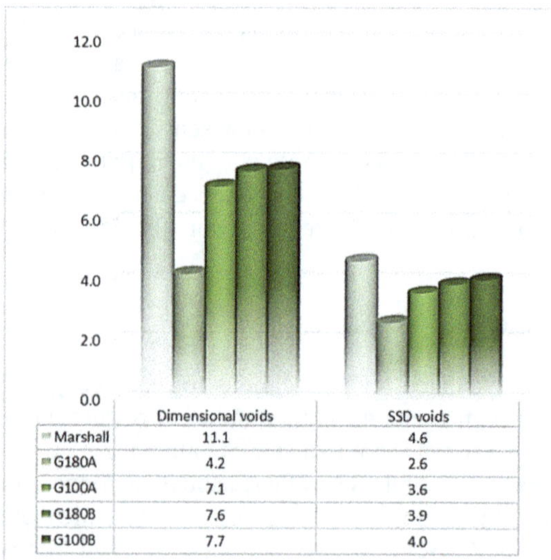

Figure 5: Relationships among geometric and imbibed voids in average values.

At the same time, on the same specimens adopted for the ITS test air voids have been evaluated using SSD and dimensional procedures. The results show a difference about of 6.9% in terms of dimensional air voids between Marshall specimens and G180A. Instead the mixture G100A, G180B and G100B show approximately the same dimensional voids value. Nevertheless, the SSD voids are also different between specimens compacted through Marshall compactor and G180A with a difference of 2%, instead G100A, G180B and G100B are the same.

3.2 Statistical analysis validation

As well as to clearly compare the properties of the various specimen types, a statistical analysis was performed to evaluate the significance of the differences among the properties. In the first place to verify that the data are distributed normally, the Kolmogorov–Smirnov (KS) test has been adopted. This type of test is also used to check the assumption of normality in Analysis of Variance. It compares the cumulative distribution function for a variable with a specified distribution. The test was carried out on ITS value of each type of combination described above. Table 6 shows the results of KS test and as it is possible that the critical value $Dn_{,critical,\alpha}$ equals to 0.48342 is always greater than D_n. In order to demonstrate the normal distribution of the date, the analysis of Variance (ANOVA) was carried out.

Table 6: KS test_RTI Normal distribution.

Type	Dn value	Dn critical	Normal distribution
Marshall	0.25629	0.48	Verified
G100A	0.22159		Verified
G180A	0.30967		Verified
G100B	0.14908		Verified
G180B	0.30754		Verified

The ANOVA was conducted at 5% significance level (α=0.05), which means that for each comparison reported as being significantly different or not, there is only a 5% chance that it is not true. The results of a one-way ANOVA, to establish the difference between the group in analysis, considering only one variable, are reported in Table 7. It is shown that there is no statistical difference in terms of ITS before the soaked in water bath between the same size specimens compacted at different number of revolutions (GB180=GB100 and G100A=G180A).

Table 7: One-way ANOVA results.

	LSD test; variable ITS (N/mm²) Homogenous Groups, alpha= 0.05000 ERROR: Between MS = 0.01087 df =65,000			
Compaction Method	Mean ITS (N/mm²)	1	2	3
G180B	0.967143	****		
G100B	1.021429	****		
MARSHALL	1.164286			****
G100A	1.375714		****	
G180A	1.432857		****	

4 CONCLUSION

The principal characteristic of a road superstructure is to guarantee a mixture that should have high strength, to be capable of supporting heavy traffic loads without cracking or rutting. It is possibly highlighted that the duration of the useful life of a road pavement is precisely the resistance that the mixture is able to exert if subjected to strong stresses such as those due to road traffic.

For this reason, the evaluation of indirect tensile strength for each layer of a road flexible pavement, is one of the fundamental tasks that must be carried out in order to guarantee better performance in terms of stiffness and strength resistance.
Based on the experimental analysis, the following main conclusions could be drawn:

- Varying the type of compaction, the dimensional voids, SSD voids and ITS value changed. In particular, fixing the size of the specimen and varying the number of revolutions does not change the ITS, instead changing the size and fixing the number of revolutions changes the ITS.
- The ITS value is proportional to the air voids percentage. In particular if the ITS value grows, voids percentage (evaluated with both procedures) decrease. This correlation could be due to the different specimen densification; in particular the bulk density of specimens packaged according to UNI EN 12697-31 is higher than bulk density of specimens packaged according to UNI EN 12697-23.
- Statistically it is confirmed that varying the number of revolutions but fixing the size of the specimen does not make significant difference.

This present study had as purpose to achieve the optimum specimen size for the evaluating of the ITS parameter obtaining the same results in terms of voids for the optimum of the mixtures, obtained using Marshall test, with the gyratory compactor. From the results it was concluded that, for the determination of ITS value, there can be an assessment on mixture compacted according to ITS standard (UNI EN 12697-23) with a N_{design} = 100 revolutions.

REFERENCES

[1] Bocci, M. & Giuliani, F., Caratterizzazione di filler per conglomerati bituminosi. *XXIII AIPCR Convegno Nazionale Stradale*, pp. 18–21, 1998.

[2] Veropalumbo, R., Viscione, N. & Formisano, A., Hot mix asphalt with fly ashes for dense-graded surface Layers of rural roads. *WIT Transactions on Ecology and the Environment*, vol. 215, WIT Press, pp. 93–105, 2018.

[3] Sheng, Y., Zhang, B., Yan, Y., Chen, H., Xiong, R. & Geng, J., Effects of phosphorus slag powder and polyester fiber on performance characteristics of asphalt binders and resultant mixtures. *Construction and Building Materials*, **141**, pp. 289–295, 2017.

[4] Sangiorgi, C., Tataranni, P., Simone, A., Vignali, V., Lantieri, C. & Dondi, G., Assessment of waste bleaching clay as alternative filler for the production of porous asphalts. *Construction and Building Materials*, **109**, pp. 1–7, 2016.

[5] Sangiorgi, C., Tataranni, P., Simone, A., Vignali, V. Lantieri, C. & Dondi, G., Assessment of waste bleaching clay as alternative filler for the production of porous asphalts. *Construction and Building Materials*, **109**, pp. 1–7, 2016.

[6] Chen, M., Lin, J., Wua, S. & Liu, C., Utilization of recycled brick powder as alternative filler in asphalt mixture. *Construction and Building Materials*, **25**, pp. 1532–1536, 2011.

[7] Pasetto, M. & Baldo, N., Recycling of waste aggregate in cement bound mixtures for road pavement bases and sub-bases. *Construction and Building Materials*, **108**, pp. 112–118, 2016.

[8] Hu, J., Liu, P., Wang, D., Oeser, M. & Tan, Y., Investigation on fatigue damage of asphalt mixture with different air-voids using microstructural analysis. *Construction and Building Materials*, **125**, pp. 936–945, 2016.

[9] Simone, A. et al., Experimental application of waste glass powder filler in recycled dense-graded asphalt mixtures. *Road Materials and Pavement Design*, pp. 1–16, 2017.

[10] Russo, F., Biancardo, S.A., Formisano, A. & Dell'Acqua, G., Predicting percent air voids content in compacted bituminous hot mixture specimens by varying the energy laboratory compaction and the bulk density assessment method. *Construction and Building Materials*, **164**, pp. 508–524, 2018.

Author index

Computational and Experimental Methods in Multiphase and Complex Flow X

Edited by: ***S. HERNÁNDEZ****, University of A Coruña, Spain, and* ***P. VOROBIEFF****, University of New Mexico, USA*

Composed of papers presented at the 10th conference on Multiphase flow this book presents the latest research on the subject. The research included in this volume focuses on using synergies between experimental and computational techniques to gain a better understanding of all classes of multiphase and complex flow.

The presented papers illustrate the close interaction between numerical modellers and researchers working to gradually resolve the many outstanding issues in our understanding of multiphase flow.

Recently multiphase fluid dynamics have generated a great deal of attention, leading to many notable advances in experimental, analytical and numerical studies. Progress in numerical methods has permitted the solution of many practical problems, helping to improve our understanding of the physics involved.

Multiphase flows are found in all areas of technology and the range of related problems of interest is vast, including astrophysics, biology, geophysics, atmospheric process, and many areas of engineering.

The papers in the book cover a number of topics, including: Experimental measurements; Numerical methods; Multiphase flows; Flow in porous media.

ISBN: 978-1-78466-329-2 **eISBN: 978-1-78466-330-8**
Published 2019 / 250pp

www.ingramcontent.com/pod-product-compliance
Lightning Source LLC
LaVergne TN
LVHW062226200926
848484LV00006B/1486